Accommodation and Vergence Mechanisms in the Visual System

Edited by
Ove Franzén
Hans Richter
Lawrence Stark

Birkhäuser Verlag
Basel · Boston · Berlin

Editors:

Prof. Dr. Ove Franzén
Mid Sweden University
Rödhakevägen 24
S-75652 Uppsala
Sweden

Dr. Lawrence Stark
University of California
School of Optometry
485 Minor Hall
Berkeley, CA 94720
USA

Dr. Hans Richter
Department of Ophthalmology
Karolinska Institute
Huddinge Hospital
S-14186 Huddinge
Sweden

Library of Congress Cataloging-in-Publication Data

Deutsche Bibliothek Cataloging-in-Publication Data
Accommodation and vergence mechanisms in the visual system / ed. by Ove Franzén
  ... - Basel ; Boston ; Berlin : Birkhäuser, 2000
   ISBN 3-7643-6073-9

ISBN 3-7643-6073-9 Birkhäuser Verlag, Basel – Boston – Berlin

© 2000 Birkhäuser Verlag, P.O. Box 133, CH-4010 Basel, Switzerland
Printed on acid-free paper produced from chlorine-free pulp. TCF ∞
Printed in Germany
ISBN 3-7643-6073-9

9 8 7 6 5 4 3 2 1

Contents

Part I CNS pathways, single unit, neural population and structural correlates of the accommodative system

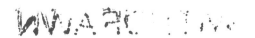

Part IV Health and clinical aspects of accommodation/vergence

Epilogue

Subject Index

This volume is dedicated
to the memory of
Professor Ivar Lie and Professor Robert von Sandor.

PREFACE

This conference was instigated by a combination of factors: The nature of the problem, the widespread occupational epidemiology reported on eye symptoms and eye fatigue in the workplace, and the organizers' awareness of the complexity of the scientific and clinical bases of knowledge that might be usefully applied. The introduction of new methods into system neurobiology provides new insights into how we receive and process information from the external world, and act upon it. New, non-invasive methods have opened the way to direct observation of the human brain in action.

Due particularly to the interaction between the visual and oculomotor requirements involved, several clinical and scientific fields intersect when these issues are considered. To provide clear vision the accommodative and pupillary mechanisms are used. To maintain binocularity, the vergence oculomotor system, sensitive to fatigue, must attain congruence with accommodative levels. This accommodation-vergence linkage was a focus of our symposium.

Many scientific disciplines could make a contribution through their own specialized knowledge --- the basic anatomy, physiology, and pharmacology of these neuromuscular systems; the development of careful experimental studies that can approach subclinical aspects of oculomotor vision stressed by artificial displays; appropriate animal and bioengineering models to prove or falsify theories put forward in a clinical setting and to help elucidate the complexity of interactions. Theoretical neurobiologists have developed methods and concepts for testing models of neuronal networks and whole brain functions that have led to heuristic interactions between · theory and experiment and also to an appreciation of the complex sensory mechanisms of the visual cortex for processing blur. Thus, the thrust of the symposium was to attempt to unravel these interconnected problems at many levels of analysis and to understand function in adults, developing children and the aged.

Indeed, the issues raised by the papers and the discussions they inspired during the meeting were very fruitful, which can be illustrated by just one example. In landmark research on primate models for myopia development, Torsten Wiesel and Elio Raviola, from the Rockefeller University and Harvard Medical School, showed two different outcomes to an operatively produced neonatal defect in two closely related species. Their results in terms of gene expression were exciting, and clinicians at the symposium could immediately understand that associated problems in man might not be the consequence of a single syndrome.

The symposium ended with the participants filled with enthusiasm for future work in their own laboratories, and even more importantly, with a consciousness of the potential productivity that is realized through informal contacts and collaboration.

Ove Franzén   Hans Richter   Lawrence Stark

Acknowledgements

This symposium was generously supported by grants from the Swedish Medical Research Council, The Swedish Natural Science Research Council, The Wenner-Gren Center Foundation for Scientific Research, the Swedish Confederation of Professional Employees (TCO) and the Mid Sweden University.

Scientific Committee

Ove Franzén, Uppsala, Sweden
Ivars Lacis, Riga, Latvia
Gunnar Lennerstrand, Stockholm, Sweden
Bruno Piccoli, Milan, Italy
Hans Richter, Uppsala, Sweden
Robert von Sandor, Stockholm, Sweden
Clifton Schor, Berkeley, CA, USA
Yasuyoshi Watanabe, Osaka, Japan

The First International Symposium on
"Accommodation/Vergence Mechanisms in the Visual System"
Stockholm, September 16, 1996

University Chancellor Stig Hagström

National Agency for Higher Education

Ladies and Gentlemen. Dear Research Colleagues!

You are coming from all over the world. Therefore, it might not be appropriate to start by wishing you a good morning since your biological clock may say something different.

It is a great pleasure for me to welcome you to Stockholm and to this wonderful day. I hope that you will find some time to look around the city and its beautiful setting and surroundings. Those of us who live here confess that we are biased when we say that Stockholm at this time of the year is one of the most beautiful cities in the world. And we want to continue to be biased in this respect.

Your theme for the symposium is certainly very adequate for this since the beauty we experience of Stockholm comes through our visual sense.

However, your theme is also very adequate in many other ways. We all know what an impact that modern information technology is having on our daily lives. Household words are now Internet, World Wide Web, multimedia etc. Much of the progress in this field hinges on our ability to create, manipulate, transmit and store pictures and images. In contrast to the word as the carrier of information which can be understood by several senses, the image as the carrier of information presupposes vision.

The possibility of using images in a large number of fields has the potential of revolutionising areas like learning, pedagogy, creative arts, and many others. However, in order for us to fully utilise and develop the image technologies we must have a better understanding how the visual system is working and how we are making use of visual information.

With the development of modern high speed electronic networks we know that we can transmit images so that the difference between virtual space and the real space is nearly zero. There has been some suggestions that this would lead to a situation where we don't have to meet physically person to person. I have no fear that this will happen. Sometimes such predictions have a tendency to get the opposite effect. I don't think that the telephone decreased the travelling. And I don't think that anybody has the experience that we are now living in the paperless society, despite all our electronic ways of handling information. No, there is no substitute for the personal meeting and therefore we are very glad that you came here in person.

Large numbers of persons are now spending many hours reading textual material from such electronic information networks in the workplace. The aim of "ergo-ophthalmology" is to supply scientific knowledge to enable the creation of displays with which every employee can be assured of maximum comfort and visual efficiency as well as the prevention of any impairment related to visual work. Our question is not whether there are various accommodative and other impairments of the eyes, but whether these impairments are caused by or in some way connected to operator tasks. Our immediate goals are the development of objective tests for such impairments so that careful epidemiological studies can ferret out such possible connections and remediate visual comforts for persons involved.

However, the eye is not the only sensor for our vision. There is also a type of vision that is created in our heads and when we interact together like in a symposium as you are having. I want to encourage you to also use and develop that vision because that is what we all need in a world of change and a world, not only filled with problems, but perhaps even more filled with opportunities. Opportunities that we only see when we have the special vision of discovering them.

Once again, let me add my warm welcome to you all to this symposium. We are glad and honoured that you have chosen to join us here. Have a wonderful, interesting, rewarding and fruitful symposium and don't forget to use all your senses and visions to look around.

1. Satoshi Ishikawa
2. Ove Franzén
3. José Pardo
4. Richard Held
5. Hans Richter
6. Birgitta Neikter
7. Bernhard Gilmartin
8. Kenneth Ciuffreda
9. Louise Hainline
10. Kazuhiko Ukai
11. Hiroshi Oyamada
12. Robert Owens
13. Clifton Schoor
14. Lawrence Stark
15. Tsunehiro Takeda
16. Lawrence Mays
17. Kenji Ohtsuka
18. Han Ying
19. Ivar Lie
20. Jan Ygge

21. Gunnar Lennerstrand
22. Ivars Lacis
23. Richard Tyrell
24. Jeffrey Andre
25. Jonathan Erichsen
26. Jim Gnadt
27. Torsten Wiesel
28. Neville McBrien
29. Philip Kruger
30. Agneta Rydberg
31. Takehiko Bando
32. Tetsuto Yamada
33. Waldemar Rojna
34. Roberto Bolzani
35. Barry Winn
36. Stephen Morse
37. Horika Otsuka
38. Kunihiko Tsuchiya
39. Neil Charman

Participants of the International Symposium at the Wenner-Gren Center, Stockholm

The Contributors

K.R. Aggarwala
Schnurmacher Institute for Vision
Research
State University of New York
New York, N.Y. 10010, USA

Jeffrey T. Andre
Department of Psychology
Franklin & Marshall College
P.O. Box 3003
Lancaster, PA 17604, USA

Takehiko Bando
Department of Physiology and
Ophthalmology
Niigata University
School of Medicine
Asahi-machi, Niigata 951, Japan

S. Bean
Schnurmacher Institute for Vision
Research
State University of New York
New York, N.Y. 10010, USA

Neil Charman
Department of Optometry and Vision
Sciences
University of Manchester
Institute of Science and Technology
Manchester, M60 1QD, England

Hai-Wen Chen
Department of Vision Sciences
State University of New York
State College of Optometry
N.Y. 10010, USA

Ken Ciuffreda
Department of Vision Sciences
State University of New York
State College of Optometry
N.Y. 10010, USA

S. Cohen
Schnurmacher Institute for Vision
Research
State University of New York
New York, N.Y. 10010, USA

Patricia Costello
Cognitive Neuroimaging Unit
Psychiatry Service
Veterans Affairs Medical Center
Minneapolis, MN, USA
and
Division of Neuroscience Research
Department of Psychiatry
University of Minnesota
Minneapolis, MN, USA

H. Endo
National Institute of Bioscience and
Human Technology
Human Informatics Department
1-1, Higashi, Tsukuba City
Ibaraki 305, Japan
Jonathan Erichsen
Deptartment of Optometry and Vision
Sciences
University of Wales, Cardiff
Cardiff CF1 3XF, U.K.

C. Evinger
Departments of Neurobiology and
Behavior and Ophthalmology
State University of New York
Stony Brook, New York, USA

Knut Inge Fostervold
Vision Laboratory
Institute of Psychology
University of Oslo
P.O. Box 1094, Blindern
N-0317 Oslo, Norway

Ove Franzén
Department of Clinical Science
Division of Ophthalmology
Karolinska Institute
Huddinge University Hospital
141 86 Huddinge, Sweden
and
Mid Sweden University
Sundsvall, Sweden

P.D.R. Gamlin
Department of Physiological Optics
Vision Science Research Center
University of Alabama
Birmingham AL 35294-4390, USA

Bernard Gilmartin
Department of Vision Sciences
Aston University
Birmingham B4 7ET, U.K.

Jim W. Gnadt
Department of Neurobiology and Behavior
State University of New York
Stony Brook, New York 11794, USA

J. Gwiazda
New England College of Optometry
Myopia Research Center
Boston, Massachusetts 02115, USA

Louise Hainline
Infant Study Center
Department of Psychology
Brooklyn College
Brooklyn, New York 11210, USA

Ying Han
Department of Clinical Science
Division of Ophthalmology
Karolinska Institute
Huddinge University Hospital
141 86 Huddinge, Sweden

N. Hara
Department of Physiology and
Ophthalmology
Niigata University
School of Medicine
Asahi-machi, Niigata 951, Japan

H. Hasebe
Department of Physiology and
Ophthalmology
Niigata University
School of Medicine
Asahi-machi, Niigata 951, Japan

K. Hashimoto
National Institute of Bioscience and
Human Technology
Human Informatics Department
1-1, Higashi, Tsukuba City
Ibaraki 305, Japan

Richard Held
New England College of Optometry
Myopia Research Center
Boston, Massachusetts 02115, USA

W. Hodos
Department of Psychology
University of Maryland
College Park, Maryland, USA
Satoshi Ishikawa
Department of Ophthalmology
School of Medicine
Kitasato University
Sagamihara Kanagawa 228, Japan

Bai-Chuan Jiang
University of Houston
College of Optometry
Houston, TX 77204-6052, USA

J.C. Kotulak
U.S. Army Aeromedical Research
Laboratory
Ft. Rucker, AL 36362, USA

V.V. Krishnan
University of California
School of Optometry
Berkeley, CA 94720-2020, USA

Philip Kruger
Schnurmacher Institute for Vision
Research
State University of New York
New York, N.Y. 10010, USA

J. Lee
Schnurmacher Institute for Vision
Research
State University of New York
New York, N.Y. 10010, USA

Joel T. Lee
Division of Neuroscience Research
Department of Psychiatry
University of Minnesota
Minneapolis, MN, USA

Gunnar Lennerstrand
Department of Clinical Science
Division of Ophthalmology
Karolinska Institute
Huddinge University Hospital
141 86 Huddinge, Sweden

Ivar Lie
Vision Laboratory
Institute of Psychology
University of Oslo
P.O. Box 1094, Blindern
N-0317 Oslo, Norway

L. Marran
University of California
School of Optometry
Berkeley, CA 94720-2020, USA

S. Mathews
Department of Ophthalmology
Texas Tech. U.H.S.C.
Lubbock, Texas 79430, USA

Lawrence Mays
Department of Physiological Optics
Vision Science Research Center
University of Alabama
Birmingham AL 35294-4390, USA

Neville McBrien
Department of Optometry and Vision
Sciences
Cardiff University of Wales
Cardiff CF1-3XF, Wales, U.K.

John Mordi
Department of Vision Sciences
State University of New York
State College of Optometry
N.Y. 10010, USA

Stephen Morse
University of Houston
College of Optometry
Houston, TX 77204-6052, USA

Tatsuto Namba
Department of Ophthalmology
School of Medicine
Kitasato University
Sagamihara Kanagawa 228, Japan

Charles Neveu
University of California
School of Optometry
Berkeley, CA 94720-2020, USA

Kenji Ohtsuka
Department of Ophthalmology
Sapporo Medical University
Sapporo 060, Hokkaido, Japan

Norika Otsuka
Department of Ophthalmology
School of Medicine
Kitasato University
Sagamihara Kanagawa 228, Japan

Alfred Owens
Whitely Psychology Laboratories
Franklin and Marshall College
Lancaster, PA 17604, USA

Robert L. Owens
Dr. Robert L. Owens & Associates
Optometrists
New Holland, Pennsylvania, USA

Hiroshi Oyamada
Department of Physiology
School of Medicine
Niigata University
Niigata 951, Japan

José V. Pardo
Cognitive Neuroimaging Unit
Psychiatry Service
Veterans Affairs Medical Center
Minneapolis, MN, USA
and
Division of Neuroscience Research
Department of Psychiatry
University of Minnesota
Minneapolis, MN, USA

M.A. Pearson
Department of Psychology
Clemson University
Clemson, SC 29634-1511, USA

Bruno Piccoli
Department of Occupational Health
ICP Hospital
University of Milan
Milan, Italy

J.C. Rabin
U.S. Army Aeromedical Research
Laboratory
Ft. Rucker, AL 36362, USA

Hans Richter
Department of Clinical Science
Division of Ophthalmology
Karolinska Institute
Huddinge University Hospital
141 86 Huddinge, Sweden
and
PET Center, Uppsala University
Uppsala, Sweden

Mark Rosenfield
Department of Vision Sciences
State University of New York
State College of Optometry
N.Y. 10010, USA

A. Sato
Department of Ophthalmology
Sapporo Medical University
Sapporo 060, Hokkaido, Japan

Clifton M. Schor
University of California
School of Optometry
Berkeley, CA 94720-2020, USA

James Sheedy
University of California
School of Optometry
Berkeley, CA 94720, USA

Scott R. Sponheim
Department of Psychology
University of Minnesota
and
Psychology Service
Veterans Affairs Medical Center
Minneapolis, MN, USA

Lawrence Stark
University of California
School of Optometry
Berkeley, CA 94720-2020, USA

R. Takada
Department of Physiology and
Ophthalmology
Niigata University
School of Medicine
Asahi-machi, Niigata 951, Japan

M. Takagi
Department of Physiology and
Ophthalmology
Niigata University
School of Medicine
Asahi-machi, Niigata 951, Japan

Tsunehiro Takeda
National Institute of Bioscience and
Human Technology
Human Informatics Department
1-1, Higashi, Tsukuba City
Ibaraki 305, Japan

J.F. Thayer
Department of Psychology
University of Missouri
Columbia, MO 65211, USA

H. Toda
Department of Physiology and
Ophthalmology
Niigata University
School of Medicine
Asahi-machi, Niigata 951, Japan

Kunihiko Tsuchiya
Department of Ophthalmology
School of Medicine
Kitasato University
Sagamihara Kanagawa 228, Japan

Richard A. Tyrrell
Department of Psychology
Clemson University
Clemson, SC 29634-1511, USA

Kazuhiko Ukai
Faculty of Social and Information Sciences
Nihon Fukushi University
Handa, Aichi 475, Japan

Reidulf Watten
Vision Laboratory
Institute of Psychology
University of Oslo
P.O. Box 1094, Blindern
N-0317 Oslo, Norway

Torsten Wiesel
Office of the President
Rockefeller University
New York, 10021-6399 N.Y., USA

Barry Winn
Department of Optometry
University of Bradford
West Yorkshire BD7 1DP, England

Tetsuto Yamada
Department of Ophthalmology
School of Medicine
Kitasato University
Sagamihara Kanagawa 228, Japan

Jan Ygge
Department of Clinical Science
Division of Ophthalmology
Karolinska Institute
Huddinge University Hospital
141 86 Huddinge, Sweden

Han Ying
Department of Clinical Science
Division of Ophthalmology
Karolinska Institute
Huddinge Hospital
141 86 Huddinge, Sweden

Accommodation and Vergence Mechanisms
in the Visual System
ed. by O. Franzén, H. Richter and L. Stark
© 2000 Birkhäuser Verlag Basel/Switzerland

Neuronal circuits for accommodation and vergence in the primate

L. E. Mays and P. D. R. Gamlin

Department of Physiological Optics and the Vision Science Research Center, University of Alabama at Birmingham, Birmingham AL 35294 USA

Summary: Single-unit recording studies of the activity of the motor and premotor neurons controlling vergence and accommodation in primates have provided significant insights into the putative accommodation and vergence controllers and the mechanisms by which they interact. Studies of midbrain near-response neurons in general, and of those that project to medial rectus motoneurons in particular, have revealed that these cells receive input both from the blur-accommodation and the disparity-vergence controller and that their activity is generally related to both vergence and to accommodation. These results indicated that, at least at the level of the midbrain, near-response neurons are not confined to independent channels for accommodation and for vergence control. These results are explained by a modification to the dual-interaction model in which each midbrain cell receives inputs from both the blur-accommodation and the disparity-vergence controllers but these inputs vary in relative strength. Additional studies have revealed that neural signals on medial rectus motoneurons and Edinger-Westphal neurons are related to both the static and dynamic components of vergence and accommodation. The existence of these signals suggests that, similar to the saccadic control system, the brain may control the velocity of these movements and integrate the velocity signal to obtain a position signal. This theory is provided with further support by the observation that while some premotor midbrain near-response neurons are predominately related to vergence angle and to accommodation, others are much more closely related to the velocity of these movements.

Introduction

An analysis of the signals found on neurons within the brain stem circuits that control accommodation and vergence is critical for understanding the way in which these behaviors function and interact. The current study focuses on the output neurons for accommodation and vergence control in alert, trained monkeys, and the premotor circuits which are presumed to drive them. The activity patterns of horizontally-acting extraocular motoneurons and identified Edinger-Westphal preganglionic neurons were recorded during accommodation and vergence eye movements. In addition, midbrain near response neurons have been identified which presumably provide these output neurons with the signals they require. An investigation of these signals, and the transformations they undergo, provides new insights into the putative accommodation and vergence controllers, and the mechanism by which accommodation and vergence are cross-coupled.

Materials and Methods

These studies were conducted on juvenile rhesus monkeys (*Macaca mulatta*) which had undergone surgical procedures for the implantation of a head holder, two eye coils, and craniotomies with the attachment of two recording chambers. All surgical procedures were carried out under general anesthesia and were followed by the use of appropriate analgesics. The animals were cared for in accordance with the National Institutes of Health *Guide for the Care and Use of Animals*, and the guidelines of the Institutional Animal Care and Use Committee of the University. The monkeys were trained to look at small visual targets for a fruit juice reward. The visual display incorporated moveable lenses which allowed the accommodative demand of the targets (light vergence) to be varied independently of binocular disparity. The horizontal and vertical positions of both eyes were measured using the search coil technique, and the accommodative state of one eye was measured using a recording infrared optometer. Single-unit recordings were made from brain stem sites using parylene-coated metal microelectrodes.

Results

Midbrain near response (NR) cells were recorded in an area just dorsal or lateral to the oculomotor nucleus. Many NR cells are located within 2 mm of this nucleus in the mesencephalic reticular formation. The activity of most NR cells increases for near viewing, but is unchanged during either vertical or horizontal conjugate eye movements. Earlier studies (Mays, 1984; Judge and Cumming, 1986) have shown that the behavior of most NR cells is unaffected when either the accommodative or disparity feedback loop is opened. This implies that these cells are driven by both the putative blur-accommodation and disparity-vergence controllers.

The activity of a midbrain NR cell during symmetrical convergence is shown in Fig. 1. Control trials (not shown) demonstrated that there was no change in activity for conjugate eye movements. Since there was an increase in both accommodation and vergence, it is impossible to determine whether the activity of this neuron is more closely related to accommodation, convergence, or to some combination of these. The relationship of a cell's activity to accommodation or vergence can only be determined by dissociating, at least partially,

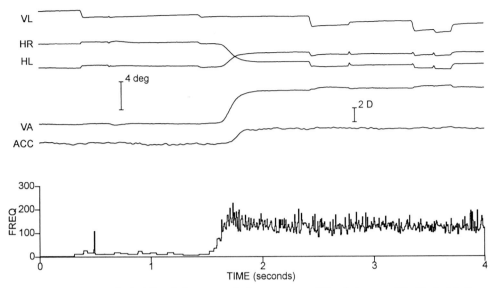

Figure 1. Activity of midbrain NR cell for symmetrical convergence. Abbreviations are VL, Vertical Left eye position, HR and HL, Horizontal Left and Right eye position; VA, Vergence Angle; ACC, Accommodation. Neural activity is represented as instantaneous firing rate.

accommodation from vergence. The results of dissociation experiments are best interpreted in the context of the "Dual-Interaction" model of accommodation and vergence, described by Semmlow and Heerema (1979) and Schor (1979). According to this model (shown in Fig. 2), accommodation and disparity vergence are controlled by their own feedback controllers and by cross-links between the controllers. The cross-links are responsible for accommodative convergence (AC) when the disparity feedback loop is open (lower switch in Fig. 2), and for convergence accommodation (CA) when the blur feedback loop is open (upper switch in Fig. 2). Consider the effects of partially dissociating accommodation from convergence by reducing the accommodative demand below that needed for a given change in viewing distance. In this case, with the loops closed, the disparity controller would be required to produce a greater than normal output to compensate for the lower AC input, and the blur-accommodation controller would produce a lower output to counter the increased CA input. What effect should this manipulation have on the activity of NR cells? If NR cells merely reflect the output of the two controllers, one should see two categories of NR cells emerge: one class, associated with the accommodative output, will have activity associated with the accommodation level, while the other, vergence-related cells, will have activity associated with vergence, not accommodation.

4

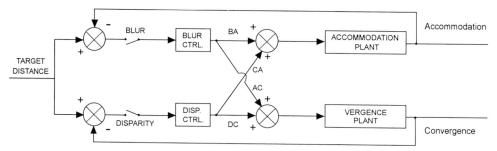

Figure 2. Dual-Interaction Model of accommodation and vergence. Abbreviations are BA, Blur-driven Accommodation; DC, Disparity-driven Convergence; CA, Convergence Accommodation; AC, Accommodative Convergence.

This dissociation experiment has been done for NR cells by Judge and Cumming (1986) and by Zhang, Mays, and Gamlin (1992). While it has been possible to find "vergence" and "accommodation" NR cells, many, if not most NR cells, do not fall readily into either classification. This is illustrated by the activity of a NR cell shown in Fig. 3. Under normal viewing conditions (Normal Accommodation), this cell increases its firing rate with decreased viewing distance (increased vergence angle). When accommodation was restricted (Near Zero Accommodation), the firing rate is still related to vergence angle, but is even higher. Other NR cells displayed a decrease in activity with reduced accommodation. For most of these cells, the firing rate was not strictly associated with vergence angle or accommodation alone.

The most parsimonious explanation of these effects is provided by a modification of the Dual-Interaction model, shown in Fig. 4. In this version of the model, NR cells (labeled NR_1 ... NR_6) receive both direct and cross-link inputs, but the relative strengths of these inputs vary as indicated by the gain blocks (G_1 ... G_6). In this scheme, gains are such that $G_3 > G_2 > G_1$ and $G_4 > G_5 > G_6$. The cell in Fig. 3 could be represented by NR_1, a cell which receives a relatively weak cross-link input from the blur-driven accommodation controller.

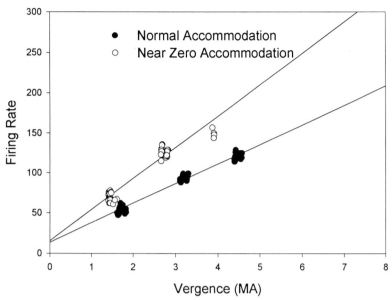

Figure 3. Firing rate of a NR cell as a function of vergence angle (in Meter-Angles) for convergence with and without accommodation.

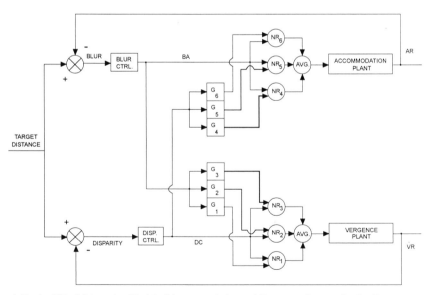

Figure 4. Revised Dual-Interaction Model of Accommodation and Vergence. See text for details.

6

Although it could theoretically also occupy position NR_4 in Fig. 4, this cell was antidromically activated from the medial rectus subdivision of the oculomotor nucleus, and so is more likely to be in the vergence output path.

In addition to being involved in accommodation and vergence interactions, NR cells have other characteristics which may clarify the way the accommodation and vergence controllers function. A substantial number of NR cells have activity patterns which match vergence velocity profiles when gaze is shifted between targets at different distances (Mays, Porter, Gamlin and Tello, 1986). Figure 5 shows an example of a NR cell which bursts for convergence. The profile of the burst matches the vergence velocity (VV) trace for this and other trials, including ramp and sine-wave tracking trials. It is likely that the NR burst cell activity is related to accommodation as well as vergence, as in the case of tonic NR cells. In addition to NR cells with predominantly tonic (Fig. 1) and burst (Fig. 5) firing patterns, many burst-tonic NR cells have been observed, suggesting that the burst and tonic signals properties are continuously distributed.

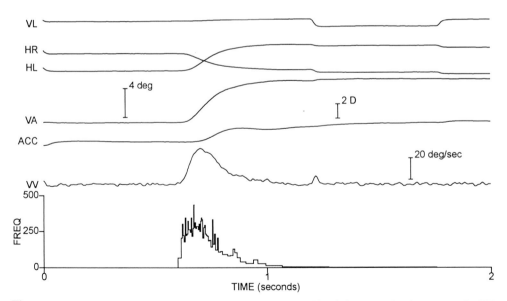

Figure 5. Activity of a NR burst neuron for symmetrical convergence. Abbreviations are as in Fig. 1, except for VV, which is the vergence velocity.

The activity pattern of putative medial rectus motoneurons appears to have much in common with NR cells. This should not be surprising, since it has been possible to antidromically identify NR cells projecting to the medial rectus subdivisions. A convergence velocity signal, in addition to a convergence signal, has been seen on nearly all medial rectus motoneurons (Gamlin and Mays, 1992), suggesting that the NR cells may be the source of both the vergence position and velocity signals. Thus far, there have been no systematic attempts to record from medial rectus motoneurons during accommodation-vergence dissociation, but it is likely that some accommodation-related signals might be found on these extraocular motoneurons.

The behavior of antidromically-identified Edinger-Westphal (E-W) preganglionic neurons has been studied as well (Gamlin,. Zhang, Clendaniel and Mays, 1994). Many of these neurons in the accommodation output path have a velocity as well as a position signal. Under conditions of accommodation-vergence dissociation, some identified E-W neurons also show some evidence of a vergence-related signal. A major difference among E-W neurons, medial rectus motoneurons, and NR cells is the overall gain of the vergence and accommodation signal. This is shown in Fig. 6. Overall, E-W cells have the lowest gains, medial rectus motoneurons have intermediate gains, and NR neurons have the highest gains (measured in equivalent units, meter-angles or diopters). The observation that some NR cells have low gains (<10) may be due to the incorrect identification of some E-W cells as NR cells. This could occur because many NR cells were not tested to determine if they could be activated by third nerve stimulation.

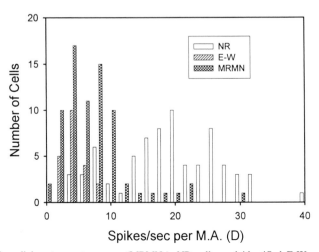

Figure 6. Gains of medial rectus motoneurons (MRMNs), NR cells, and identified E-W neurons, measured in equivalent units (Meter-Angles or Diopters).

8

Discussion

The results of these, and other related studies, strongly suggest that a revised version of the Dual-Interaction model of accommodation and vergence (such as that presented in Fig. 4) can account for most of the tonic firing seen in midbrain NR cells, and implies that these neurons are important for accommodation-vergence interactions. It appears that NR cells get inputs from both the blur-driven accommodation controller and the disparity vergence controller. The balance of these inputs, over the entire population of NR cells, determines the relative strengths of the cross-links (i.e., the AC/A and CA/C ratios). Indeed, residuals of these signals have appeared on some identified E-W accommodation neurons, but medial rectus motoneurons have not been so examined. Results of antidromic activation studies (Zhang, Gamlin, and Mays, 1991; Zhang, Mays, and Gamlin, 1992) strongly imply that NR neurons provide medial rectus motoneurons with the vergence signals they require. Identified E-W neurons have not been tested in this way. Current evidence suggests that accommodation and vergence are highly intertwined at the premotor level, and that is may be very difficult to unravel these subsystems by looking at the outputs alone.

Interestingly, velocity signals, as well as position signals, appear on E-W neurons as well as NR cells and medial rectus motoneurons, suggesting that NR cells may provide E-W neurons with both velocity and position signals, even though velocity signals would appear to be unnecessary for accommodation control. The prevalence of velocity signals suggests that vergence and accommodation systems may rely on a velocity control mechanism, not unlike that employed by the saccadic eye movement system (Robinson, 1975).

Acknowledgments
These studies were supported by National Eye Institute Grants EY03463, EY07558, and CORE Grant EY03039, and by the McKnight Fund for Neuroscience Research.

References

Gamlin, P. D. R. and Mays, L. E. (1992) Dynamic properties of medial rectus motoneurons during vergence eye movements. *Journal of Neurophysiology* 67: 64-74.

Gamlin, P. D. R., Zhang, Y., Clendaniel, R. A. and Mays, L. E. (1994) Behavior of identified Edinger-Westphal neurons during ocular accommodation.. 72: 2368-2382.

Judge, S. J. and Cumming, B. G. (1986) Neurons in the monkey midbrain with activity related to vergence eye movement and accommodation. *Journal of Neurophysiology* 55: 915-930.

Mays, L. E. (1984) Neural control of vergence eye movements: Convergence and divergence neurons in midbrain. *Journal of Neurophysiology* 51: 1091-1108.

Mays, L. E., Porter, J. D., Gamlin, P. D. R. and Tello, C. A. (1986) Neural control of vergence eye movements: Neurons encoding vergence velocity. *Journal of Neurophysiology* 56: 1007-1021.

Robinson, D. A. (1975) Oculomotor control signals. *In:* Bach-y-Rita, P., Lennerstrand, G. (ed.): Basic mechanisms of ocular motility and their clinical implications. Pergamon Press, Oxford, pp. 337-374.

Schor, C. M. (1979) The relationship between fusional vergence eye movements and fixation disparity. *Vision Research* 19: 1359-1367.

Semmlow, J. J. and Heerema, D. (1979) The synkinetic interaction of convergence accommodation and accommodative convergence. *Vision Research* 19: 1237-1242.

Zhang, Y., Gamlin, P. D. R. and Mays, L. E. (1991) Antidromic identification of midbrain near response cells projecting to the oculomotor nucleus. *Experimental. Brain Research* 84: 525-528.

Zhang, Y., Mays, L. E. and Gamlin, P. D. R. (1992) Characteristics of near response cells projecting to the oculomotor nucleus. *Journal of Neurophysiology* 67: 944-960.

Accommodation and Vergence Mechanisms
in the Visual System
ed. by O. Franzén, H. Richter and L. Stark
© 2000 Birkhäuser Verlag Basel/Switzerland

Neural codes for three-dimensional space

J.W. Gnadt

Dept. Neurobiology and Behavior, State University of New York, New York 11794, USA

Summary. Several lines of evidence suggest that cortical areas in posterior parietal cortex of primates are involved in visual-spatial processing and in pre-motor planning and execution of eye movements. Most experimental investigations in parietal cortex have focused on issues related to two-dimensional visual function and on conjugate eye movements. However, there are several indications that posterior parietal cortex may operate in a three-dimensional coordinate system. In monkeys, we have investigated neurons in a specialized region of parietal cortex, sometimes referred to as the "parietal eye fields" or as area LIP due to its location on the lateral bank of the intraparietal sulcus. The vast majority of neurons in area LIP have eccentric response fields, most of which are tuned in three-dimensional space. There are also a large number of cells with tonic activity proportional to gaze position in three-dimensional space. We have found a small minority of neurons with activity related directly to sensory or motor parameters of the oculomotor near response. These included responses to binocular disparity, depth-related blur, vergence velocity and accommodation velocity; each of which was independent of retinal eccentricity or conjugate eye position.

Background

In considering possible sources of pre-motor cortical signals for making eye movements in depth, two candidate areas in primates are the frontal eye fields (FEF) - the prefrontal motor area representing eye movements - and the parietal eye fields - a sensorimotor association area in the posterior parietal cortex. The common nomenclature of this posterior eye field in monkeys is area LIP due to its location on the lateral bank of the intraparietal sulcus.

Little is currently known about the possible involvement of the FEF in the control of eye movements in depth, though preliminary experiments (Gamlin and Yoon, 1995) indicate that neurons in prefrontal cortex may indeed express activity related to depth. Regarding the parietal eye fields, there is considerable evidence that posterior parietal cortex in primates is involved in visual-spatial processing and in the pre-motor planning and execution of eye movements. Some of the first evidence came from studies of human brain injury early in the 20th century (Balint, 1909; Holmes, 1918; Holmes and Horrax, 1919), where it was noted that injuries to parietal cortex could cause deficits in visual-spatial perception and in guiding voluntary eye movements. Later, experimental lesions made in non-human primates have confirmed and extended these observations (Lynch and Mclaren, 1989; Lynch, 1992). Additional evidence for a role of LIP in oculomotor function comes from microstimulation studies in monkeys (Shibutani et al., 1984; Kurylo and Skavenski, 1991), which showed that conjugate saccades are elicited by microstimulation in the intraparietal sulcus.

Anatomically, the connectional hallmark of area LIP is its position as a recipient of visual inputs from the dorsal ("spatial") visual stream of cortical processing in primates, and its output projections to cortical and subcortical centers directly linked to eye movements (Fries 1984; Lynch et al. 1985; Andersen et al., 1985; Asanuma et al. 1985; Cavada and Goldman–Rakic, 1989; Blatt et al., 1990; Cavada and Goldman–Rakic, 1991; Baizer et al. 1993).

Most investigations of visual spatial function in parietal cortex have focused on issues related to two-dimensional visual function and on conjugate eye movements. However, there are several indications that posterior parietal cortex may operate in a three-dimensional coordinate system. For example, there are numerous reports of patients with parietal lesions having documented perceptual or behavioral deficits in depth (Holmes, 1918; Holmes and Horrax, 1919; Brain, 1941; Paterson and Zangwell, 1944; Cogan and Adams, 1953; Cogan, 1965; Godwin-Austin, 1965; Manor et al., 1988; Ohtsuka et al., 1988; Fowler et al., 1989). Furthermore, we have demonstrated in physiological studies of monkeys that many neurons in area LIP encode spatial parameters in depth (Gnadt, 1992; Gnadt and Mays, 1995).

Methods

A comprehensive description of many of our research methods and our three-dimensional visual display apparatus can be found in Gnadt and Mays (1995). Briefly, Rhesus monkeys (*Macaca mulatta*) were trained to fixate and follow small visual targets while isolated neurons were recorded using standard recording techniques from chronically-prepared, behaving subjects. Stimulus presentation and control of the behavioral tasks were maintained under computer control. The positions of both eyes were sampled at 500 times/sec using the scleral search coil technique and the monkeys received liquid reward for correctly completing the desired eye movement tasks. Trans-dural, single unit extracellular recordings were made using parylene-insulated tungsten microelectrodes. Visual stimuli were presented using a three-dimensional display system that can present visual stimuli at any position within ± 15 deg from straight ahead, from infinitely distant to within 10 cm.

Results

To date, we have tested many neurons in the intraparietal sulcus and superior temporal sulcus in several monkeys for responses related to depth. From these studies, I divide the neurons into three general categories of neurons with responses related to depth. The first category, which accounts for the vast majority of area LIP cells, have eccentric response fields related to the spatial location of eye movements targets. Initial experiments described these cells with respect to saccadic eye movements (Gnadt and Andersen, 1988). More recently, we tested similar neurons at different depths relative to the fixation plane. While holding the location of the target constant at the center of the frontoparallel response field, the neurons were systematically tested during eye movements to target positions proximal (i.e., near) to the plane of fixation, at the plane of fixation, and distal (i.e., far) to the plane of fixation. By necessity, the movements to these targets required a combination of version and vergence movements. Initial tests presented the stimulus with appropriately matched binocular disparity and depth-related blur. Later control tasks presented binocular disparity or monocular blur independently. Interestingly, we found that roughly 3/4 (72%) of the neurons had broad tuning curves as a function of target depth. From a variety of factors, I suggest that these neurons provide a volitional signal for oculomotor control based on a perceptual representation of three-dimensional space (Gnadt and Mays, 1995).

Additionally, we have found a second category of cells for which we can not find a measurable eccentric response field, but which express a tonic rate of activity proportional to current eye position (gaze direction). Originally, these cells were characterized for their activity related to conjugate eye position (Andersen et al., 1990). However, I have now tested this type of neuron for sensitivity to vergence position as well. Many of the cells modulate their tonic firing rate for binocular gaze direction. Fig. 1 provides an example of one cell that changes its firing rate for both version and vergence components of current gaze position. The sensitivity to conjugate position at two different values of vergence is illustrated in Figs. 1A and 1B as the height of the data points up the vertical spikes for data samples taken during visual fixation. The neuron had its greatest activity for gaze positions up and to the right. Fig. 1C shows the neuron's sensitivity to vergence during fixation at two different conjugate positions. Note that the firing rate increased when the subject gazed at converged positions. Finally, Fig. 1 D shows the neuron's activity during dynamic visual tracking of a target between the neuron's least and most preferred positions in three dimensions. Note that this cell's activity is in phase with eye position with a 200 ms delay.

All cells recorded to date that express a firing rate proportional to vergence position also responded to conjugate eye position. Since their activity confounds version and vergence eye position, these cells are not uniquely related to the oculomotor near response. Furthermore, evidence from Gamlin and Mays (1992) suggests that direct pre-motor inputs to vergence centers in the brainstem should receive velocity signals, not position. In addition, the change in the neuron's activity follows the eye movements. Because they do not lead the movements, they are not consistent with a pre-motor command signal. My interpretation of these data is that the cells express an extra-retinal signal of gaze position in three-dimensional space.

At a minimum, neurons directly related to the near response should express sensory signals or motor activity correlated with some component of the near response (Schor, 1985). Specifically, we have looked for activity related to the sensory inputs (binocular disparity or accommodative demand) or to the motor behavior (vergence or accommodation). We have not considered the third component of the near triad, pupillary constriction, in our studies. Furthermore, the depth-related responses should be independent of retinal eccentricity or conjugate eye position. Only a minority of parietal neurons have met those

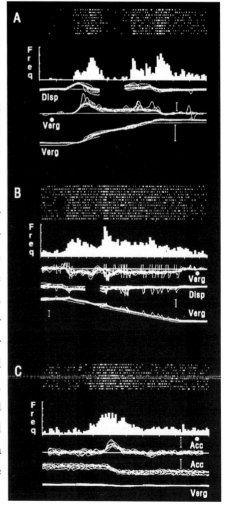

criteria. These neurons make up the third category of area LIP cells related to depth. However, it is likely that they are not a homogeneous population.

Over thirty neurons have been recorded that have some activity during visual tracking in depth, independent of conjugate eye position (Gnadt, 1992). Some neurons demonstrate a sensitivity to simple binocular disparity. One example is illustrated in Fig. 2A for a cell that was active during convergence tracking of a visual target. The neuron was not active during conjugate tracking of the visual target by saccades or by smooth pursuit (not shown). The depend-ence of the cell's activity on visual stimulation was revealed by the cessation of activity during convergence tracking when the visual target was briefly extinguished., even as the

vergence continued at constant velocity in the dark (Fig. 2A). Using visual probe trials, the disparity appeared to have a small (< 1.0 deg.) foveal receptive field with sensitivity to small (< 1.0 deg.) crossed disparity. During tracking, the neuron was stimulated due to the slight lag of the vergence response to the approaching stimulus. The latency of the response to the stimulus was 90 ms.

Perhaps more interesting are those neurons related to motor parameters, such as the one presented in Fig. 2B. Note that it continues to respond during the blank out of the visual stimulus and that the activity profile is similar to the profile of divergence velocity. The activity leads the movement velocity by less than 40 ms. We have found that for many of these neurons the activity remains constant when the accommodative stimulus (depth-related blur) is held constant during the visually guided vergence. This suggests that the activity is

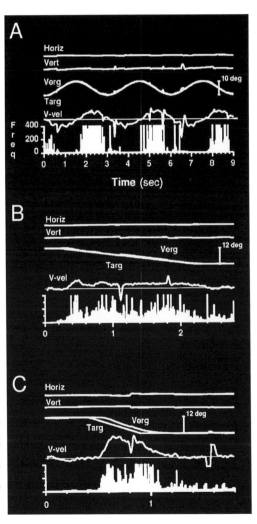

primarily related to the vergence component of the near response.

In contrast, we have found two cells in one subject that was related to an accommodation response, independent of vergence. This is illustrated in Fig. 2C where the subject made an accommodation response to a binocular presentation of depth-related blur while holding the vergence demand constant. This neuron's activity lead the velocity response by more than 160 ms.

In all cases, the neurons' firing rate was proportional to the movement velocity. This is shown in Fig. 3A for a neuron with activity in phase with divergence velocity during sinusoidal tracking. Figs. 3B and 3C show the response during two constant velocity trials of visually guided vergence. A linear fit of the firing rate to vergence velocity found the relation to be 8.1 spikes/sec per deg/sec with a linear constant of 79 spikes/sec (r = 0.63). The relationship was non-linear below 2.5 deg./sec, with a rapid decrease towards 0 spikes/sec.

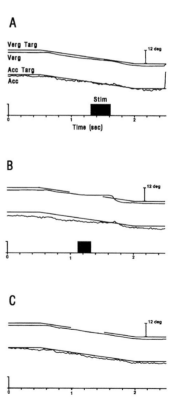

These vergence and accommodation velocity responses are present for both accommo-dative vergence and vergence accommo-dation. This suggests that they occur after the cross-linking between the two subsystems. However, there is evidence that the cross-linking occurs, at least in part, in the brainstem (Zhang et al., 1992). Thus, it is not clear how these cortical signals might relate to these lower structures. Further-more, many of the neurons do not have a substantial phase lead to movement velocity (<40 ms.) and may represent feedback or intracortical loops, not the primary driving signals for the brainstem circuits.

To further investigate the involvement of area LIP in the oculomotor near response, I have tested the effects of both stimulation and inactivation in the lateral bank of the intraparietal sulcus. Even when applied at sites with depth related activity, I was unable to produce near response movements by microstimulation. However, I did find that stimulation could disrupt ongoing vergence and accommodation (Fig. 4). The effect was subtle if applied during tracking in depth for a visible stimulus (Fig. 4A), but the stimulation completely and reliably arrested the movements if applied during a brief blanking of the stimulus (Fig. 4B). However, this disruption effect was not specific to the near response. When the stimulation was applied during tracking of a target that simultaneously required vergence and conjugate smooth pursuit (tracking a target along the line of sight of one eye), both the vergence and the version components were disrupted. These results seem to indicate that microstimulation in area LIP is not sufficient to produce specific near response behavior, though stimulation here certainly does effect oculomotor behavior at some level.

Finally, I tested the effect of reversible inactivation of LIP on the near response using multiple injections of large volumes (2 ul) of lidocaine along the intraparietal sulcus. I could find no measurable effect.

Summary

In summary, we have shown that the majority of area LIP cells have eccentric response fields that are tuned in three-dimensional space. Because the tuning in depth was not uniquely related to binocular disparity, depth-related blur or simple combination of two monocular response fields, we suggested that the tuning was due to a perceptually-based representation of depth. An unresolved question at this time is to determine how these 3-D signals in LIP are distributed and transformed to its target structures. Area LIP neurons seem to lay in the middle of the sensorimotor translation, after the brain has extracted depth from the visual centers of the brain and before the dynamic motor signals that drive the movements' are expressed.

Additionally, a large number of area LIP neurons have a tonic firing rate proportional to binocular eye position. The form of the activity is consistent with being a spatial signal of the Cyclopean position of gaze in three-dimensional space. We have known since the time of Helmholtz early in this century (1919) that the visual system must account for the current gaze position to maintain a perceptually stable, egocentric frame of reference. This activity appears to be a neural correlate of that process.

Finally, I have provided physiological data demonstrating neural activity specifically related to the near response in area LIP. These neurons are functionally different from those cell types described above in that they are related to the vergence and/or accommodation velocity and have activity independent of conjugate eye position or target eccentricity. These neurons, however, are a minority population among the other cell types described above. Moreover, evidence from microstimulation and reversible inactivation suggests that area LIP is neither necessary nor sufficient for the near response. However, the fact that area LIP appears to be neither absolutely sufficient nor necessary for the oculomotor near response does not argue that it is not involved. Indeed, there is considerable evidence from lesions in humans and from neurophysiological studies in monkeys that this parietal visual-spatial area does contribute to voluntary control of eye movements, including responses to visual targets in depth.

Thus, there is substantial evidence that neurons in area LIP provide a perceptual representation of three-dimensional space for the voluntary control of oculomotor behavior. These various multi-dimensional signals must be decomposed into appropriate signals for generating the specific submodalities of eye movements. Other experiments are now in progress to pursue this hypothesis.

Acknowledgments
The author is grateful to Ms. Janine Beyer for her valuable technical expertise. This research was supported by NIH grant EY08217, a Sloan Foundation Research Fellowship and a Dean's Research Grant (Medicine, SUNY at Stony Brook).

References

Andersen R.A., Asanuma C. and Cowen M. (1985) Callosal and prefrontal associational projecting cell populations in area 7a of the macaque monkey: a study using retrogradely transported fluorescent dyes *Journal of Comparative Neurology* 232: 443-455.

Andersen R.A., Bracewell R.M., Barash S., Gnadt J.W. and Fogassi L. (1990) Eye position effects on visual, memory, and saccade-related activity in areas LIP and 7a of macaque. *Journal of Neuroscience* 10(4):1176-96.

Asanuma C., Andersen R.A. and Cowan W.M. (1985) The thalamic relations of the caudal inferior parietal lobule and the lateral prefrontal cortex in monkeys: divergent cortical projections from cell clusters in the medial pulvinar nucleus. *Journal of Comparative Neurology* 241:357-381.

Baizer J.S., Desimone R. and Ungerleider L.G. (1993) Comparison of subcortical connections of inferior temporal and posterior parietal cortex in monkeys. *Visual Neuroscience* 10(1):59-72.

Balint R. (1909) Seelenlahmung des `schauens', optische ataxie, raumliche storung der aufmerksamkeit. Psychiatric Neurology 25:51-81.

Blatt G.J., Andersen R.A. and Stoner G.R. (1990) Visual receptive field organization and cortico-cortical connections of the lateral intraparietal area (area LIP) in the macaque. *Journal of Comparative Neurology* 299(4):421-45.

Brain W.R. (1941) Visual disorientation with special reference to lesions of the right cerebral hemisphere. *Brain.* 64: 244-272.

Cavada C. and Goldman-Rakic P.S. (1989) Posterior parietal cortex in rhesus monkey: I. Parcellation of areas based on distinctive limbic and sensory corticocortical connections. *Journal of Comparative Neurology* 287(4):393-421.

Cogan, D.G. (1965) Ophthalmic manifestations of bilateral non-occipital cerebral lesions. *British Journal of Ophthalmology* 49: 281-297.

Cogan, D.G. and Adams, R.D. (1953) A type of paralysis of conjugate gaze (ocular motor apraxia). A.M.A Archives of Opthalmology 50: 434-442.

Fowler S., Munro N., Richardson A., and Stein J. (1989) Vergence control in patients with lesions of the posterior parietal cortex. *Journal of Physiology* 417: 92P.

Fries W. (1984) Cortical projections to the superior colliculus in the macaque monkey: a retrograde study using horseradish peroxidase. *Journal of Comparative Neurology* 230:55-76.

Gamlin P.D.R. and Mays L.E. (1992) Dynamic properties of medial rectus motoneurons during vergence eye movements. *Journal of Neurophysiology* 67: 64-74.

Gamlin P.D.R. and Yoon K. (1995) Single-unit activity relate to near response in area 8 of the primate frontal cortex. **Soc. Neurosci. Abstr.**, 1918.

Gnadt J.W. (1992) Area LIP: Three-dimensional space and visual to oculomotor transformation. Behavioural Brain Research 15: 745-746.

Gnadt J.W. and Andersen R.A. (1988) Memory related motor planning activity in posterior parietal cortex of macaque. *Experimental Brain Research* 70(1):216-20.

Gnadt J.W. and Mays L.E. (1995) Neurons in monkey parietal area LIP are tuned for eye-movement parameters in three-dimensional space. *Journal of Neurophysiology* 73(1):280-297.

Godwin-Austin R.B. (1965) A case of visual disorientation. *Journal of Neurology Neurosurgery and Psychiatry* 28: 453-458.

Holmes G. (1918) Disturbances of visual orientation. *British Journal of Ophtalmology* 2:506-516.

Holmes G. and Horrax G. (1919) Disturbances of spatial orientation and visual attention with loss of stereoscopic vision. Archives of Neurology and Psychiatry 1:385-407.

Kurylo D.D. and Skavenski A.A. (1991) Eye movements elicited by electrical stimulation of area PG in the monkey. Journal of Neurophysiology 65(6):1243-53.

Lynch J.C. (1992) Saccade initiation and latency deficits after combined lesions of the frontal and posterior eye fields in monkey. *Journal of Neurophysiology* 68:1913-1916.

Lynch J.C., Graybiel A.M. and Lobeck L.J. (1985) The differential projection of two cytoarchitectonic subregions of the inferior parietal lobule of macaque upon the deep layers of the superior colliculus. *Journal of Comparative Neurology* 235(2):241-54.

Lynch J.C. and McLaren J.W. (1980) Deficits of visual attention and saccadic eye movements after lesions of parietooccipital cortex in monkeys. *Journal of Comparative Neurology* 61:74-90.

Manor R.S., Heilbronn Y.D., Sherf I., and Ben-Sira I. (1988) Loss of accommodation produced by peristriate lesion in man? *Journal of Clinical Neurology-Ophthalmology* 8(1): 19-23.

Ohtsuka K., Maekawa H., Takeda M., Uede N., and Chiba S. (1988) Accommodation and convergence insufficiency with left middle cerebral artery occlusion. *American Journal of Ophthalmology.* 106: 60-64.

Paterson A. and Zangwell O.L. (1944) Disorders of visual space perception associated with lesions of the right cerebral hemisphere. *Brain.* 67:331-358.

Schor C.M. (1985) Models of mutual interactions between accommodation and convergence. *American Journal of Optometry & Physiological Optics* 62: 369-374.

Shibutani H., Sakata H. and Hyvarinen J. (1984) Saccade and blinking evoked by microstimulation of the posterior parietal association cortex of the monkey. *Experimental Brain Research* 55:1-8.

Accommodation and Vergence Mechanisms
in the Visual System
ed. by O. Franzén, H. Richter and L. Stark

Neuronal connectivity between accommodative and active fixation systems

K. Ohtsuka and A. Sato

Sapporo Medical University, Dept. of Ophthalmology, S-1, W-16, Chuo-ku, Sapporo, 060, Hokkaido, Japan

Summary: The superior colliculus (SC) has long been recognized as an important structure in the generation of saccadic eye movements. The SC of the cat and the monkey has recently been shown to be involved in the control of visual fixation. A subpopulation of collicular cells exhibits tonic discharge when the animal fixates on a target of interest. These cells are located in the rostral SC where the central visual field is represented. Active fixation is thought to be important for the ocular near response; accommodation, vergence and pupillo-constriction, and these systems are functionally linked. Therefore, it is possible that the rostral SC is also involved in the control of accommodation or vergence. The results of several recent studies have suggested that the rostral SC is also involved in the control of accommodation. The accommodation-related area in the rostral SC also corresponds to the area of representation of the central visual field. The accommodation-related area in the SC receives heavy projections from the accommodation area in the lateral suprasylvian (LS) area of the cat cortex. It is well known that the LS area is also involved in the control of vergence and pupillo-constriction. The rostral SC projects to the pretectal nuclei (PT) where accommodative responses are evoked by microstimulation, the raphe interpositus (RIP), in which omnipause neurons are located, and the dorsomedial portion of the nucleus reticularis tegmenti pontis. Many neurons in the intermediate layers of the rostral SC have divergent axon collaterals to the PT and the RIP. The rostral SC is likely a key structure involved in the near response and visual fixation.

Introduction

Accommodation and visual fixation systems are assumed to be functionally linked, allowing maintenance of clear images on the fovea. The accommodation system provides a clear image of a target of interest on the fovea, and the visual fixation system maintains the image on the fovea by inhibiting saccadic intrusions. However, the neural substrate for this linkage is not known. The results of recent neurophysiological and neuroanatomical studies suggest that the rostral superior colliculus (SC) where the central visual field is represented is involved in both systems in the cat and the monkey (Munoz and Guitton, 1989; Munoz and Guitton, 1991; Munoz, et al., 1991; Munoz and Wurtz, 1993; Maekawa and Ohtsuka, 1993; Sawa and Ohtsuka, 1994).

The SC has long been recognized as an important structure in the generation of saccadic eye movements (Sparks, 1986). Neurons in the intermediate layers of the SC exhibit burst discharge in relation to saccades. These neurons are organized into a retinotopically coded motor map with amplitude and direction of a saccade determined by the spatial distribution of neural activity on this map (Lee, et al., 1988; Sparks, et al., 1976). Saccades evoked by microstimulation of the caudal SC have larger amplitude, and the amplitude can be decreased by shifting the stimulating electrode rostrally (Robinson, 1972). The rostral SC represents the central visual field on the motor map, and a subpopulation of neurons in this area exhibits tonic discharge when the animal looks at the fixation target and pause in the rate of tonic discharge during saccades (Munoz and Guitton, 1989; Munoz and Wurtz, 1993). Microstimulation of the rostral SC delivered during a saccade produces an interruption of the saccade in the monkey (Munoz and Wurtz, 1993) and the cat (Pare and Guitton, 1994). Microstimulation of the rostral SC of the cat also elicits accommodative responses with low currents < 20 μA (Sawa and Ohtsuka, 1994). Both accommodation- and visual fixation-related areas are located in the rostral SC where the central visual field is represented (Munoz and Wurtz, 1993; Sawa and Ohtsuka, 1994; Ohtsuka and Sato, 1996b).

Accommodation-Related Area in the Rostral SC

The results of previous studies have suggested that the lateral suprasylvian (LS) area, the cortical area surrounding the middle suprasylvian sulcus (Mss) of the cat, is involved in the control of accommodation (Bando, et al., 1981; Bando, et al., 1984b; Sawa, et al., 1992). The LS area receives visual inputs. Some neurons in this area respond to changes in ocular disparity and target size and to motion in depth, which are important visual cues for accommodation (Toyama and Kozasa, 1982; Toyama, et al., 1986a,b). Some LS neurons also exhibit burst discharges preceding the onset of spontaneous accommodation (Bando, et al., 1984b). It is likely that these neurons have an important role in the control of accommodation. Microstimulation of the LS area evokes accommodative responses (Bando, et al., 1981; Sawa, et al., 1992). Results of systematic microstimulation of the LS area in the cat showed that low-threshold areas for evoking accommodation are located in the lower parts of the medial banks of the Mss from A1 to A4, and at A8 in the stereotaxic coordinates (Sawa, et al., 1992). The latency of accommodative responses evoked by stimulation of the caudal area (A1-A4) is shorter than that of accommodative responses evoked by stimulation of the rostral area (A8). It is likely that the caudal accommodation area provides output motor signals for the subcortical system to control accommodation.

The LS area projects to many areas, such as other cortical areas, the thalamus, the pulvinar, the striatum, the pretectum, the SC and the pontine nuclei (Kawamura, et al., 1974; Norita, et al., 1991; Segal and Beckstead, 1984; Updyke, 1981). Bando et al. (1984b) reported that about 70% of accommodation-related neurons in the LS area were antidromically activated by stimulation via electrodes placed in the pretectum and/or the SC with average latencies of 2.4-2.5 ms. Maekawa and Ohtsuka (1993) reported that following injections of WGA-HRP into the accommodation-related area in the caudal LS area of the cat, dense labeling of axon terminals was observed in the rostral SC where the central visual field is represented, while labeled terminals in the pretectal area were not clearly evident. Systematic mapping with microstimulation revealed that accommodative responses could be evoked with weak currents < 20 µA from a circumscribed area in the superficial and intermediate layers of the rostral SC (Sawa and Ohtsuka, 1994). Following injection of WGA-HRP into the area in the rostral SC where accommodative responses were elicited with low currents < 20µA, retrogradely labeled ganglion cells were observed mainly in the area centralis of the retinae (Ohtsuka and Sato, 1996b). The low-threshold area for evoking lens accommodation in the rostral SC was almost compatible with the terminal area from the accommodation-related area in the LS area (Sawa and Ohtsuka, 1994).

In addition, following injections of WGA-HRP into the accommodation-related area in the rostral SC, retrogradely labeled cells were observed in the lower part of the medial bank of the Mss, which corresponded to the accommodation-related area in the LS area of the cat (Ohtsuka and Sato, 1996a). On the other hand, following injections of WGA-HRP into the pretectum, retrogradely labeled cells were seen in the upper part of the medial bank of the Mss, which did not correspond to the accommodation-related area in the LS area of the cat (Ohtsuka and Sato, 1996a). Therefore, neurons in the cortical accommodation area may project to the rostral SC, but not to the pretectal area. Accommodative responses evoked by stimulation of the LS area were abolished by the injection of muscimol (inhibitory neurotransmitter, GABA agonist) into the accommodation-related area in the rostral SC of the cat (Ohtsuka and Sato, 1996a). It is possible that descending signals from the cortex which are involved in the control of accommodation from the cortex finally converge upon the rostral SC in the brainstem. The rostral SC is assumed to be part of the final common pathway from the cortical accommodation area to the brainstem premotor circuit.

Visual Fixation Area in the Rostral SC

The premotor circuit for controlling saccadic eye movements includes neurons exhibiting a high-frequency burst of activity during saccades (burst neurons), which are crucial for the generation of saccades. The burst neurons are inhibited tonically by inhibitory neurons in the raphe interpositus (RIP) and the omnipause neurons (OPNs), which discharge tonically during fixation (Evinger, et al., 1982; Fuchs, et al., 1985). The saccade system is thought to be reciprocally related to the visual fixation system (saccade suppression system). For eliciting a saccade, the burst neurons are driven by a motor error signal while tonic discharge of the OPNs is turned off by an inhibitory trigger signal (Fuchs, et al. 1985). On the other hand, the saccade system is inhibited by the activity of OPNs during fixation for maintenance of stable foveal images. Neurons in the SC exhibit burst discharge in relation to saccades and project to the premotor area (Sparks, 1986). Therefore, the SC provides the motor error signal to generate saccades. The SC also projects to the OPNs (Buttner-Ennever et al. 1988; Raybourn and Keller, 1977; King, et al., 1980; Kaneko and Fuchs, 1982; Pare and Guitton, 1994). The direct projection from the SC to OPNs was found to be excitatory (Raybourn and Keller, 1977). Therefore, the SC must also facilitate the tonic activity of OPNs. These findings suggest that the SC has two different roles in the control of saccades, saccade generation and saccade suppression.

Recently, a subpopulation of collicular cells (the fixation cells) which exhibit tonic discharge similar to that of OPNs have been found in the area centralis representation in the cat (Munoz and Guitton, 1989; Peck, 1989) and in the foveal representation in the monkey (Munoz and Wurtz, 1993). The fixation cells exhibit tonic discharge when the animal looks at a fixation target and a pause in the rate of tonic discharge during either saccadic eye movements or saccadic gaze shifts. The pause in fixation cell discharge always begin before the onset of saccades. Pause duration is highly correlated with both ipsiversive and contraversive saccade duration (Munoz and Wurtz, 1993). The activity of these fixation cells contributes to active visual fixation through maintenance of the tonic activity of OPNs.

The visual fixation area in the rostral SC is part of a larger system for visual fixation within the central nervous system that interacts with the saccadic system. In cortical areas, visual fixation cells have been found in area 7a and the lateral intraparietal area (LIP) of the posterior parietal cortex (Lynch, et al., 1977; Mountcastle, et al., 1975; Sakata, et al., 1980; Ben Hamed, et al., 1995). Visual fixation cells in area 7a are sporadically active when the animal spontaneously looks around at its surroundings but increase their activity abruptly with fixation on a target to obtain a reward. They pause in their activity during saccades but not in all directions and many cells do not discharge during fixation at all orbital positions of the eye. A subset of LIP neurons activated by visual stimulation of the central visual field exhibit three different patterns of discharge (Ben Hamed, et al., 1995). Some neurons exhibiting the first pattern have sustained discharges during periods of fixation and are inhibited by saccades. The second pattern is a mirror image of the first pattern. Neurons exhibiting the second pattern only respond during saccades. Neurons exhibiting the third pattern respond immediately after saccades to memorized targets, until the reappearance of the saccade target. LIP neurons carry signals related to various stages of fixation engagement and disengagement and to the underlying attention process. In the frontal eye fields, one population of cells projecting to the SC responded to foveal visual stimulation but did not exhibit a change in the discharge rate in association with active fixation (Segraves and Goldberg, 1987). However, stimulation of the frontal eye field representing the foveal area delayed the onset of saccades (Azuma et al., 1986). The posterior parietal cortex and the frontal eye fields project to the SC (Aizawa, et al., 1995; Lynch, et al., 1985). The fixation cells in the rostral SC may be part of the final common pathway from cortical fixation areas to the brainstem premotor circuit (Munoz and Wurtz, 1993).

The Efferent Projection from the Rostral SC

The accommodation-related area in the rostral SC projects to many parts of the brain as follows (Sato and Ohtsuka, 1996); (1) in the ascending projections, the caudal portion of the nucleus of the optic tract (NOT), the nucleus of the posterior commissure (NPC), the posterior pretectal nucleus (PPN), the olivary pretectal nucleus (OPN), the mesencephalic reticular formation (MRF) at the level of the oculomotor nucleus, the lateral posterior nucleus of the thalamus, the anterior pretectal nucleus (APN), the zona incerta, and the centromedian nucleus of the thalamus on the ipsilateral side, and (2) in the descending projections; the paramedian pontine reticular formation (PPRF), the RIP, the dorsolateral pontine nucleus (DLPN) and the dorsomedial portion of the nucleus reticularis tegmenti pontis (NRTP) on the contralateral side, and the nucleus of the brachium of the inferior colliculus, the cuneiform nucleus, the medial part of the paralemniscal tegmental field and the dorsolateral division of the pontine nuclei on the ipsilateral side. In these parts of the brain, the NOT, the NPC, and the NRTP are assumed to be involved in the control of accommodation and/or vergence (Konno and Ohtsuka, 1996; Hultborn, et al.,

1978a; Gamlin and Clarke, 1995). The NOT and the NPC project to parasympathetic neurons in the oculomotor nucleus (Breen, et al., 1983; Buttner-Ennever, et al., 1996). Therefore, the projection from the rostral SC to the oculomotor nucleus via the PT may be an important pathway for controlling accommodation.

The DLPN and the dorsomedial portion of the NRTP are primary sources of afferents to the flocculus and vermal lobules VI and VII (Brodal, 1979; Langer, et al., 1985), which are involved in the control of saccades, smooth-pursuit eye movements and fixation

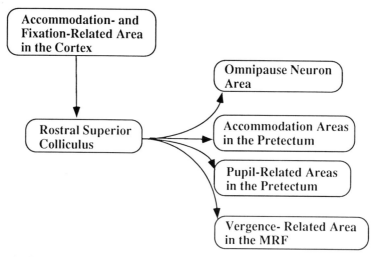

Fig. 1 Schematic diagram of afferent and efferent pathways of the rostral SC indicating its involvement in the control of the ocular near response and active fixation. The rostral SC receives descending projections from the cortical accommodation-related area and the fixation-related area in the posterior parietal cortex, and projects to the omnipause neuron area, the accommodation areas in the pretectum, pupil-related areas in the pretectum and the vergence-related area in the mesencephalic reticular formation (MRF).

(Lisberger and Fuchs, 1978; Noda and Suzuki, 1979; Ohtsuka and Noda, 1991; Ohtsuka and Noda, 1995). In addition, accommodation was evoked by stimulation of the cerebellar nuclei, i.e., the interpositus nucleus and the fastigial nucleus (Hosoba, et al., 1978; Hultborn, et al., 1978b). This evoked accommodation was suppressed when stimulation of the cerebellar nuclei was preceded by stimulation of the vermal area (lobules VI and VII) (Hultborn, et al., 1978b). It is known that the interpositus nucleus projects to the parasympathetic oculomotor neurons with a short (mono- or disynaptic) latency (Hultborn, et al., 1978b). Therefore, the rostral SC also projects to the parasympathetic oculomotor nucleus via the cerebellum.

In the MRF, neurons which exhibit discharge preceding accommodation, convergence, or both were found (Bando, et al., 1984a; Mays, 1984; Judge and Cumming, 1986). These cells were confined to a small region dorsolateral to the oculomotor nucleus in both the cat and the monkey. The rostral SC projects to the portion dorsolateral to the oculomotor nucleus, which must be comparable to the accommodation-vergence area in the MRF. Accommodation and vergence eye movement are closely linked as the ocular near response. Results of a previous study suggest that the SC is also involved in the control of vergence eye movements (Cowey, et al., 1984). It is likely that the connection between the rostral SC and the MRF provides the basis for the link between accommodation and vergence in the ocular near response.

The OPN and PPN are also involved in the control of pupillary movement (Gamlin, et al., 1995; Clarke and Ikeda, 1985). Accommodation and pupillary movement are also closely linked as the ocular near response. Whether the rostral SC is involved in the control of pupillary movement is not known. It is likely that the rostral SC provides signals related to accommodation for the pupillary control system in the brainstem.

The accommodation-related area in the rostral SC also projects to the omnipause neuron area in the RIP. About 85% of primate OPNs and 66% of cat OPNs are activated by electrical stimulation of the SC (Raybourn and Keller, 1977; Kaneko and Fuchs, 1982). Pare and Guitton reported that 61% of OPNs were activated by stimulation of the rostral SC, whereas only 29% responded to caudal stimulation. In a study performed using a fluorescent double-labeling technique, about 60% of neurons in the rostral SC which projected to the RIP also projected to the low-threshold area for evoking accommodation in the pretectum (Nagasaka and Ohtsuka, unpublished data). The rostral SC is a neural substrate for linkage between the accommodation and visual fixation systems. Figure 1 shows a schematic diagram of afferent and efferent pathways of the rostral SC. The rostral SC may be part of the final common pathway from the ocular near response and fixation systems in the cortex to the brainstem premotor circuit.

Concluding Remarks

The SC has long been recognized as an important structure in the generation of saccadic eye movements. Results of recent studies suggest that the rostral SC where the central visual field is represented is involved in the control of accommodation and visual fixation. In addition, convergence eye movements are evoked by microstimulation of the rostral SC, indicating that the rostral SC is also involved in the control of vergence (Ohtsuka and Sawa, unpublished data). The rostral SC receives descending projections from the cortical accommodation-related area and the fixation-related area in the posterior parietal cortex, and projects to the accommodation-, vergence- and pupilloconstriction-related areas in the brainstem and the omnipause neuron area in the RIP. A subgroup of collicullar cells in the rostral SC has divergent axon collaterals to the accommodation-related area and to the omnipause neuron area. These findings suggest that the rostral SC is a key structure for controlling the ocular near response, although the function of the SC in the control of vergence remains to be elucidated.

References

Aizawa, H., Gerfen, C.R. and Wurtz R.H. (1995) Afferent connection to fixation zone of monkey superior colliculus from frontal cortex and basal ganglia. *Society of Neuroscience Abstract.* 21: 1194.

Azuma, M., Nakayama, H. and Suzuki, H. (1986) Suppression of visually triggered saccades by electrical stimulation of the monkey frontal eye field. *Journal of Physiology (Japanese Society of Physiology)* 48: 266

Bando, T., Tsukuda, K., Yamamoto, N., Maeda, J. and Tsukahara, N. (1981) Cortical neurons in and around the Clare-Bishop area related with lens accommodation in the cat. *Brain Research* 225: 195-199.

Bando, T., Tsukuda, K., Yamamoto, N., Maeda, J. and Tsukahara, N. (1984a) Physiological identification of midbrain neurons related to lens accommodation in cats. *Journal of Neurophysiology* 52: 870-878.

Bando, T., Yamamoto, N. and Tsukahara, N. (1984b) Cortical neurons related to lens accommodation in posterior lateral suprasylvian area in cats. *Journal of Neurophysiology.* 52: 879-891.

Ben Hamed, S., Duhamel, J-R., Bremmer, F. and Graf, W. (1995) Representation of the central visual field and fixation-related activity in the lateral intraparietal area of macaque monkeys. *Society of Neuroscience, Abstract* 21:665.

Breen, L., Burde, R.M. and Loewy, A.D. (1983) Brainstem connections to the Edinger-Westphal nucleus of the cat: a retrograde tracer study. *Brain Research* 261: 303-306.

Brodal, P. (1979) The pontocerebellar projection in the rhesus monkey: an experimental study with retrograde axonal transport of horseradish peroxidase. *Neuroscience* 4: 193-208.

Buttner-Ennever, J.A., Cohen, B., Horn, A.K.E. and Reisine, H. (1996) Pretectal projections to the oculomotor complex of the monkey and their role in eye movements. *Journal of Comparative Neurology* 366: 348-359.

Buttner-Ennever, J.A., Cohen, B., Pause, M. and Fries, W. (1988) Raphe nucleus of the pons containing omnipause neurons of the oculomotor system in the monkey. *Journal of Comparative Neurology* 267: 307-321.

Clarke, R.J. and Ikeda, H. (1985) Luminance and darkness detectors in the olivary and posterior pretectal nuclei and their relationship to the pupillary light reflex in the rat. I. Studies with steady luminance levels. *Experimental Brain Research* 57: 224-232.

Cowey, A., Smith, B. and Butter, C.M. (1984) Effects of damage to superior colliculi and pretectum on movement discrimination in rhesus monkeys. *Experimental Brain Research* 56: 79-91.

Evinger, C., Kaneko, C.R.S. and Fuchs, A.F. (1982) Activity of omnipause neurons in alert cats during saccadic eye movements and visual stimuli. *Journal of Neurophysiology* 47: 827-844.

Fuchs, A.F., Kaneko, C.R.S. and Scudder, C.A. (1985) Brainstem control of saccadic eye movements. *Annual Review of Neuroscience* 8: 307-337.

Gamlin, P.D.R. and Clarke, R.J. (1995) Single-unit activity in the promate nucleus reticularis tegmenti pontis related to vergence and ocular accommodation. *Journal of Neurophysiology* 73: 2115-2119.

Gamlin P.D.R., Zhang, H. and Clarke, R.J. (1995) Luminance neurons in the pretectal olivary nucleus mediate the pupillary light reflex in the rhesus monkey. *Experimental Brain Research* 106: 169-176.

Hosoba, M., Bando, T. and Tsukahara, N. (1978) The cerebellar control of accommodation of the eye in the cat. *Brain Research* 153: 495-505.

Hultborn, H., Mori, K. and Tsukahara, N. (1978a) The neuronal pathway subserving the pupillary light reflex. *Brain Research* 159: 255-267.

Hultborn, H., Mori, K. and Tsukahara, N. (1978b) Cerebellar influence on parasympathetic neurons innervating intra-ocular muscles. *Brain Research.* 159: 269-278.

Judge, S.J. and Cumming, B.G. (1986) Neurons in the monkey midbrain with activity related to vergence eye movement and accommodation. *Journal of Neurophysiology.* 55: 915-930.

Kaneko, C.R.S. and Fuchs, A.F. (1982) Connections of cat omnipause neurons. *Brain Research* 241: 166-170.

Kawamura, S., Spraugue, J.M. and Niimi, K. (1974) Corticofugal projections from the visual cortices to thalamus, pretectum and superior colliculus in the cat. *Journal of Comparative Neurology* 158: 339-362.

Konno, S. and Ohtsuka, K. (1996) Accommodation- and pupilloconstriction-related areas in the midbrain of the cat. *Japanese Journal of Ophthalmology* (in press).

King, W.M., Precht, W. and Dieringer, W. (1980) Afferent and efferent connections of cat omnipause neurons. *Experimental Brain Research* 38: 395-403.

Langer, T.P., Fuchs, A.F., Scudder, C.A. and Chubb, M.C. (1985) Afferents to the flocculus of the cerebellum in the rhesus macaque as revealed by retrograde transport of horseradish peroxidase. *Journal of Comparative Neurology* 235: 1-25.

Lee, C., Rohrer, W.H. and Sparks, D.L. (1988) Population coding of saccadic eye movements by neurons in the superior colliculus. *Nature London* 332: 357-360.

Lisberger, S.G. and Fuchs, A.F. (1978) A role of primate flocculus during rapid behavioral modification of vestibuloocular reflex. I. Purkinje cell activity during visually guided horizontal smooth-pursuit eye movement and passive head rotation. *Journal of Neurophysiology* 41: 733-763.

Lynch, T.P., Graybiel, A.M. and Lobeck. (1985) The differential projection of two cytoarchitectonic subregions of the inferior parietal lobule of macaque upon the deep layers of the superior colliculus. *Journal of Comparative Neurology* 235: 241-254.

Lynch, T.P., Moutcastle, V.B., Talbot, W.H. and Yin, T.C.T (1977) Parietal lobe mechanisms for directed visual attention. *Journal of Neurophysiology* 40: 362-389.

Maekawa, H. and Ohtsuka, K. (1993) Afferent and efferent connections of the cortical accommodation area in the cat. *Neuroscience Research* 17: 315-323.

Mays, L.E. (1984) Neural control of vergence eye movements: Convergence and divergence neurons in midbrain. *Journal of Neurophysiology* 51: 1091-1108.

Mountcastle, V.B., Lynch, J.C., Georgopoulos, A., Sakata, H. and Acuna, C. (1975) Posterior parietal association cortex of the monkey: command functions for operations within extrapersonal space. *Journal of Neurophysiology* 38: 871-908.

Munoz, D.P. and Guitton, D. (1989) Fixation and orientation control by the tecto-reticulo-spinal system in the cat whose head is unrestrained. *Revue Neurologique Paris* 145: 567-579.

Munoz, D.P. and Guitton, D. (1991) Control of orienting gaze shifts by the tectoreticulospinal system in the head-free cat. II. Sustained discharges during motor preparation and fixation. *Journal of Neurophysiology* 66: 1624-1641.

Munoz, D.P., Guitton, D. and Pelisson, D. (1991) Control of orienting gaze shifts by the tectoreticulospinal system in the head-free cat. III. Spatiotemporal characteristics of phasic motor dicharge. *Journal of Neurophysiology* 66: 1642-1666.

Munoz, D.P. and Wurtz, R.H. (1993) Fixation cells in monkey superior colliculus. I. Characteristics of cell discharge. *Journal of Neurophysiology* 70: 559-575.

Noda, H. and Suzuki, D. (1979) The role of the flocculus of the monkey in fixation and smooth pursuit eye movements. *Journal of Physiology London* 294: 335-348.

Norita, M., McHaffie, J.G., Shimizu, H. and Stein, B.E. (1991) The corticostriatal and corticotectal projections of the feline lateral suprasylvian cortex demonstrated with anterograde biocytin and retrograde fluorescent techniques. *Neuroscience Research* 10: 149-155.

Ohtsuka, K. and Noda, H. (1991) Saccadic burst neurons in the oculomotor region of the fastigial nucleus of macaque monkeys. *Journal of Neurophysiology* 65: 1422-1434.

Ohtsuka, K. and Noda, H. (1995) Discharge properties of Purkinje cells in the oculomotor vermis during visually guided saccades in the macaque monkey. *Journal of Neurophysiology* 74: 1828-1840.

Ohtsuka, K. and Sato, A. (1996a) Descending projections from the cortical accommodation area in the cat. *Investigative Ophthalmology & Visual Science* 37: 1429-1436.

Ohtsuka, K. and Sato, A. (1996b) Retinal projections to the accommodation-related area in the rostral superior colliculus of the cat. *Experimental Brain Research* (in press).

Pare, M. and Guitton, D. (1994) The fixation area of the cat superior colliculus: effects of electrical stimulation and direct connection with brainstem omnipause neurons. *Experimental Brain Research* 101: 109-122.

Peck, C.K. (1989) Visual responses of neurons in cat superior colliculus in relation to fixation of targets. *Journal of Physiology, London* 414: 301-315.

Raybourn, M.S. and Keller, E.L. (1977) Colliculo-reticular organization in primate oculomotor system. *Journal of Neurophysiology* 40: 861-878.

Robinson, D.A. (1972) Eye movements evoked by collicular stimulation in the alert monkey. *Vision Research* 12: 1795-1808.

Sakata, H., Shibutani, H. and Kawano, K. (1980) Spatial properties of visual fixation neurons in posterior parietal association cortex of the monkey. *Journal of Neurophysiology* 43: 1654-1672.

Sato, A. and Ohtsuka, K. (1996) Projection from the accommodation-related area in the superior colliculus of the cat. *Journal of Comparative Neurology* 367:465-476

Sawa, M., Maekawa, H. and Ohtsuka, K. (1992) Cortical area related to lens accommodation in cat Japanese *Journal of Ophthalmology* 36: 371-379.

Sawa, M. and Ohtsuka, K. (1994) Lens accommodation evoked by microstimulation of the superior colliculus in the cat. *Vision Research* 34: 975-981.

Segal, R.L. and Beckstead, R.M. (1984) The lateral suprasylvian corticotectal projection in cats. *Journal of Comparative Neurology* 225: 259-275.

Segraves, M.A. and Goldberg, M.E. (1987) Functional properties of corticotectal neurons in the monkey's frontal eye field. *Journal of Neurophysiology* 58: 1387-1419.

Sparks, D.L. (1986) Transition of sensory signals into commands for control of saccadic eye movements: role of the superior colliculus. *Physiological Reviews.* 66: 118-171.

Sparks, D.L., Holland, R. and Guthrie, B.L. (1976) Size and distribution of movement fields in the monkey superior colliculus. *Brain Research* 113: 21-34.

Toyama, K. and Kozasa, T. (1982) Responses of Clare-Bishop neurons to three-dimensional movement of a light stimulus. *Vision Research* 22: 571-574.

Toyama, K., Fujii, K. Kasai, S. and Maeda, K. (1986a) The responsiveness of Clare-Bishop neurons to size cues for motion stereopsis. *Neuroscience Research* 4: 110-128.

Toyama, K., Komatsu, Y. and Kozasa, T. (1986b) The responsiveness of Clare-Bishop neurons to motion cues for motion stereopsis. *Neuroscience Research* 4: 83-109.

Updyke, B.V. (1981) Projections from visual areas of the middle suprasylvian sulcus onto the lateral posterior complex and adjacent thalamic nuclei in cat. *Journal of Comparative Neurology* 201: 477-506.

Accommodation and Vergence Mechanisms
in the Visual System
ed. by O. Franzén, H. Richter and L. Stark

The pupillary light reflex, accommodation and convergence: Comparative considerations

J.T. Erichsen, W. Hodos[1] and C. Evinger[2]

University of Wales, Cardiff, Dept. of Optometry and Vision Sciences, P.O. Box 905, Cardiff, United Kingdom
[1]*University of Maryland, Dept. of Psychology, Maryland, USA,* [2]*State University of New York at Stony Brook, Depts. of Neurobiology & Behavior and Ophthalmology, Stony Brook, New York, USA*

Summary. In most vertebrates, preganglionic neurons of the midbrain parasympathetic nucleus of Edinger-Westphal [nEW] mediate positive accommodation and pupilloconstriction via their projection to the ciliary ganglion. Our extensive anatomical studies in the pigeon (*Columba livia*) show that nEW is organized into distinct subdivisions. Microstimulation, as well as retrograde transsynaptic transport of wheat germ agglutinin [WGA] from the iris or ciliary muscle to nEW, has helped us to confirm the functional specificity of two of these subdivisions. The anatomical basis for the vergence component of the near response has also been investigated. Using immunohistochemistry, as well as transneuronal transport methods, we have identified a small dorsal subpopulation of medial rectus motoneurons [nMRd] in the pigeon. Our results suggest that nMRd in the pigeon is similar to the C subdivision of the medial rectus nucleus in the monkey and thus may play a role in vergence. Acting in concert with nEW neurons, nMRd might also be involved in the near response.

Introduction

With few exceptions, vertebrates possess functional intraocular muscles (iris and ciliary muscle) and share the ability to use these muscles to alter the size of the pupil and the shape of the lens. In response to relevant visual feedback, these changes allow the animal to control the amount of light entering the eye, the refractive state of the eye and to some extent its depth of focus. In most of the animals investigated thus far, the final motor pathway mediating pupilloconstriction and accommodation appears to be highly conserved from an evolutionary perspective. Thus, preganglionic neurons of the midbrain parasympathetic nucleus of Edinger-Westphal [nEW] project to cells in the ciliary ganglion which in turn innervate either the iris or ciliary muscle. This suggests that the more detailed levels of anatomical organization of this system may also be similar in different species. Indeed, such parallels could even extend to the inputs to nEW. One of us (JTE) was involved in a earlier investigation of the pupillary light reflex of the pigeon (*Columba livia*) (Gamlin et al., 1984). For the first time in any vertebrate, we were able to identify the entire pathway from the retinal ganglion cells mediating the luminance response via synapses in a specific pretectal nucleus, nEW and the ciliary ganglion to the iris. Subsequent reports involving various mammals have strongly suggested striking similarities in the organization of this pathway across vertebrates (e.g., Gamlin and Clarke, 1995).

In our studies over the last fifteen years, we have used birds as the model for our anatomical studies of the pupillary light reflex and accommodation pathways. Birds, as highly visual animals, afford several distinct advantages as models for vision. First, their eyes are generally very large in proportion to the size of the rest of the nervous system. For example, the diameter of the pigeon eye is half that of an adult human and its two eyes easily outweigh the entire brain (Marshall et al., 1973). One consequence of this is that many visual structures within the brain are usually relatively large and more distinct than in many mammals. Second, the retinofugal pathways are entirely crossed, which means that in the course of pathway tracing experiments, one of each pair of visual structures in the brain can often serve as a control for the other (Cowan et al., 1961; Karten et al., 1973). Such within animal comparisons greatly facilitate the interpretation of labeled cells and terminals after injection of a tracer. Third, all birds possess a central fovea in each eye which is used, as in primates, for the visual grasp reflex and for high acuity vision (Fite and Rosenfield-Wessels, 1975; Erichsen, 1979). Although other mammals have central areas in the retina with higher acuity, only the eyes of primates have foveae. Finally, birds have been shown to display some of the common pathologies of vision that are found in humans. For example, numerous studies have demonstrated the ability to induce myopia in the chick, and in many ground foraging birds, a localized lower frontal field myopia may serve as an important, naturally occurring adaptation for feeding (Hodos and Erichsen, 1990).

For largely historical reasons, the final motor pathway from nEW to the intraocular muscles is especially well described in the pigeon and chick (see Figure 1 below). Located in the midbrain just dorsal and somewhat lateral to the main oculomotor complex, nEW is a small ovoid nucleus that is approximately 750 µm in diameter. The parasympathetic preganglionic cells of nEW provide the exclusive input to the ciliary ganglion. The postganglionic cells consist of two morphologically distinct subpopulations. The ciliary cells, which are innervated by a single large calyx-like ending, project either to the iris or the ciliary muscle. The choroid cells, which receive numerous small boutonal terminations, project to the choroid underlying the retina (see below).

In the context of investigating pupilloconstriction or accommodation, it is also necessary to point out the differences that sharply distinguish birds from primates. First, as a consequence of the laterally directed placement of the eyes in the head, as much as the totally crossed visual system, the pupillary and accommodative responses of the two eyes can function entirely independently of one another (Schaeffel et al., 1986). Obviously, this means that, although birds are certainly able to converge their eyes, vergence and accommodation can be completely

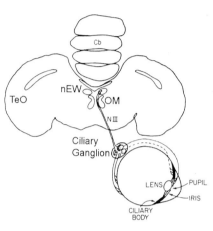

Figure 1. Pathway from nEW to the eye in bird

uncoupled. Second, in many birds, eye movements are not sufficient to allow the central foveae to be used for forward or binocular vision (Nalbach et al., 1990). Such vision must be restricted to an area of the temporal peripheral retina which often contains a specialized area or even a second fovea (Galifret, 1968). The need to look obliquely through the optical system of the eye may explain the fact that regional myopias can be induced in the chick eye. Indeed, depending on the visual manipulation performed on the eye, the equatorial and axial diameters can be affected independently (Diether and Schaeffel, 1997). It also raises the interesting possibility that there are differences between axial (central) and oblique (peripheral) accommodation in birds and that such differences may be reflected in the central nervous system. Finally, as pointed out above, it has been demonstrated in the pigeon and chick that the cells of the ciliary ganglion not only project to the iris and ciliary muscle but some cells also innervate the choroid vasculature of the eye (Landmesser and Pilar, 1970; Reiner et al., 1990). The source of such a projection in mammals has yet to be determined.

Our own extensive anatomical studies of the pigeon, as well as work that we have carried out in collaboration with others, has largely centered on the anatomical organization of the nucleus of Edinger-Westphal [nEW]. Using conventional and transsynaptic pathway tracing techniques, as well as selective lesions, immunohistochemistry and microstimulation, we have found clear evidence that nEW is organized into distinct functional subdivisions (Reiner et al., 1983; 1991). Our hope is that a better understanding of these subdivisions will facilitate the identification of the afferent pathway(s) mediating accommodation. In addition, we have also begun investigating the possibility that, as in primates, pigeons possess a specialized portion of the medial rectus nucleus [nMR] that controls vergence (Erichsen and Evinger, 1989). Although it is clear that accommodation in the pigeon can and does occur in the absence of vergence, this does not rule out the intriguing possibility that something analogous to a "near response" might occur when a pigeon fixates objects in front of its beak.

Material and Methods

Histology, pathway tracing and immunohistochemistry. In all of the experiments, White Carneaux pigeons (*Columba livia*) of undetermined sex and aged 4 months to 2 years were used. Most of the individual techniques employed in the series of experiments reviewed here have already been described. Specifically, conventional pathway tracing methods are as detailed in Gamlin et al. (1984). The procedures for using immunohistochemistry to localize the transsynaptic tracers, wheat germ agglutinin [WGA] and tetanus toxin C fragment, after injection can be found in Collins et al. (1991). For the analysis of immunohistochemical deficits after electrolytic lesions see Reiner et al. (1991). Our most recent protocols for the immunohistochemical localization of various neurotransmitters and related enzymes have been published (Erichsen et al., 1991; Bagnoli et al., 1992).

Image processing and analysis. Mounted and coverslipped brain sections were examined on a Leitz Diaplan brightfield microscope equipped with Nomarski optics. A Dage CCD-72S camera connected to a Scion PCI LG-3 frame digitizing board was used to capture images of the relevant brain sections and/or labeled cells. These image files were stored on a Macintosh 8500/150 and then analyzed with NIH Image software. The analysis consisted of tracing lesions or the relevant neural structures, such as nEW, and then marking the location of labeled and unlabeled cells within each of those structures. Subsequent to this stage of analysis, the program could automatically use the calibrated tracings to measure the location of the lesion, the volume of a neural structure, or the number and distribution of the labeled and/or unlabeled cells.

Microstimulation. In this ongoing study, we attempted to place an electrode stereotaxically in nEW. As the electrode was lowered into the brain, current was passed in brief trains while monitoring the pupil and lens for any observable changes. When the maximal accommodative and/or pupillary response was found, a retinoscopic assessment of the current response of the eye's refractive state as well as the threshold for pupilloconstriction was recorded for each animal. At the end of each experiment, a marker lesion was made to allow subsequent determination of the electrode's location. The bird's head was always carefully positioned in the stereotaxic apparatus to match its normal head orientation. The initial measurements of refractive state changes were always made along the eye's axis, which in the pigeon is located at an azimuth of 75° and an elevation of 0° (Clarke and Whitteridge, 1976; Erichsen, 1979). The parameters for the microstimulation of the midbrain were originally established in Erichsen et al. (1981) and continue to be followed in our more recent studies (see Reiner et al., 1991). In view of the naturally occurring lower field myopia in the pigeon (Hodos and Erichsen, 1990), we also monitored accommodative responses at an azimuth of 30-45° and an elevation of -60°.

Results

Subdivisions of the nucleus of Edinger-Westphal [nEW]

Pathway tracing: Iris. Gamlin's et al. (1984) earlier study of the pupillary light reflex pathway of the pigeon involved the use of conventional direct retrograde and anterograde tracers. One of the more interesting findings arose from the distribution of nEW afferents visualized after an injection of the pretectal pupillary nucleus (area pretectalis) with tritiated proline. The labeled terminal field was restricted to the caudal pole as well as the lateral portion of nEW. In an experiment by Gamlin et al. (1982), involving a parallel injection of the suprachiasmatic nucleus [SCN], the labeled afferents were now restricted to the medial aspect of nEW. This suggested that nEW consisted of distinct subdivisions. The putative pupillary subdivision has been named nEWcl and the region receiving input from SCN is called nEWm. Selective lesions of nEWcl and nEWm result in the loss of terminal endings (immunoreactive for Substance P; see Erichsen et al., 1982a,b) on ciliary and choroid cells, respectively (Reiner et al., 1991). The delineation of a putative accommodative subdivision was prevented by the fact that the direct accommodative input(s) to nEW are as yet unknown in the pigeon or any other vertebrate. More recently, a pathway tracing study by Gamlin and Reiner (1991) suggests that cells in both the medial and lateral mesencephalic reticular formations (MRF, LRF) of the midbrain project to nEW and may be involved in the mediation of accommodation.

Any attempt to determine the specific cells of nEW controlling either the iris or ciliary muscle would ordinarily require the use of direct retrograde tracers. However, the iris and ciliary muscle cells of the ciliary ganglion are intermingled and thus an injection restricted to one or the other subpopulation is not possible. However, we have succeeded in using the transsynaptic retrograde tracer, WGA, to overcome this problem (Erichsen et al., 1995). Within one day of an injection of WGA into the anterior chamber of the pigeon, a relatively small number of ciliary neurons are retrogradely labeled in the ciliary ganglion. More significantly, after four days, transsynaptically labeled cells appear in a restricted portion of nEW. No such cells were ever observed in less than 2-3 days after injection, indicating the absence of direct retrograde transport. A detailed analysis of the distribution of these presumably pupillary preganglionic neurons indicates that they are found in the same region of nEW which, as Gamlin et al. (1984) demonstrated, receives projections from area pretectalis. Moreover, the number of transsynaptically labeled cells in nEW (see Table 1) is relatively constant from case to case and consistent with our previous estimates of pupillary neurons in nEW (Reiner et al., 1991).

Pathway tracing: Ciliary muscle. The injections of the anterior chamber never resulted in anterograde labeling of central visual structures, indicating that the WGA did not pass from the aqueous into the vitreous. On the other hand, injections of the vitreous, which strongly labeled primary visual areas, always produced the same pattern of labeled cells in nEW as the anterior chamber injections (i.e., EWcl). This suggests that the ciliary body and muscle are normally inaccessible to substances such as WGA that are contained in the surrounding vitreous, in marked contrast to the iris which seems to readily take up the WGA from the aqueous and/or vitreous.

In light of these findings, we attempted to inject the ciliary muscle directly through the sclera just outside the limbus. In two cases, such injections resulted in much more extensive transsynaptic labeling of cells in nEW. Anterograde labeling of primary visual areas was consistent with leakage of WGA into the vitreous. Thus, labeling of the pupillary subdivision (EWcl) would be expected. A detailed analysis of the number and distribution of

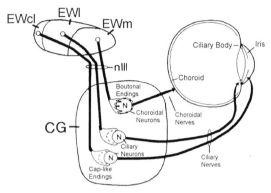

Figure 2. Function-specific subdivisions of nEW in pigeon

these labeled cells in one case indicates that the entire pupillary subdivision is labeled, as well as numerous cells located more medially (Table 1). In fact, a reconstruction of these cells indicates that the putative accommodative subdivision occupies the middle portion of nEW throughout most of its rostral-caudal extent (i.e., EWl) as well as the dorsal portion of what had been named nEWm (see Figure 2). The ventromedial region of nEW, which innervates the choroid cells of the ciliary ganglion, is clearly devoid of labeled cells.

Table 1: Transsynaptic labeling of pupillary and accommodative subdivisions of nEW

Injection site:	Iris (N=9)		Ciliary Muscle (N=1)	
Number of WGA-labeled cells	123 ($\pm$ 3.6)	(11%)	625	(55%)
Number of unlabeled cells	1005 ($\pm$ 12.2)	(89%)	504	(45%)
Total number of cells in nEW	1128 ($\pm$ 12.8)		1129	

Microstimulation. Microstimulation of nEW in anesthetized adult pigeons produced a current-dependent change in the refractive state of the ipsilateral eye. Increasing current made the eye progressively more myopic until an asymptotic value was reached. Depending on the location of the electrode within or near nEW and on the strength of the current, constriction of the pupil also occurred. Correlating electrode location with recorded responses, we have found that excitation of caudolateral nEW cells tended to produce exclusively pupilloconstriction, whereas increased accommodative changes were evoked more medially and/or rostrally. These results are in broad agreement with the anatomical findings presented in the previous section.

The maximum range of accommodation that we observed in the central visual field was 5-7 diopters, which agrees well with previous reports of the normal range found in awake behaving pigeons. One very interesting result obtained in the course of our experiments is that, at any given current, a consistently greater change in refractive state was always observed in the lower visual field than in the central visual field. This was particularly striking in one bird where there was no pupillary response to stimulation which made it relatively easy to measure accommodative responses at fairly high currents. Subsequent analysis revealed that the electrode tip was located within the ipsilateral MRF, as described by Gamlin and Reiner (1991).

Subdivision of the medial rectus nucleus [nMR]

In all vertebrates, motoneurons of the medial rectus nucleus [nMR] innervate the medial rectus muscle and thus control vergence movements of the eye. As in primates, nMR of the pigeon is located near and just ventral to nEW. We have now accumulated considerable evidence suggesting that a small subpopulation of nMR cells in the pigeon comprises an immunologically distinct and perhaps functionally unique class of motoneurons. Forming a dorsal crescent in nMR just medial to nEW, these cells [nMRd] are generally smaller than the other medial rectus motoneurons, but are retrogradely labeled by injection of HRP into the medial rectus muscle. A similar injection of a transsynaptic tracer, tetanus toxin C fragment, also results in direct retrograde labeling of the entire ipsilateral nMR, but in addition the contralateral nMRd contains labeled terminals. In view of our previous experience with this tracer, the latter result suggests the possibility of an afferent shared by both nMRd. Immunohistochemical studies have revealed both vasoactive intestinal polypeptide- and substance P-labeled fibers that apparently terminate in nMRd but not in the rest of nMR. In the developing chick embryo, cells within both nEW and nMRd express enkephalin but not other cells within nMR. Taken together, the neurons of nMRd share many of the same characteristics found in the C subdivision of nMR in the primate (Büttner-Ennever and

Akert, 1981), i.e., small size, dorsal position within nMR, proximity to nEW, etc. These cells in the pigeon also appear to receive a specific afferentation. As the C subdivision is believed to innervate the special fibers of the medial rectus muscle that are involved in vergence, our results raise the interesting possibility that nMRd subserves a similar role in the pigeon.

Discussion

In vertebrates, the final motor pathway for pupilloconstriction and accommodation is the projection from the nucleus of Edinger-Westphal [nEW] to the cells of the ciliary ganglion, which in turn innervate the iris and ciliary muscles. Studies in some mammals suggest that not all preganglionic neurons are located within nEW and that nEW may also contain cells that do not project to the ciliary ganglion (e.g., Warwick, 1954; Loewy et al., 1978; Burde, 1988; Ishikawa et al., 1990). This uncertainty has complicated studies of the central pathways mediating pupilloconstriction and accommodation in mammals. Evidence of the connection between nEW and the ciliary ganglion in birds is much clearer than in mammals. All preganglionic neurons are found in nEW, and they project exclusively to the ciliary ganglion (e.g., Narayanan and Narayanan, 1976; Erichsen et al., 1981; Reiner et al., 1991). In fact, the entire oculomotor complex of birds is probably the most discretely organized of any vertebrate (Evinger, 1988), and this anatomical clarity has facilitated our studies of the afferents to specific motoneuron populations within nEW and the medial rectus nucleus [nMR].

In the pigeon, there are three subdivisions mediating accommodation, pupilloconstriction and the choroidal blood flow. Using conventional and transsynaptic pathway tracing techniques, we and others have delineated the location of these subpopulations, and microstimulation studies have confirmed the function of each. As a consequence, we have also been better able to estimate the proportion of preganglionic neurons involved in the control of the ciliary muscle and iris. This represents a necessary first step towards the future identification of afferents to nEW that mediate positive accommodation. Our previous work on the pupillary pathway suggests that these systems are highly conserved across vertebrates and thus, the results of such studies will readily be extended to the mammal. The central control of accommodation has also been studied in a number of vertebrates. As yet, however, there is no real consensus about its general pattern of organization, much less about the specific components or the number of pathways involved in this clinically important ocular function (see review by Ruskell, 1990). Most previous studies have been electrophysiological rather than anatomical in nature (Mays and Gamlin, 1995; Gamlin et al., 1996). The recent work by Gamlin and Reiner (1991) utilized the well-defined cytoarchitecture of nEW in the pigeon to reveal at least two possible sources of accommodative input in the midbrain. Future anatomical and microstimulation studies will be necessary to confirm the functional role of one or both of these brain centers.

Together with pupilloconstriction and accommodation, convergence of the eyes completes the triad of the near response. Although considerable information is available about the electrophysiology of this system in primates, little is known about the anatomical pathways subserving vergence in any animal. Together with the lateral rectus muscle, the medial rectus [MR] muscle controls horizontal movements of the eye and, in particular, mediates the convergence of the two eyes (i.e., disconjugate eye movements). In the course of vertebrate evolution, the organization of extraocular motoneurons has been conserved to a remarkable extent. Thus, motoneurons of nMR innervate the ipsilateral MR muscle of the eye in all vertebrates studied except the elasmobranchs (Evinger, 1988). Some species differences have been noted. For example, three anatomically distinct subpopulations of MR motoneurons have been described in the primate (i.e., cell groups A-C; Büttner-Ennever and Akert, 1981). Of these cell groups, the dorsal-most C subdivision, consisting of smaller-diameter MR motoneurons, innervates the small orbital fibers of the MR muscle and is hypothesized to be important for vergence movements (Büttner-Ennever and Akert, 1981). This C subdivision is immediately ventral to nEW and extends throughout the rostral two-thirds of the oculomotor nucleus. However, such a subdivisional organization may not be unique to the primate. For example, more recent studies have revealed the existence of a small group of dorsolateral MR motoneurons in the cat that is distinct from the major population of ventrolateral MR motoneurons (Miyazaki, 1985). This smaller group may be homologous to the C subdivision of the primate (Evinger, 1988). Our finding that there is a clearly defined subpopulation of medial rectus motoneurons in the pigeon with similar characteristics strongly suggests that a number of vertebrates may possess the anatomical substrate for a near response that coordinates vergence with the activity of preganglionic neurons of nEW. If so, future anatomical and microstimulation studies of the inputs to nMRd in the pigeon may provide the basis for determining the brain center(s) mediating such a response in other vertebrates.

One of the essential properties of the near response in humans and other primates is that the three parts of the near triad are inextricably linked, as neither accommodation nor vergence can occur without the other in addition to pupilloconstriction. Presumably because of the lateral placement of the eyes, accommodation can and does occur without vergence in the pigeon. As we have described, there are both pupillary and accommodative subdivisions in nEW, and there is evidence that nMR has a subdivision that may be specialized for vergence, as in the primate. If the pigeon also turns out to possess a near response, future comparative studies may provide a useful model for investigating the relationship between the accommodation and vergence systems when operating independently and/or together.

Acknowledgements
We thank Angela Levine and Alex Adler for their technical assistance, and Nick Wright for his help with preparation of the manuscript. Supported by grants EY04587 (JTE), EY00735 (WH), and EY07391 (CE) from the National Eye Institute.

References

Bagnoli, P., Fontanesi, G., Alesci, R. and Erichsen, J.T. (1992) Distribution of neuropeptide Y, substance P, and choline acetyltransferase in the developing visual system of the pigeon and effects of unilateral retina removal. *Journal of Comparative Neurology* 318: 392-414.

Burde, R.M. (1988) Direct parasympathetic pathway to the eye: revisited. *Brain Research* 463: 158-162.

Büttner-Ennever, J.A. and Akert, K. (1981) Medial rectus subgroups of the oculomotor nucleus and their abducens internuclear input in the monkey. *Journal of Comparative Neurology* 197: 17-27.

Clarke, P.G.H. and Whitteridge, D. (1976) The projection of the retina, including the 'Red Area' on to the optic tectum of the pigeon. *Quarterly Journal of Experimental Physiology* 61: 351-358.

Collins, W.F., Erichsen, J.T. and Rose, R.D. (1991) Pudendal motor and premotor neurons in the male rat: A WGA transneuronal study. *Journal of Comparative Neurology* 308: 28-41.

Cowan, W.M., Adamson, L. and Powell, T.P.S. (1961) An experimental study of the avian visual system. *Journal of Anatomy* 95: 545-563.

Diether, S. and Schaeffel, F. (1997) Local changes in eye growth induced by local refractive error despite active accommodation. *Vision Research* 37: 659-668.

Erichsen, J.T. (1979) How birds look at objects. D.Phil. dissertation, University of Oxford.

Erichsen, J.T., Reiner, A. and Karten, H.J. (1981) Neurons of the nucleus of Edinger-Westphal are a source of enkephalinergic and substanceP-containing terminals in the ciliary ganglion. *Society for Neuroscience Abstracts* 7: 777.

Erichsen, J.T., Reiner, A. and Karten, H.J. (1982a) Co-occurrence of substance P-like and Leu-enkephalin-like immunoreactivities in neurones and fibres of avian nervous system. *Nature* 295: 407-410.

Erichsen, J.T., Karten, H.J., Eldred, W.D. and Brecha, N.C. (1982b) Localization of substance P-like and enkephalin-like immunoreactivity within preganglionic terminals of the avian ciliary ganglion: light and electron microscopy. *Journal of Neuroscience* 2:994-1003.

Erichsen, J.T., Bingman, V.P. and Krebs, J.R. (1991) The distribution of neuropeptides in the dorsomedial telencephalon of the pigeon (*Columba livia*): A basis for regional subdivisions. *Journal of Comparative Neurology* 314: 478-492.

Erichsen, J.T., Evinger, C., Hodos, W. and Adler, A.P. (1995) Function-specific subdivisions of the nucleus of Edinger-Westphal as revealed by transneuronal transport of WGA. *Society for Neuroscience Abstracts* 21: 1918.

Erichsen, J.T. and Evinger, C. (1989) A unique subpopulation of medial rectus motoneurons and its relationship with the nucleus of Edinger-Westphal. *Society for Neuroscience Abstracts* 15: 240.

Evinger, C. (1988) Extraocular motor nuclei: location, morphology and afferents. *In:* J.A. Büttner-Ennever (ed.) *Neuroanatomy of the oculomotor system.* Elsevier Science Publishers, Amsterdam, pp. 81-118.

Fite, K.V. and Rosenfield-Wessels,, S. (1975) A comparative study of deep avian foveas. *Brain, Behavior and Evolution* 12: 97-115.

Galifret, Y. (1968) Les diverses aires fonctionelles de la rétine du pigeon. *Zeitschrift fur Zellforschung* 86: 535-545.

Gamlin, P.D., Reiner, A. and Karten, H.J. (1982) Substance P-containing neurons of the avian suprachiasmatic nucleus project directly to the nucleus of Edinger-Westphal. *Proceedings of the National Academy of Sciences, USA* 79: 3891-3895.

Gamlin, P.D., Reiner, A., Erichsen, J.T., Karten, H.J. and Cohen, D.H. (1984) The neural substrate for the pupillary light reflex in the pigeon (*Columba livia*). *Journal of Comparative Neurology* 226: 523-543.

Gamlin, P.D. and Reiner, A. (1991) The Edinger-Westphal nucleus: Sources of input influencing accommodation, pupilloconstriction, and choroidal blood flow. *Journal of Comparative Neurology* 306: 425-438.

Gamlin, P.D. and Clarke, R.J. (1995) The pupillary light reflex pathway of the primate. *Journal of the American Optometry Association* 66: 415-418.

Gamlin, P.D., Yoon, K. and Zhang, H. (1996) The role of cerebro-ponto-cerebellar pathways in the control of vergence eye movements. *Eye* 10: 167-171.

Hodos, W. and Erichsen, J.T. (1990) Lower-field myopia in birds: An adaptation that keeps the ground in focus. *Vision Research* 30: 653-657.

Ishikawa, S., Sekiya, H. and Kondo, Y. (1990) The center for controlling the near reflex in the midbrain of the monkey: a double labelling study. *Barin Research* 519: 217-222.

Karten, H.J., Hodos, W., Nauta, W.J. and Revzin, A.M. (1973) Neural connections of the visual wulst of the avian telencephalon. Experimental studies in the pigeon and owl. *Journal of Comparative Neurology* 150: 253-278.

Landmesser, L. and Pilar, G. (1970) Selective reinnervation of two cell populations in the adult pigeon ciliary ganglion. *Journal of Physiology (London)* 211: 203-216.

Loewy, A.D., Saper, C.B. and Yamodis, N.D. (1978) Re-evaluation of the efferent projections of the Edinger-Westphal nucleus in the cat. *Brain Research* 141: 153-159.

Marshall, J., Mellerio, J. and Palmer, D.A. (1973) A schematic eye for the pigeon. *Vision Research* 13: 2449-2453.

Mays, L.E. and Gamlin, P.D. (1995) Neuronal circuitry controlling the near response. *Current Opinions in Neurobiology* 5: 763-768.

Miyazaki, S. (1985) Location of motoneurons in the oculomotor nucleus and the course of their axons in the oculomotor nerve. *Brain Research* 348: 57-63.

Nalbach, H.O., Wolf-Oberhollenzer, F. and Kirschfeld, K. (1990) The pigeon's eye viewed through an ophthalmoscopic microscope: orientation of retinal landmarks and significance of eye movements. *Vision Research* 30: 529-540.

Narayanan, C.H. and Narayanan, Y. (1976) An experimental inquiry into the central source of preganglionic fibers to the chick ciliary ganglion. *Journal of Comparative Neurology* 166: 101-109.

Reiner, A., Karten, H.J., Gamlin, P.D. and Erichsen, J.T. (1983) Parasympathetic ocular control: Functional subdivisions and circuitry of the avian nucleus of Edinger-Westphal. *Trends in Neuroscience* 6: 140-145.

Reiner, A., Fitzgerald, M.E.C and Gamlin, P.D. (1990) Neurotransmitter organization of the nucleus of Edinger-Westphal and its projection to the avian ciliary ganglion. *Visual Neuroscience* 6: 451-472.

Reiner, A., Erichsen, J.T., Cabot, J.B., Evinger, C., Fitzgerald, M.E.C and Karten, H.J. (1991) Neurotransmitter organization of the nucleus of Edinger-Westphal and its projection to the avian ciliary ganglion. *Visual Neuroscience* 6: 451-472.

Ruskell, G.L. (1990) Accommodation and the nerve pathway to the ciliary muscle: a review. *Ophthalmic and Physiological Optics* 10: 239-242.

Schaeffel, F., Howland, H.C. and Farkas, L. (1986) Natural accommodation in the growing chicken. *Vision Research* 26: 1977-1993.

Warwick, R. (1954) The ocular parasympathetic nerve supply and its mesencephalic sources. *Journal of Anatomy (London)* 88: 71-93.

Accommodation and Vergence Mechanisms
in the Visual System
ed. by O. Franzén, H. Richter and L. Stark
© 2000 Birkhäuser Verlag Basel/Switzerland

Vergence eye movement and lens accommodation: cortical processing and neuronal pathway

T. Bando, N. Hara, M. Takagi, H. Hasebe, R. Takada and H. Toda

Departments of Physiology and Ophthalmology, Niigata University School of Medicine, Asahi-machi, Niigata, Niigata 951, Japan

Summary: The contribution of the extrastriate visual cortex to ocular convergence and lens accommodation were studied by means of lesion, microstimulation, and recording of neuronal activities in the cat. The results support the hypothesis that the lateral suprasylvian cortex plays an essential role in these ocular movements.

Introduction

Vergence eye movement and lens accommoda-tion are triggered by visual depth cues (Fincham, 1951; Westheimer et al., 1969), and then improve the visual image in the brain by reducing binocular disparity and retinal blur, which are themselves visual depth cues. These visual cues are analyzed early in the cortical processing (Poggio and Poggio, 1964).

The lateral suprasylvian area (LS cortex) is the extrastriate visual area in the cat corresponding to V5 in the monkey (Zeki, 1974). Neurons in this area are selectively activated by motion in the visual field, and some of them are related to motion disparity (Toyama et al., 1986). This area may well provide continuing neuronal signals for the control of vergence eye movement and lens accommodation. The LS cortex receives visual cortical and collicular inputs (Spear, 1991) (Fig. 1). It projects directly to the brainstem nuclei, including the superior colliculus and the pretectum, which are superior to the motor nuclei of eye movements. In the primate, mesencephalic reticular neurons are related to lens accommodation and/or ocular convergence (Mays et al., 1986; Judge et al., 1986). The avaible data concerning activity in the brainstem in cats are limited. (Bando et al., 1984a; Sawa et al., 1994).

44

We have proposed that the LS cortex plays an important role in controlling vergence eye movement and lens accommodation (Bando et al., 1991, 1992, 1996) based on a series of experiments in the cat (Bando et al., 1984b; Toda et al., 1991; Takagi et al., 1993). In this paper, the results of our study are reviewed.

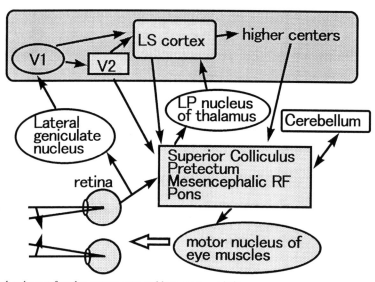

Fig. 1. Neuronal pathway of ocular convergence and lens accommodation

Materials and Methods

Cats were trained to look at an approaching target. Eye coils, a head holder and a chamber for electrode penetration were implanted under nembutal anesthesia (35 mg/kg wt). A tungsten-in-glass microelectrode was used for microstimulation and for the recording of neuronal activities. Microstimulation was performed using a current of 50 µA or less. Electrolytic lesions were made by passing a 0.5 mA DC current for 5 min through a microelectrode at 6-12 sites, covering the caudal two-thirds of the postero-medial LS cortex. Eye movement was monitored by the scleral search coil method, and lens accommodation was monitored using an infrared optometer.

The onset of convergence eye movement was determined by polynomial curve fitting (Bando et al., 1984a). Symmetry of right and left eye components of vergence eye movement was analyzed by the laterality factor. It is defined as (Ai-Ac)/(Ai+Ac), where Ai and Ac are the amplitudes of ipsilateral and contralateral horizontal eye components, respectively.

Results and Discussions

Lesion study

It is known that discrimination of moving objects (Pasternak et al., 1989; Kruger et al., 1993; Lomber et al., 1994) as well as orientating behavior (Hardy et al., 1988) are impaired by inactivation of the LS cortex. In our study, significant deficiencies in ocular convergence were found following either unilateral or bilateral lesions in the LS cortex.

In Fig. 2A, the time course of changes in parameters of ocular convergence after bilateral electrolytic lesion in the LS cortex is shown. Vergence eye movements were reduced significantly both in amplitude and velocity after lesion (Hara et al., 1992). A similar reduction in amplitudes and velocities was found after unilateral lesion (Fig. 2B).

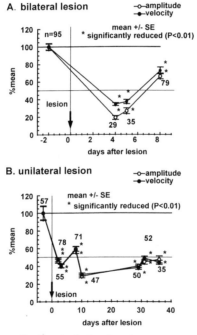

Fig. 2. Time course of changes in amplitude and velocity of ocular convergence

Differences in effects of bilateral and unilateral lesions were found in the degree of symmetry of vergence eye movements. Distribution of the laterality factor was not changed after bilateral lesions (Fig. 3A), but the component ipsilateral to the side of the lesion became dominant following a unilateral lesion (Fig. 3B). Symmetry was restored after one month.

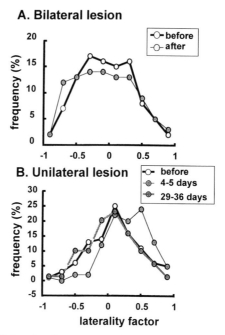

Fig. 3. Changes in the laterality factor of ocular convergence

Small and slow convergence eye movement was still elicited by visual stimulation even after bilateral lesions. One possible explanation for this is that the lesions were incomplete and that the areas of the LS cortex that were still active might be responsible for small and slow vergence eye movements. Another possibility is that the LS cortex plays a principal role in controlling vergence eye movements with higher velocities, and that other areas such as area 18 were responsible for slow vergence eye movements. Neurons related to positional and motion disparities have been found in V2 (Regan et al., 1979; Ferster, 1981).

Effects of microstimulation in the LS cortex

By microstimulation of the LS cortex, lens accommodation has been evoked in anesthetized (Bando et al., 1984; Sawa et al., 1992) cats. Disjunctive eye movement was, however, never evoked in the anesthetized animals. In conscious animals, disjunctive eye movement as well as lens accommodation were evoked by microstimulation in the LS cortex (Toda et al., 1991). The disjunctive eye movement and lens accommodation were evoked in partly overlapping regions.

Evoked disjunctive eye movements were much smaller than those elicited by visual stimulation, but both had a similar amplitude-vs.-velocity relationship. In view of this similarity, we assume that the disjunctive eye movement evoked by microstimulation in the LS cortex shared common neuronal circuitry with ocular convergence evoked by visual stimulation.

The areas effective in evoking disjunctive eye movements were systematically mapped by microstimulation with a microelectrode in the banks and the fundus of the middle suprasylvian sulcus. They were found in the rostral and caudal parts of the postero-medial LS cortex (PMLS) and the rostral postero-lateral LS cortex (PLLS), in agreement with the duplicated retinotopic organization of the LS cortex (Palmer et al., 1978).

Disjunctive eye movements have rather complex properties (Toda et al., 1996). Two peaks were found in the frequency histogram of the latencies of the disjunctive eye movements evoked by microstimulation in the caudal PMLS (Fig. 4A, abscissa): the short-latency component (SLC), and the long-latency component (LLC). As shown in Fig. 4A, the SLC and LLC had different distributions of the laterality factor. The SLC has a symmetric distribution of the laterality factor, while the LLC has a significantly dominant component contralateral to the side of stimulation.

A similar but less prominent tendency is found in disjunctive eye movements evoked from the rostral PMLS (Fig.4B). The laterality factors of SLCs showed a symmetrical distribution, and those of LLCs were significantly deviated to the contralateral side. Disjunctive eye movements evoked from the rostral PLLS were asymmetric, irrespective of the latencies (Fig. 4C).

48

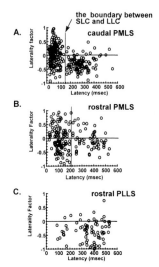

Fig. 4. The relationship between the laterality factor and the latency of evoked disjunctive eye movement

The other regional difference is that the mean velocities of disjunctive eye movements evoked from the rostral PMLS are higher than those evoked from the caudal PMLS. The mean latency of the SLC evoked from the caudal PMLS was 60 msec, the shortest of all. The latencies of the SLC evoked from the rostral PMLS were 50 msec longer than those of the SLC from the caudal PMLS. It is presumed that neuronal signals flow from the rostal to the caudal PMLS.

LS neuronal activities related to ocular convergence and lens accommodation

Neuronal activity related to lens accommodation has been reported in the LS cortex in chloralose-anesthetized cats (Bando et al., 1984b, 1988). Recently, neuronal activities in the LS cortex related to ocular convergence as well as lens accommodation were recorded in conscious cats. Activities of nineteen out of 168 LS neurons were related to convergence eye movements. They were divided into two groups: Activities of type I neurons were correlated with velocities of ocular convergence both at higher and lower target speeds, while those of type II neurons were correlated with ocular convergence only at higher target speeds. Type II neurons were also different from type I neurons in that their activities tended to correlate with lens accommodation. It is suggested that type I neurons are vergence-related, and type II neurons are accommodation-related: Type II neurons are possibly co-activated with type I neurons at higher target speeds

(Takagi et al., 1993). The activities of convergence-related LS neurons did not precede the onset of ocular convergence, but preceded the time at which ocular convergence reached its peak velocity.

Conclusion

The results of lesions, microstimulation and recordings of neuronal activities in the LS cortex support the hypothesis that the LS cortex plays a role in the control of ocular convergence as well as lens accommodation. Regional differences in the roles of the LS cortex are also suggested.

Acknowledgments
This work was supported by grants-in-aid for special projects (08279213) from the Japanese Ministry of Education, Science and Culture, and by a Grant from the Nissan Science Foundation.

References

Bando,T., Hara,N., Takagi,M., Yamamoto,K. and Toda,H. (1996) Roles of the lateral suprasylvian cortex In convergence eye movement In cats. *In: Progress in Brain Research*, vol. 112, Elsevier, pp143-156.

Bando,T., Takagi,M., Toda,H. and Yoshizawa,T. (1992) Functional roles of the lateral suprasylvian cortex in ocular near response in the cat. *Neuroscience Research* 15: 162-178.

Bando,T and Toda,H. (1991) Cerebral cortical and brainstem areas related to the central control of lens accommodation in cat and monkey. *Comparative Biochemistry and Physiology* 98C: 229-237.

Bando,T., Toda,H. and Awaji,T. (1988) Lens accommodation-related and pupil- related units in the lateral suprasylvian area in cats. In *Progress in Brain Research*, vol.75, Extrageniculo-striate Visual Mechanisms, , Elsevier, Amsterdam, pp. 231-236.

Bando,T., Tsukuda.K., Yamamoto,N., Maeda,J. and Tsukahara,N. (1984a) Physiological identification of midbrain neurons related to lens accommodation in cats. *Journal of Neurophysiology* 52: 870-878.

Bando,T., Yamamoto,N. and Tsukahara,N. (1984b) Cortical neurons related to lens accommodation in posterior lateral suprasylvian area in cats. *Journal of Neurophysiology* 52: 879-891.

Ferster,D. (1979) A comparison of binocular depth mechanisms in areas 17 and 18 of the cat visual cortex. *Journal of Physiology* 311: 623-655.

Fincham, E.F. (1951) The accommodation reflex and its stimulus. *British Journal of Ophthalmology* 35: 381-393.

Hara,N., Ando,T., Toda,H., Takagi,M., Yoshizawa,T. and Bando,T. (1992) Effect of lesion in an extrastriate cortex on convergence eye movement in the cat. *Neuroscience Research* S17: S217.

Hardy,S.C. and Stein,B.E. (1988) Small lateral suprasylvian cortex lesions produce visual neglect and derides visual activity in the superior colliculus. *Journal of Comparative Neurology* 273: 527-542.

Judge,S.J. and Cumming,B.G. (1986) Neurons in the monkey midbrain with activity related to vergence eye movement and accommodation. *Journal of Neurophysiology*, 55: 915-930.

Kruger,K., Kiefer,W. and Groh,A. (1993) Lesion of the suprasylvian cortex Impairs depth perception of cats. *NeuroReport* 4: 883-886.

Lomber,S.G., Cornwell,P., Sun,J., MacNeil,M.A. and Payne,B.R. (1994) Reversible Inactivation of visual processing operations In middle suprasylvian cortex of the behaving cat. *Proceedings of the National Academy of Science* 91:2999-3003.

Mays.L.E., Porter,J.D., Gamlin,P.D.R. and Tello,C.A. (1986) Neural control of vergence eye movements: Neurons encoding vergence velocity. *Journal of Neurophysiology* 56: 1007-1021.

Palmer,L.A., Rosenquist,A.C. and Tusa,R.J. (1978) The retinotopic organization of lateral suprasylvian visual areas in the cat. *Journal of Comparative Neurology* 177: 237-256.

Pasternak,T., Horn,K.M. and Maunsell,J.H.R. (1989) Deficits In speed discrimination following lesions of the lateral suprasylvian cortex In the cat. *Visual Neuroscience* 3: 365-375.

Poggio,G.F. and Poggio,T. (1984) The analysis of stereopsis. *Annual Review of Neuroscience* 7: 379-412.

Regan, D. , Beverley, K.I. and Cynader, M.(1979) Stereoscopic subsystems for position in depth and for motion in depth. *Proceedings of the Royal Society*, London B 204:465-501..

Sawa,M., Maekawa,H. and Ohtsuka,K. (1992) Cortical area related to lens accommodation in cat. *Japanese Journal of Ophthalmology* 36: 371-379.

Sawa, M. and Ohtsuka, K. (1994) Lens accommodation evoked by microstimulation of the superior colliculus in the cat, *Vision Research* 34: 975-981

Spear,P.D. (1991) Functions of extrastriate visual cortex in non-primate species, In A.V. Leventhal (ed.) *Vision and visual Dysfunction*. Vol. 4, The Neural Basis of Visual Function, Macmillan, Basingstroke, pp. 339-370.

Takagi,M., Toda,H. and Bando,T. (1993) Extrastriate cortical neurons correlated with ocular convergence in the cat. *Neuroscience Research* 17: 141-158

Toda,H., Takagi,M., Yoshizawa,T. and Bando,T. (1991) Disjunctive eye movement evoked by microstimulation in an extrastriate cortical area of the cat. *Neuroscience Research* 12: 300-306.

Toda,H., Yamamoto,K., Hasebe,H., Bando,T.(1996) Regional differences in the lateral suprasylvian area and the evoked potentiation of ocular convergence in cats. *Neuroscience Research*, S20: 184.

Toyama, K., Komatsu, Y. and Kozasa, T. (1986) The responsiveness of Clare-Bishop neurons to motion cues for motion stereopsis. *Neuroscience Research* 4:83-109.

Westheimer, G. and Mitchell, G. (1969) The sensory stimulus for disjunctive eye movements. *Vision Research* 9: 749-755.

Zeki S.M. (1974) Functional organization of a visual area in the posterior bank of the superior temporal sulcus of the rhesus monkey. *Journal of Physiology* 236:549-573

Research on dynamic accommodation using TDO III (Three Dimensional Optometer III) and MEG (Magnetoencephalography)

T. Takeda, H. Endo and K. Hashimoto

National Institute of Bioscience and Human-Technology (NIBH), Human Informatics Department, 1-1 Higashi, Tsukuba City, Ibaraki, 305, JAPAN

Summary: We have developed a TDO III (Three Dimensional Optometer III), which can measure dynamic accommodation, eye movement and pupil diameter simultaneously, while subjects are looking at visual objects by moving their eyes. We have also developed a TVS (Three-dimensional Visual Stimulator) using a similar optical system, which can present independent binocular targets of different distance, direction and image size.

Using the TDO III and the TVS, simultaneous accommodation and vergence responses toward a real image and/or a stereoscopic image created by the TVS were measured. It was found that the accommodation toward the stereoscopic image showed peculiar responses that have less amplitude compared with the response toward the real image, and a receding movement of accommodation from the initial peak in the response.

Recently we have installed a 64-channel whole-cortex MEG (Magnetoencephalography) system to study the control mechanism of the accommodation/vergence by the central nervous system. A special relay lens system has been developed both to impose visual stimuli and to measure accommodation responses with a dynamic refractometer. Using these systems, MEG and accommodation responses were measured simultaneously for the first time.

An accommodation response toward stepwise stimuli was found after about 300 ms from the onset of the stimuli. Two highly synchronized MEG responses probably related to the accommodative control were found at about 100 and 200 ms from the onset of the stimuli prior to the accommodative response. This line of research was deemed to be of help in elucidating the complex accom-modation/vergence mechanisms.

Introduction

It has been difficult to measure dynamic accommodation together with vergence, which is essential for the elucidation of accommodation/vergence mechanisms. Hence simultaneous measurement of three major ocular functions – accommodation, eye movement and pupil diameter – in a real working environment, has been a method demanded in many fields. As a first attempt, we developed a dynamic refractometer that could measure dynamic accommodation easily by modifying a commercially available auto refractometer (Takeda, Fukui and Iida, 1985). However, it required subjects to fix their eye direction to coincide with the optical axis. We then developed a three-dimensional optometer III (TDO III), to measure dynamic accommodation together with eye movement and pupil diameter, while subjects were moving their eyes and heads to watch visual objects (Takeda, Fukui and Iida, 1988, 1993).

It is indispensable to have a visual stimulator which can stimulate the subject's eyes by changing each visual function independently or by combining some of them freely. As there was no such stimulator, we developed a three-dimensional visual stimulator (TVS), which can move the targets within a 15° x 16° (height x width) rectangular visual field, changing their distance, direction and target size with the aid of the relay lens system (Takeda et al., 1995).

Several new non-invasive measurement methods for human brain functions such as PET, fMRI and MEG have recently been developed, As the MEG system is the most promising for the elucidation of dynamic features of the brain, we have recently introduced a 64-channel whole-cortex MEG system. The combination of an optometer with the MEG system allows a more in-depth study of accommodation by analyzing both ocular responses and brain activities. In order to carry out such research, a special relay lens system has been developed to deliver visual stimuli while measuring the accommodation responses using a dynamic refractometer from the outside of a magnetically shielded room. With these devices, MEG and accommodation responses have been simultaneously measured for the first time. This new approach should be highly effective for acquiring new insights concerning the cerebral visual functions.

TDO III (Three Dimensional Optometer III)

Figure 1 shows the TDO III system. The subjects can move their heads and eyes freely while the head movement and the response of the three major ocular functions are being measured simultaneously. Alignment is easily performed with a three-axis movable chin rest that is powered by small motors. The alignment can be performed within several minutes with the aid of a CCD camera located on the optical axis and a CCD camera located just below the eye. It can be performed within 10s for an immediate repetition on the same subjects.

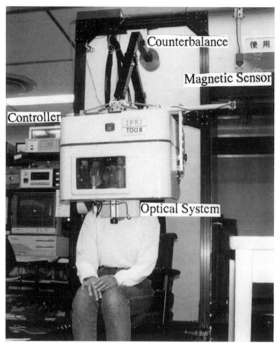

Figure 1. The three-dimensional optometeráV system. Optical components are located in a box on a subject's head which are counter balanced to allow free head movement. Controllers are located on the left.

Figure 2 depicts the system configuration of the TDO III. The measurement of accommodation is performed by the optical system of the auto refractometer (modified from NIDEC, AR-1100). With the original AR-1100, subjects are required to gaze into the apparatus and fix their eye position but in the modified model they are allowed to move their eye freely to watch objects with the aid of a relay lens system.

The X-Y tracker in the percept Scope (the first function of C3160, Hamamatsu Photonics) measures eye movement, and the galvanomirrors are driven by a servo controller of the TDO III to maintain the monitored eye in the center of a CRT using the measured data. Infrared measurement light is directed into the eye by use of the servo control mechanism, irrespective of eye movement. Pupil diameter is measured by the second function of the Percept Scope (Area Analyzer). It slices the monitored image of the eye at a given threshold level on a CRT, and counts pixels brighter than the threshold level. Then the pupil diameter is calculated using the assumption that the pupil is a circle. The microcomputer controls the measurement timing and takes in the data of the three ocular functions and the head movement.

54

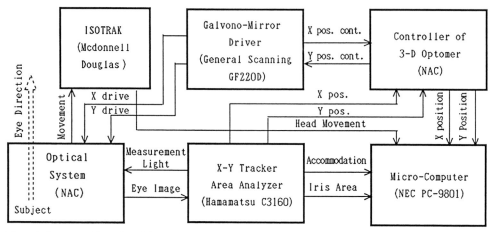

Figure 2. System configuration of the TDO III.

Figure 3 depicts the top view of the realized relay lens system of the TDO III. The dichroic mirror, that transmits almost all visible light and reflects almost all infrared measurement light, is placed as close to the eye as possible to reduce the size. Two spherical mirrors, B and E, and two convex lenses, I and K, constitue a simple relay lens system that transfers images of the eye with identical magnification at the point AR. Galvanomirrors C and G are used to compensate for eye movements to form a stabilized eye image irrespective of eye movement. The mirror F can be rotated around two axes by electric motors to adjust distance of eyes. The L O mirrors serve to adjust optical length.

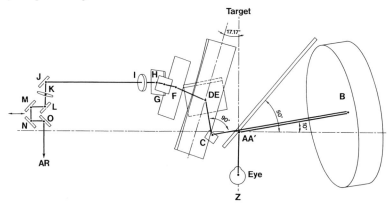

Figure 3. Relay lens system of the TDO III (top view). A,A': dichroic mirror; B,E: spherical mirrors; C,G: galvanomirrors; I,K: convex lenses; F: optical axis conversion mirror; L³O: optical length modulator mirrors; D,H,J: relay mirrors.

The optical system for the measurement of accommodation is shown in Figure 4. The infrared measurement light is emitted by IR LEDs located at the center. As its central frequency is 880 nm, the subjects sense no measurement light. The reflected image of the diaphragm on the retina is again reflected by a beam splitter and makes a conjugate image on the photo detectors located on the upper right. The principle of the measurement is the same as that of the Campbell optometer (Campbell and Robson, 1959). The TV monitor shown at the lower part of the figure makes the measurement much easier, saving the time required for alignment. The final characteristics of the TDO III are summarized in Table 1.

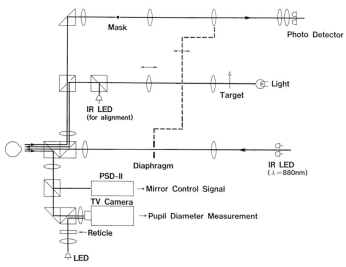

Figure 4. Optical system for the measurement of dynamic accommodation.

Table 1 Characteristics of the TDO III.

Range:	Accommodation	-12.7D [a] +26.6D
	Eye movement	
	Horizontal	±20°
	Vertical	-25° [a] +5°
	Pupil diameter	0 [a] 100 %
Speed:	Accommodation	4.7 Hz
	Eye movement	6.4 Hz
	Pupil diameter	6.4 Hz
Resolution:	Accommodation	±0.02D
	Eye movement	
	Horizontal	±0.18°
	Vertical	±0.25°
	Pupil diameter	±0.11 mm
Accuracy:	Accommodation	±0.25 D
	Eye movement	±0.5°
	Pupil diameter	±0.3 mm

Accommodation response toward actual artwork

To show the performance of the TDOáV, photographs of paintings with ample depth cues were used, such as., Christina's World 1948 by A. Wyeth (Figure 5). The photograph was 31 by 23 cm in size and presented at a distance of 33 cm (-3 D) from the subjects. As the depth of the TDO was 25 cm, there was sufficient space to permit the required lighting (8 cm). N and F on the figure denote near and far fixation points, which are not actually displayed but given as verbal instructions. The measurement was performed under normal lighting conditions, using binocular vision. However, the illuminance was rather low (178 lux), because the paintings were held upright and no special lighting was used. This condition was chosen intentionally to prevent any effect by reflected light on the measurement.

Figure 5. Reproduction of Christina's World, 1948, by A. Wyeth. The subject gazed alternately at N and F, which were not really displayed in the picture.

The measurement was performed 5 times. Each measurement consisted of four fixation point shifts (two of FÆN and two of NÆF). General trends were the same but the magnitude of responses differed. Figure 6 shows an example when the subject OR scanned the picture shown in the picture. Eye movements, especially vertical ones (Ver.), clearly indicated fixation-point shift. Arrows indicate the change in time of fixation points. N and F denote that the visual attention of the subject was directed at a near or a far fixation point. The time span of the gaze was not fixed so as to prevent subjects' anticipation.

The subjects showed clear accommodative responses to the apparent depth cues in the paintings. The steady-state portions of the response curves were averaged for 3 s and the result was defined as diopters of accommodation for near (N) or far (F) fixation. The difference between the two responses was defined as the extent of accommodation induced by the stimulation. The average amount of accommodation was 0.77 D with SD of 0.19 D (Takeda, Iida and Fukui). It was later determined that the induced accommodation was mainly a response to the perceived depth sensation and was not caused by the lowering of the eye position (Takeda, Neveu and Stark, 1992)

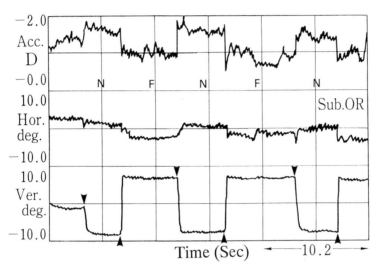

Figure 6. The response of subject OR to monocular vision while looking at the painting (clear accommodative response is shown by perceiving apparent depth).

TVS (Three-dimensional Visual Stimulator)

The TVS (Figure 7) can change the directions, distances, sizes, luminance and varieties of two sets of targets for both eyes independently (Takeda et al., 1995). Figure 8 depicts the schematic optical system of the TVS, which consists of (1) liquid crystal projectors (LCPs) that generate flexible visual targets for both eyes, (2) Badal optometers that change target distances while maintaining constant visual angles, and (3) an optical relay lens system. The ranges of stimulation are as follows; distances: 0^a -20 D, directions: ±16° horizontally and ±15° vertically, sizes: $0°^a 2°$ of visual angle and luminance: 10^{-2} ^{a} 10^2 cd/m2. The target images are refreshed at 60 Hz and speeds of smooth change of the target (ramp stimuli) are as follows: distance: 5D/s, direction: 30°/s, size: 10°/s.

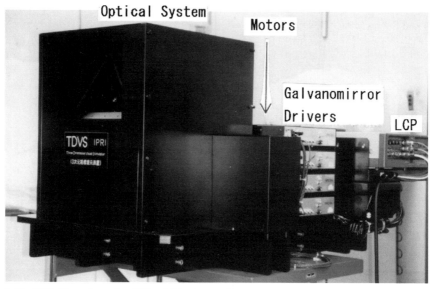

Figure 7. The TVS (Three dimensional Visual Stimulator) which can binocularly present a visual target and change its distance, direction and size.

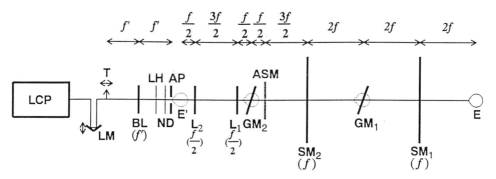

Figure 8. Optical systems of the TVS.

Accommodation response toward real and stereoscopic images

A maltese cross was presented binocularly by the TVS under the following 2 conditions: (Real): the accommodative distance, vergence angle and image size were changed as if a real target moved from-1 D (1m) to -4 D (.25m); (Stereoscopic):the distance was kept constant at -1 D while the vergence angle and the size were changed as if a stereoscopic TV image was shown. Figure 9 shows typical accommodation (Acc.), vergence (Ver.) and pupil (Pup.) responses of subject HN for the two conditions.

Measurements clearly show that the accommodation is evoked in both conditions. Very interestingly, accommodative response to a stereoscopic image reached approximately the same magnitude as that evoked by the image and then fell off sharply. There have been many subjectively examined papers reporting that severe visual fatigue is induced by using the stereoscopic displays. It seems likely that the difference in accommodative response found when real and stereoscopic images were compared, are closely linked to the phenomenon of visual fatigue reported in this study.

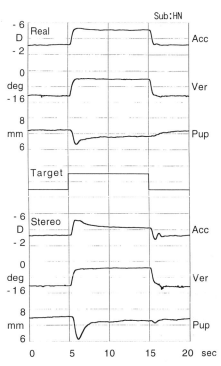

Figure 9. Responses of subject HN toward a real image (upper three graphs) and a stereoscopic image (lower three graphs) together with the target movement (middle). Accommodation toward the stereoscopic image changed to almost the same extent as that of the real image, then receded considerably.

MEG (Magnetoencephalography)

Figure 10 depicts the schematic structure of the MEG system. The sensor coils with a 5 cm base line distance are equally distributed over the head with a mean distance of 4 cm. There are reference coils to measure environmental noise to reduce zero effect of noise, which is the most important characteristic of the system (Vrba et al., 1982). The whole system is kept cool with liquid helium for the four days the experiment lasts. The measured data are pre-amplified and digitized using an A/D converter and stored in a micro-computer.

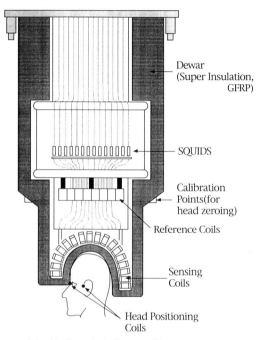

Figure 10. Schematic structure of the 64-channel whole-cortex biomagnetometer.

A phantom head (75 mm in diameter) was used to evaluate the noise rejection algorithm. The noise reduction was evaluated in two conditions varying the number of recordings. The door of the magnetically shielded room was closed in condition C and was left open with an active CRT placed at the entrance of the magnetically shielded room in condition O. The results (Figure 11) show that the noise rejection algorithm was effective for condition O, especially for fewer average numbers. The data confirm that the noise was reduced to an acceptable level with the door open after averaging 2^5 recordings or more (Takeda et al., 1996a). The main characteristics of the whole-head MEG system are shown in Table 2.

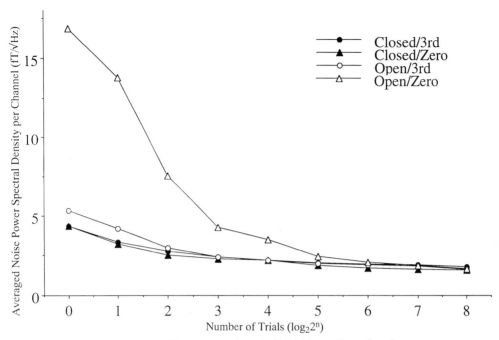

Figure 11. Reduction in the average noise power per channel with the increase in number of averages.

Table 2 Characteristics of the MEG system.

Dewar	
Weight	[a] 200 kg
Volume of liquid He	[a] 70 l
Holding time	[a] 4 days
Helmet Shape	Over 95% of Japanese can be measured.
Sensor Coils	
Type	1st Gradient
Diameter	2 cm
Baseline	5 cm
Number	64 channels
Distribution	Equally distributed over whole cortex.
Software	1-3 order noise reduction
	Filters up to 20th order with no phase shift
	DC offset reduction

The accommodative amplitude in this paper were measured objectively with the TDO III and were about 1D less than the diopters measured subjectively.

Accommodation stimulus and recording system

In order to measure the accommodation together with MEG, we made a special relay lens system which optically transferred the subject's eye image to the outside of the magnetically shielded room (Figure 12). It consisted of 4 identical spherical lenses with 400 mm focal lengths, moving the eye by 3200 mm to form a real image just in front of the dynamic refractometer where the subject's eye was placed in normal measurements. The accommodation measurement was performed using the real image of the eye. The dynamic changes of the focus point of the eye were measured by the optical system shown in Figure 4.

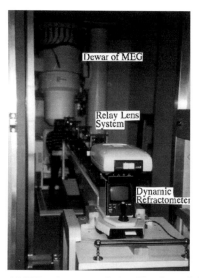

Figure 12. Measurement systems of MEG and accommodation. The door of the shielded room was kept open allowing the presence of a relay lens system which transfers a real eye image outside the room to measure accommodation.

MEG and accommodation responses

Subjects and Procedure

 Subjects were 3 right-handed volunteers with adequate accommodative power and no ocular problems other than myopia. The accommodative power was measured with the dynamic refractometer by moving the target from a position located beyond the far point (the farthest point on which the subject could focus) to another position nearer than the near point (Takeda, Fukui and Iida., 1985). The amplitude of a stepwise stimulius was set to about 60 % of each subject's accommodative power to prevent excessive visual fatigue. Subjects were instructed to try to see the target as clearly as possible. The target was changed in a stepwise manner with a random time interval of 5±0.5 s.

The measurements were performed on the right eye, occluding the left eye with an eye patch. The experiments consisted of more than 64 recordings; the collected data were averaged after discarding those contaminated by eye blinks. The sampling rate was set to 250 Hz. The recording was started 1.5 s before the onset of the stimulus and lasted 2.5 s. The evoked fields, the accommodation response and the target position were simultaneously recorded on a personal computer. The trigger signal for MEG measurements was set at a predetermined threshold of the target. The MEG signals were band-pass filtered between 0.5-40 Hz and the DC offset was removed using a 1.5 s pre-recording.

Response example

Only the responses of the subject HK (36, male with 4D of accommodative power) are shown in this paper to save space. Figure 13 depicts the response of HK, when he gazed at a stationary target at the farthest distance?? (-5D) in the refractometer. MEG recordings of all the 64 channels were superimposed, and showed a constant noise-like baseline activity of the brain with the intensity of about 50 fT.

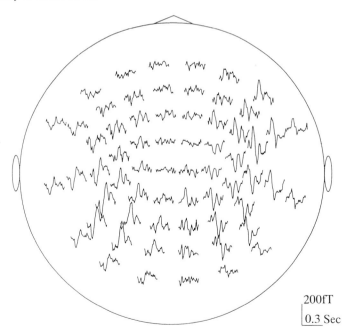

200fT

0.3 Sec

Figure 13. MEG response of all channels. Active sensors are shown mainly over the temporal and occipital regions.

The target was then moved between -5 and -8 D, because HK was -5 D myopic. Figure 14 depicts whole channels' responses for the target movement from far to near, which is shown on a circular projection of the head. Clear responses are found on the sensors located over occipital and lateral lobes.

Figure 14. MEG responses without accommodative stimulus.

Figure 15 shows the accommodation and superimposed MEG responses toward the same stimulus together with the target movement (Takeda et al., 1996b). The time lag of the accommodative response was 288±50 ms. It was calculated by averaging the time lags of the 64 recordings. The beginning of accommodative responses is indicated by an arrow in the figure. Although the amplitudes do not match the stimulus, this is normal and is called the accommodation lag (Tucker and Charman, 1979). The mean time lag and the shape of the accommodation responses correlate well with the literature (Campbell and Westheimer, 1959).

Superimposed MEG responses indicated well synchronized activation of brain around 100 and 200 ms. Both peaks showed the phase reversal phenomenon. On the other hand, the responses were rather quiet around 300 ms when the average accommodative response began to rise. Several additional control experiments strongly imply that the second synchronized activity in the MEG comes from the brain activity involved in controlling accommodation. The control center of the accommodation, about which little is known, even though electrophysiological studies in mammals (Bando and Toda, 1991), is now under careful examination and more experimental data are being gathered.

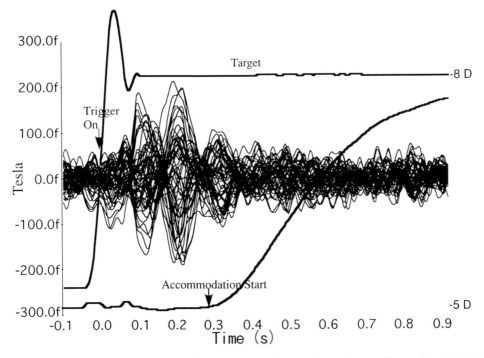

Figure 15. Typical MEG responses, accommodation responses and target change. Two peculiar synchronized MEG responses were found prior to the accommodation response.

Conclusions

This paper shows the basic functions of the TDO and the TVS when measuring accommodation and vergence objectively and simultaneously, and of the MEG for measuring the magnetic response of brain activity. It also introduces interesting accommodative responses when subjects are looking at 2-dimensional paintings, and also the time relationship between MEG and the accommodative responses. The control mechanism of accommodation/vergence is now under intensive research using these new devices. We believe that the TDO, TVS and the MEG will open the door to new insights into the complicated accommodation/vergence control mechanism.

References

Bando, T. and Toda, H.. (1991) Cerebral cortical and brainstem areas related to the central control of lens accommodation in cat and monkey, *Comparative Biochemistry and Physiology* 98C, 1, 229-237.

Campbell, F.W. and Robson, J.G. (1959) High-Speed Infrared Optometer, *Journal of the Optical Society of America*, 49, 268-272.

Campbell, F.W. and Westheimer, G. (1959) Factors influencing accommodation responses of the human eye, *Journal of the Optical Society of America*, 49, 568-571.

Takeda, T. Fukui,Y. and Iida, T. (1985) An Objective measurement apparatus for accommodation ability change cause by VDT work, *Journal of Ophthalmology Optical Society of Japan*, 6, 59-66

Takeda, T. Fukui, Y. and Iida, T. (1988) Three-dimensional optometer, *Applied Optics.*, 27, 12, 2595-2602.

Takeda,T. Iida,T. and Fukui,Y. (1990) Dynamic eye accommodation evoked by apparent distances, *Optometry and Vision Science.*, 67, 450-455.

Takeda,T. Neveu,C. and Stark,L. Accommodation on downward gaze, *Optometry and Vision Science*, 69, 556-561 (1992).

Takeda, T. Fukui, Y. and Iida, T. (1993) Three-dimensional optometerV, *Applied Optics*, 32, 22, 4155-4168.

Takeda,T., Fukui,T., Hashimoto,K., Hiruma,N. (1995) Three-dimensional visual stimulator, *Applied Optics*, 34, 4, 732-738.

Takeda, T., Morabito, M., Kumagai, T. and Endo, H. (1996a) Use of CRT as a visual stimulator in MEG measurements, in Recent advances in event-related brain potential research, eds. Ogura, C., Koga, Y. and Shimokochi, M., Elsevier Amsterdam, 510-516.

Takeda,T., Morabito,M., Xiao,R., Hashimoto,K., Endo,H. (1996b) Cerebral activity related to accommodation : *A neuromagnetic study, EEG Supplement.* 47, 283-291.

Tucker,J. and Charman,W.N. (1979) Reaction and response times for accommodation, *American Journal of Optometry & Physiological Optics*, 56, 490-503.

Vrba J. Fife, A. A. and Burbank, M. B. (1982) Spatial discrimination in SQUID gradiometers and 3rd order gradiometer performance, *Canadian Journal of Physics*, 60, 1060-1073.

Neuroanatomical correlates of the near response:
Voluntary modulation of accommodation in the human visual system

Hans Richter,[1,2] Joel T. Lee,[3] and José V. Pardo[3,4]

[1]*Department of Clinical Science, Division of Ophthalmology, Karolinska Institute, Stockholm, Sweden.* [2]*PET Centre, Uppsala University, Uppsala, Sweden.* [3]*Division of Neuroscience Research, Department of Psychiatry, University of Minnesota, Minneapolis, Minnesota, USA.* [4]*Psychiatry PET Unit, Psychiatry Service, Veterans Affairs Medical Center, Minneapolis, Minnesota, USA.*

Summary. This study attempts at identifying brain regions concerned with the execution, during monocular viewing, of eye movements for voluntary positive accommodation (VPA) to a blurred checkerboard pattern. Neuronal activity was estimated in thirteen normal volunteers by measurement of changes in regional cerebral blood flow (rCBF) with positron emission tomography (PET) and the autoradiographic ^{15}O-water technique. Stereotactic brain atlases, originally used in surgery, permit cerebral localization in a 3-D coordinate grid based on cerebral landmarks. VPA requires attentional capacity and involves an integrated network of changes in neural activity distributed through occipital, frontal, parietal, and temporal cortices as well as the cerebellum. Cerebral blood flow (CBF) along the dorsal pathway underlying spatial vision was systematically inhibited or reduced, whereas activity along the ventral pathway, more concerned with object vision, was increased during processing and execution of repetitive, large-amplitude, VPA responses.

Introduction

The principal stimulus triggering a visual accommodative response (AR) under typical viewing conditions is dioptric blur. By synergistic coupling, an AR is normally accompanied by vergence movements of the eyes and pupillary constriction/dilatation. The human central nervous system (CNS) processes that transform the blurred sensory representation into an appropriately scaled motor-command signal, to be executed by the ciliary muscles, are poorly understood (Richter and Franzén, 1994). The present experiment tests the hypothesis that monocular VPA movements of the crystalline eye-lens involve neural brain circuits which can be visualised in healthy, alert, humans with PET neuroimaging.

Methods

Subjects. Thirteen naive subject (7 females and 6 males; mean age of 27 yrs., S.D. $\pm$ 6; range 20 - 43; all but one right-handed) participated in the PET-study. All participants had normal or normal, corrected vision. The volunteers gave written informed consent according to guidelines of the institutional review board and the radioactive drug research committee.

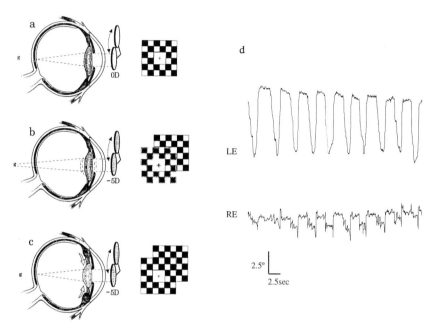

Figure 1. Experimental design for neuroimaging of VPA. (**a**) The *passive viewing* condition, involved maintaining fixation while a 0.0 D lens was intermittently placed in front of the dominant eye. (**b**) The *blurred viewing* condition required the subject to avoid focusing upon the checkerboard whenever a -5.0 D lens intermittently interrupted the line of sight. The AR is inhibited volitionally and the retinal image remains blurred. A "breakdown" in the AR during large (>3.0 D) dioptric defocusing can nullify the normal, closed-loop, negative feedback mechanisms (Richter et al., 1994). The AR may not posture precisely onto the checkerboard in this blurred condition because of subthreshold contrast. The AR would instead approximate the intermediate resting state of the accommodative system (Owens and Higgins, 1983). Consequently, a small amplitude negative (or positive depending on magnitude of individual resting level) reflexive AR may result upon removal of the -5.0 D lens. (**c**) Condition *voluntary positive accommodation* (VPA) required the subject to maintain maximal focus upon the checkerboard whenever the -5.0 D lens interrupted the line of sight. High frequency (0.3 Hz), large amplitude (~+5.0 D), voluntary AR must occur for successful performance. A compatible distribution of negative AR results after the removal of the -5.0 D lens, after each successful VPA response. Two additional controls (not shown) included passive viewing of a -5.0 D *degaussed checkerboard* and *eyes closed rest* (ECR). (**d**) Typical amplitudes of eye-movement response during VPA to a -5.0 D blurred checkerboard. Co-activation of vergence in left (occluded eye) during VPA is obvious (M: 8.26 SD:$\pm$1.19). Although most vergence in fixating RE is inhibited, small order conjunctional divergence movements to the right can be observed (M: 0.74, SD: $\pm$0.27). RE: right eye position. LE: left eye position. Downward movements deflection denotes eye-movement to the right.

The monochromatic high-contrast checkerboard (1 c/deg) subtended 11° (horizontally/vertically) and contained a central fixation cross (1.5°) so as to reduce or inhibit saccadic eye-movements and smooth pursuit (Fig.1). Space average luminance of the screen and surround was 37.5 cd/m^2 and 1.0 cd/m^2, respectively (Quantum Instrument Inc., Photometer LX, Garden City, NJ). The absence of saccadic eye-movements was confirmed by electro-oculogram (EOG) recordings.

PET scans were acquired on an ECAT 953B camera (Siemens, Knoxville, TN) in 2D mode for 60 sec following an intravenous bolus injection of 50 mCi H$_2$^{15}O (for details on scan-acquisition, scan-realignment and stereotactic transformation, see: Pardo et al., this volume; Zald and Pardo, 1997). Potentially, the tasks selected in this protocol afford visualisation of the retinal image (blurred checkerboard), of the changes in the retinal image induced by VPA, and of the sensory-motor processes in VPA/vergence.

Data analysis

To explore the functional anatomy of voluntary accommodation processes across the whole brain, paired image subtraction analysis were completed for the following two contrasts: 1) VPA minus *passive viewing* (c-a) and 2) VPA minus *blurred viewing* (c-b). The results from the remaining two contrasts 3) VPA minus *degaussed* (c-d) and 4) VPA minus *eyes closed rest* (c-e) will be reported elsewhere.

Spherical 7 mm radius region of interest (ROI) were centred on the coordinates of peak subtracted rCBF (ΔrCBF) resulting from VPA minus either *passive* or *blurred viewing*. The Taliarach coordinates that guided these ROIs are listed in Table 1. Pearson product moment correlations were calculated to provide a common metric for displaying the relation between activity in the different ROIs. Preliminary findings will be reported for this analysis. By cancelling sources of variance unrelated to VPA, the between-task ΔrCBF matrix shows a commonality in how structures become activated or deactivated during the production of VPA (for details of methodology, see Zald, Donndelinger and Pardo, 1998).

Since the blurred (or unblurred) sensory representation plays a crucial role in VA processing, unsubtracted rCBF in visual cortex was assessed across the different scan conditions and repeated-measures ANOVAs were performed on individual activity in calcarine or occipital ROIs (7 respectively 16 mm radius). The ROIs were centred on the Taliarach coordinates (x, y, z) ±17, -94, -7 bilaterally in BA 17 or similarly on ±17, -87, and z: 0, so that in the second case extrastriate areas (BA18 and 19) also were included.

Result and discussion

Occipital activation during VPA

An analysis of mean rCBF (AVERAGE) in the calcarine ROIs (7 mm radius) involved a 5 x 2 design with the first variable, SCAN CONDITION, consisting of five levels: 1) *Passive viewing*, 2) *Blurred viewing*, 3) *VPA*, 4) *Degaussed checkerboard*, 5) *Eyes closed rest*, and a second variable HEMISPHERE, consisting of two levels (left, right). No significant effects surfaced involving HEMISPHERE (P=0.32). The main effect of SCAN CONDITION was significant (P=0.01). Since no significant effect of HEMISPHERE occurred, left and right hemispheric values were averaged. Group mean rCBF for the ROIs in BA17 are displayed as a function of scan condition in Figure 2.

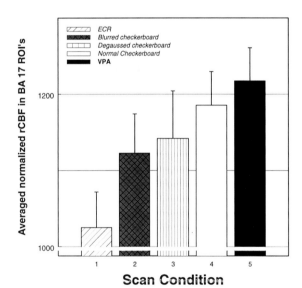

Figure 2. Calcarine sulcus ROIs. Group mean unsubtracted activity (rCBF) within bilateral 7 mm diameter spherical ROIs centred on the foveal portion of BA 17 during controls and during VPA. Stereotactic co-ordinates: (x, y, z): ±17, -94, -7.

Data from the occipital ROIs (16 mm radius) were used to calculate product moment correlation coefficients, r_{xy} as a measure of hemispheric coupling (x = ipsilateral rCBF during *passive viewing*; y = contralateral rCBF during *passive viewing*). A significant correlation was obtained during VPA (r_{xy} = 0.62, P<0.05) and left individual hemispheres were associate with right hemispheres of similar magnitude. The hemispheric dependence on the degree of activation was somewhat less evident during during *passive viewing* (r_{xy} = 0.54, P=0.06). These correlations may be explained by the fact that individual vascular supply represents a stable feature of occipital functional responses. Nine out of the thirteen subjects manifested a contralateral dominance in rCBF, i.e., the contralateral values tended to have a higher magnitude then the ipsilateral values (P<0.05 for both *passive viewing* and VPA). This asymmetry [(dominant hemisphere/none-dominant hemisphere)-1] remained unchanged during *passive viewing* or during VPA. The degree to contralateral asymmetry averaged 8% (range: 3.9% - 11%, S.D. $\pm$ 2.72) whereas averaged ipsilateral asymmetry equalled 4.26% (range: 3.7% - 5.6%, S.D. $\pm$ 0.98).

A distributed set of activation foci appeared within the occipital lobule (BA 17/18) when VPA was contrasted to any one of the four controls in the paired image subtraction analysis. These results closely mirrored the outcome from the preceding ROI analysis. For example, the z-scores in the VPA minus *blurred viewing* subtraction (c-b) were higher relative to the z-scores obtained in the VPA minus *passive viewing* comparison (c-a. See Table 1).

The visual activation in BA17 from the VPA vs. *blurred viewing* comparison c-b (Table 1c) probably resulted from an increased retinal contrast in condition VPA from successful accommodation. The increased visual activation in the VPA minus *passive viewing* comparison c-a (Table 1a and Fig. 4) can not be explained so readily because even if the subjects had been completely accurate in their accommodation, no activation in condition VPA would be expected over that in the normal checkerboard. Degradation of the retinal image by dioptric defocus is known to decrease the amplitude of the contrast dependent component in visual evoked potentials recorded over the occiput (Franzén, Lennerstrand and Richter, 1994). In the *passive viewing* condition the visual target was always in focus.

Cortical cells are know to rapidly decrease their firing rate in response to continuous stimulation. Hence, contrast adaptation may have occurred selectively in only the *passive viewing* condition while the intermittent lens presentations at 0.3 Hz might have acted to prevent contrast adaptation in both the *blurred viewing* and the VPA conditions, respectively. If contrast adaptation did not occur in the *blurred viewing* condition, then rCBF in this condition should be higher than the rCBF evoked by the *degaussed checkerboard*. In the later case, the visual target was constantly blurred and some adaptation could have occurred. However, the resulting rCBF did not change across these conditions (corrolated students *t*-test; P=0.82. Fig. 2).

An intriguing possibility is that part of the striate activation (Fig. 4) was caused by re-entrance. Geniculocortical terminations in the monkey constitute only about 30% of the total number of synapses, even in layer 4. The remaining inputs derive from a variety of intrinsic and extrinsic sources. Rockland and Van Hoesen (1994) detailed the morphology of feedback connections from area V2 to V1. The laminal termination of these connections is mainly in layer 1, 2, 3A, 5 or infrequently 6. The widespread, divergent arborization of feedback axons, especially in layer 1, implies that one axon may contact a large number of neurones. One aspect of feedback connections that has attracted interest, arises from their relationship with feedforward connections (reciprocity of connections). Many feedback connections from cortical areas do not 'reciprocate' any identified feedforward pathway from V1 (Rockland and Van Hoesen, 1994). Two hypothetical re-entrant mechanisms operating within the occipital lobule during accommodative processing are briefly suggested.

First, the occipital activations (Table 1c and Fig.4) arise from hardwired, local re-entrance within micro-circuitry analogous to servo-mechanisms. Even-error and odd-error focusing signals may form at an early stage within visual cortex; the occipital activations reported here may in part be a consequence of ascending and descending streams of sensory information (Ullman, 1995). For example, the descending stream may represent a preset value of focusing power to achieve the intended contrast, while the ascending stream relays the updated retinal contrast information.

Second, modulation related to perceptual, cognitive, and attentional processing during VPA provide another source for visual activations in the VPA vs. the *passive viewing* comparison (c-a). The behavioral relevance of the visual target in condition VPA probably is much greater than during the unblurred checkerboard condition (e.g. Fulton, 1928; Kosslyn et al., 1993; Heinze et al., 1994; Motter, 1993).

Temporal activation during VPA

Activity in the right temporal region (BA22) was raised above the baseline in the VPA vs. the *blurred viewing* comparison (c-b, Table 1c, #3). In the VPA minus the *passive viewing* comparison the same area right BA22 was activated, but failed to reach conservative levels of significance (Z = 2.78). Right BA22 ΔCBF correlated (P<0.05) with several regions that surfaced in the same comparison: right BA 17; left BA 18; thalamus; and right basal ganglia (Table 1c, #1, 2, 4 and 5).

Table 1.   Brain regions implicated in voluntary positive accommodation (VPA).

Foci	Talairach Coordinates			Magnitude of peak activation

(a) Regions activated by the VPA response (VPA minus "*passive viewing*" comparison, c-a).

	x	y	z	Z-Score
1. Right striate visual cortex (BA 17)	19	-89	-7	4.35
2. Anterior lobe of vermis	8	-67	-2	3.94
3. Left visual cortex (BA 17)	-10	-76	14	3.54
4. Anterior lobe of vermis	3	-69	-16	3.22
5. Left visual cortex (BA 18)	-6	-85	20	3.20

(b) Regions deactivated by the VPA response (Comparison of "*passive viewing*" with VPA, a-c).

1. Left superior frontal gyrus (BA10)	-15	62	25	-5.09
2. Right Frontal Lobule (BA6)	26	-4	45	-3.85
3. Left/right posterior Cingulate	-3	-55	29	-3.48
4. Left Parietal Lobule (BA 39/40)	-35	-46	32	-3.40

(c) Regions activated by VPA, as shown by the VPA minus "*blurred viewing*" comparison (c-b).

1. Right primary visual cortex (BA 17)	17	-94	-7	5.16
2. Left visual cortex (BA 18)	-17	-80	-9	4.29
3. Right Lateral temporal sulcus (BA22)	60	5	4	3.44
4. Thalamus (midbrain)	-1	-8	0	3.40
5. Right Lenticular nucleus (Basal ganglia)	33	-10	-2	3.32
6. Right lateral cerebellum	26	-73	-27	3.22

(d) Regions deactivated by the VPA response (Comparison of "*blurred viewing*" with VPA, b-c).

1. Left superior frontal gyrus (BA10)[*]	-15	62	25	-4.27

Response magnitudes and locations in coordinates of the atlas of Talairach and Tournoux (1988). x, y, and z (in mm) correspond to right-left, anterior-posterior, and superior-inferior dimensionsof the brain. Right, anterior, and superior sides have positive coordinates. BA refers to approximate Brodmann areas. Foci just outside the brain are labeled by an asterisk.

Temporal cortex contribution to processing within human visual accommodative (VA) circuitry is presently unknown. However, VA processing in the cat within the Lateral Syprasylvian sulcus (LSS) and Middle Syprasylvian sulcus (MSS), both extrastriate parietotemporal regions homologous to macaque V5, have been studied extensively by Bando and co-workers (Bando and Toda, 1991; Yoshizawa et al., 1991). Individual components of the near triad have been dissected using the microelectrode stimulation technique. The LS region contains two specialisations: 1) a rostral region (receiving visual afferents) primarily involved with visual-sensory processing (secondary, odd-error cues); 2) a caudal region showing shorter latency and threshold for the VA response, which may provide output motor signals for the subcortical system to control lens accommodation. The rostral LS areas may not send motor signals to the subcortical areas. Instead, it is likely that the rostral area plays the role of providing sensory information for the caudal area by processing the visual blur signal from the eye. Distinct brain regions relevant to vergence and pupillary changes, also appears localised to partially overlapping or to separate cortical loci in LS and MS

Jampel (1955) found ARs, in addition to vergence eye movements and pupillary constriction, evoked by faradic stimulation of the cerebral cortex surrounding the STS. Electrophysiological results obtained in the macaque temporal lobe in the vicinity of STS underscore sensory responsiveness. The posterior dorsal segment of the rhesus STS receives projections from visual cortices. These neurons respond to spatial aspects of motion (sometimes in depth) and project to the intermediate layers of the superior colliculus (Zeki, 1974). The human homologue to V5/BA 37, lies closer to the temporal lobe than the parietal lobe unlike LS area or Macaque V5 (McKeefry et al., 1997). Neurones in the anterior ventral portion of the macaque STS have large receptive fields and often respond to motion in depth with little or no preference for stimulus size, shape, orientation, or contrast (Bruce et al., 1981). Areas critical for object recognition are also known to coexist in macaque STS area.

Parietal and frontal deactivations

The activity of the right premotor area BA6 (frontal eye field) was significantly reduced during the monocular VPA responses in the *passive viewing* vs. VPA comparison a-c (Table 1b, #2). This decrease was highly correlated ($r_{xy} = 0.80$, P<0.001) with a simultaneous activity decrease in the left parietal lobule BA39/40 (Table1b, #4, Fig. 3). In a parallel PET study (Richter et al.,

1998), adjacent BA6 coordinates (x, y, z): ±35, -11, 60 were deactivated bilateraly during the execution of monocular voluntary negative accommodation.[1]

These rCBF decreases are interpreted principally as a disengagement of the saccade system during cortical processing of accommodation/vergence eye-movements. The ocular near-response could functionally interact with the fixation system to obtain saccade suppression for the purpose of providing a clear foveal image. Ohtsuka and Sato (1996) presented evidence linking omnipause activity with accommodation activity within the rostral superior colliculus (SC). Omnipaus neurons in SC are tonally active in all periods of fixation and pause during saccades (Munoz and Wurtz, 1993). Sawa and Othsuka (1994) described elicitation of AR by low-current stimuli of the rostral-portion of the SC corresponding to the representation of the central visual field.

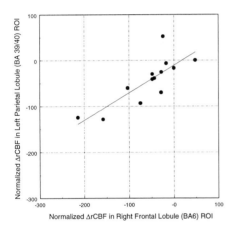

Figure. 3. Individual ΔrCBF decreases from VPA minus *passive viewing* **(i.e., a-c) in right frontal BA6 coupled with simultaneous decreases in left parietal BA39/40.** The Taliarach co-ordinates (x, y, z) for these ROIs are listed in Table 1b (activation #2 and #4).

No frontal or parietal activation was observed during both VPA minus *passive viewing* (c-a) or VPA minus *blurred viewing* (c-b). The only other frontal response appeared as a deactivation in both of these conditions, which localised to BA10 (Table 1b, #1 and 1d).

---

[1]Multiple prefrontal cortical fields involve sensorimotor processing for eye movements: frontal eye fields (FEF); supplementary motor area (SMA); supplementary eye fields (SEF); and, several adjacent prefrontal regions (Mitz and Godschalk, 1989). Many of these regions have been visualized during neuroimaging studies of oculomotor tasks. The reported coordinates from PET studies for FEF range in x from 18 to 44; in y from 11 to -3; and in z from 41-56 (Fox et al., 1985; Paus et al., 1993; Petit et al., 1993; Lang et al., 1994).

left BA31 (Table 1b, #3). The posterior parietal association cortices are critical for directed attention, visual-spatial analysis, and vigilance in the contralateral hemisphere (Mountcastle, Talbot and Yin, 1977). Inferior parietal lobule (IPL) and BA39 posterior parietal neurons in the macaque are also known to carry a premotor spatial signals relevant to changes in gaze within 3D space relative to the plane of fixation (Gnadt and Mays, 1995; Sakata et al 1980). However, the present VPA paradigm did not require any overt 3D spatial shifts in posturing of the ARs.

Cerebellar involvement in VPA

Two activations localised to the vermis (Table 1a, #2 and 4) and one to cerebellar cortex (Table 1c, #6). The cerebellum is interconnected with many regions showing changes in activity during VPA (Bando et al., 1984). For a review of cerebellar VA circuitry, see: Pardo, Lee and Richter, this volume.

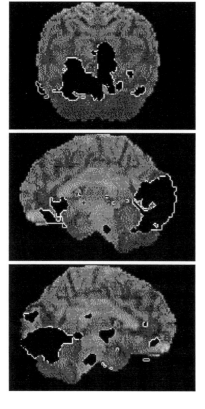

Figure 4. Cerebral activation during voluntary positive accommodation (VPA). VPA minus *passive viewing*. Surface rendering of posterior activations. Threshold for these images set at Z=2.1 to display optimally the pattern of activity; note that significance threshold, as shown in Table 1, was selected as Z=3.2. Whiter color denotes greater Z scores; light gray set to maximum. Panels show that VPA give rise to baseline increases of occipital rCBF. The visual responses show contralateral hemispheric asymmetry during monocular VPA (See discussion).

Conclusion

VPA recruited many regions already active during *passive* or *blurred* viewing. However, our data also show that VPA involves circuits dissociable in part from those participating in other oculomotor fixation tasks. Although VPA must be linked tightly to systems involved in target selection (Richter and Franzén, 1994), no changes were localised to the region of the right anterior cingulate cortex which is part of the anterior attention system (Pardo et al., 1990). An important observation is that an apparent dynamic and reciprocal functional connectivity exists between the accommodation and cortical saccade systems during VPA. The human cortical correlates to vergence remain still obscure (Mays and Gamlin, 1995), as do the rCBF effects from the partially suppressed vergence. Future studies will address these findings and further dissection of higher-order processes involved in the oculomotor near response.

Acknowledgements
This work was supported by the Department of Veterans Affairs, USA, and the Karolinska Institute/University of Minnesota Exchange Program. Dr. Hans Richter was a Postdoctoral Fellow supported in part by the Fulbright Commission, the Swedish Institute, the Sweden-America Foundation and the Swedish Crown. We thank our participants for their cooperation and patience; Dr. Satoshi Minoshima (University of Michigan) for providing software for PET data analysis; Dr. Richard Robb (Mayo Clinic) for providing ANALYZE software; Chris Fy and the Eye Clinic (VAMC, Minneapolis, MN) for performing optometric evaluations; Dr. Ying Han, Huddinge University Hospital, Huddinge, Sweden, for eye-movement recordings with the Ober-2 system; and the technical staff of the PET imaging service (VAMC, Minneapolis, MN).

References

Bando, T., and Toda, H. (1991) Cerebral cortical and brainstem areas related to the central control of lens accommodation in cat and monkey. *Comparative Biochemistry and Physiology* 98:229-237.

Bando, T., Tsukuda, K., Yamamoto, N, Maeda, J., Tsukahara, N. (1984) Physiological identification of midbrain neurons related to lens accommodation in the cat. *Journal of Neurophysiology* 52:870-878.

Bruce, C., Desimone, R., and Gross, C. (1981) Visual properties of neurons in a polysensory area in superior temporal sulcus of the macaque. *Journal of Neurophysiology* 46:369-384.

Franzén, O., Lennerstrand, G., and Richter, H. (1994) Brain potential correlates of supraliminal contrast functions and defocus. *International Journal of Human-Computer Interaction* 6:155-176.

Fulton, J.F. (1928) Observations upon the vascularity of the human occipital lobe during visual activity. *Brain* 51: 310-320.

Fox, P.T., Fox, J.M., Raichle, M.E., Burde, R.M. (1985) The role of cerebral cortex in the generation of voluntary saccades: A positron emission tomographic study. *Journal of Neurophysiology* 54:348-369.

Gnadt, J.W., and Mays, L.E. (1995) Neurons in monkey parietal area LIP are tuned for eye-movement parameters in 3D space. *Journal of Neurophysiology* 73:280-297.

Heinze, H.J., Mangun, G.R., Burchert, W., Hinrichs, H., Scholz, M., Munte, T.F., Gs, A., Scher, M., Johannes, S., Hundeshagen, H., Gazzaniga, M.S., and Hillyard, S.A. (1994) Combined spatial and temporal imaging of brain activity during visual selective attention in humans. *Nature* 372:543-546.

Jampel, R. (1955) Convergence, divergence, pupillary reactions and accommodation of the eyes from faradic stimulation of the macaque brain. *Journal of Comparative Neurology* 115:371-399.

Kosslyn, S.M., Alpert, N.M., Thompson, W.L., Maljkovic, V., Weise, S.B., Chabris, C.F., Hamilton, S.E., Rauch, S.L., and Buonanno, F. (1993) Visual mental imagery activates topographically organized visual cortex: PET investigations. *Journal of Cognitive Neuroscience* 5:263-287.

Lang, W., Petit, L., Höllinger, P., Pietrzyk, U., Tzourio, N., Mazoyer, B., and Berthoz, A. (1994) A positron emission tomography study of oculomotor imagery. *Neuroreport* 5:921-924.

Mays, L.E., and Gamlin, P.D.R. (1995) Neuronal circuitry controlling the near response. *Current Opinion in Neurobiology* 5:763-768.

McKeefry, D.J., Watson, J.D., Frackowiak, R.S., Fong, K., and Zeki, S. (1997) The activity in human areas V1/V2, V3, and V5 during the perception of coherent and incoherent motion. *Neuroimage* 5:1-12

Mitz, A.R., and Godschalk, M. (1989) Eye-movement representation in the frontal lobe of rhesus monkey. *Neuroscience Letter* 106:152-162.

Motter, B.C. (1993) Focal attention produces spatially selective processing in visual cortical areas V1, V2 and V4 in the presence of competing stimuli. *Journal of Neurophysiology* 70:909-919.

Mountcastle,V.B., Talbot, W.H., and Yin, T.C.T. (1977) Parietal lobe mechanisms for directed visual attention. *Journal of Neurophysiology* 40:362-389.

Munoz, D.P., and Wurtz, R.H. (1993) Fixation cells in monkey superior colliculus I. Characteristics of cell discharge. *Journal of Neurophysiology* 70:559-575.

Ohtsuka, K., and Sato,A. (1996) Descending projections from the cortical accommodation area in cat. *Investigtive Ophthalmology and Visual Science* 37:1429-1436.

Owens, D.A., and Higgins, K.E. (1983) Long-term stability of the dark focus of accommodation. *American Journal of Optometry and Physiological Optics* 60:32-38.

Pardo J., Pardo, P., Janer, K., and Raichle, M. (1990) The anterior cingulate cortex mediates processing selection in the Stroop attentional conflict paradigm. *Proceedings of National Academy of Science USA* 87:256-259.

Paus T., Petrides, M., Evans, A., and Meyer, E. (1993) Role of the human anterior cingulate cortex in the control of oculomotor, manual, and speech responses: A positron tomography study *Journal of Neurophysiology* 70:453-469.

Petit, L., Orssaud, C., Tzourio, N., Salamon, G., Mazoyer, B., and Berthoz, A. (1993) PET study of voluntary saccadic eye movements in humans: Basal ganglia-thalamocortical system and cingulate cortex involvement. *Journal of Neurophysiology* 69:1009-1017.

Richter, H., Franzén, O., and von Sandor R (1994) Quantitative judgements and matching of subjective speed of apparent laser speckle flow induced by refractive defocus. *Behavioural Brain Research* 62:81-91.

Richter, H., and Franzén, O. (1994) Velocity percepts of apparent laser speckle motion modulated by voluntary changes of visual accommodation: Real-time, in-vivo measurements of the accommodative response. *Behavioural Brain Research* 62:93-102.

Richter, H., Abdi, S., Han, Y., Lennerstrand,G., Franzén, O., Anderson, J., Pardo, J.V., Schneider, H., and Långström, B. (1998) Higher order control processes and plasticity in neural CNS circuitry subserving VNA. *The Association for Research in Vision and Ophthalmology (ARVO)* 4835-B586, S1047 *(Abstract)*.

Rockland, K.S., and Van Hoesen, G.W. (1994) Direct temporal-occipital feedback connections to striate cortex (V1) in the macaque monkey. *Cerebral Cortex* 4: 300-313.

Sakata, H., Shibutani, H., and Kawano, K. (1980) Spatial properties of visual fixation neurons in posterior parietal association cortex of the monkey. *Journal of Neurophysiology* 43:1654-1672.

Sawa, M., and Ohtsuka, K. (1994) Lens accommodation evoked by microstimulation of the superior colliculus in the cat. *Vision Research* 34:975-981.

Talairach, J., and Tournoux, P. (1988) *Co-Planar Stereotaxic Atlas of the Human Brain.* New York: Theime Medical Publishers.

Ullman, S. (1995) Sequence seeking and counter streams: A computational model for bidirectional information flow in the visual cortex. *Cerebral Cortex* 1:1-11.

Yoshizawa, T., Takagi, M., Toda, H., Shimizu, H., Norita, M., Hirano, T., Abe, H., Iwata,K., and Bando, T. (1991) A projection of the lens accommodation-related area to the pupillo-constrictor area in the posteromedial lateral suprasylvian area in cats. *Japanese Journal of Ophthalmology* 35:107-118.

Zald, D.H., and Pardo,J.V. (1997) Emotion, olfaction, and the human amygdala: Amygdala activation during aversive olfactory stimulation. *Proceedings of National Academy of Science* 94:4119-4124.

Zald, D.H., Donndelinger, M.J., and Pardo, J.V. (1998) Elucidating dynamic brain interactions with across-subjects correlational analysis of PET data: The functional connectivity of the amygdala and orbitofrontal cortex during olfactory tasks. *Journal of Cerebral Blood Flow Metabolism* In press

Zeki, S. (1974) Functional organization of a visual area in the posterior bank of the superior temporal sulcus of the rhesus monkey. *Journal of Physiology* 236:549-573.

Functional neuroanatomy of the human near/far response of the visual system to blur cues: Fixation to point targets at different viewing distances

José V. Pardo,[1,2] Scott R. Sponheim,[3,4] Patricia Costello,[1,2] Joel T. Lee,[2] Ying Han,[5] and Hans Richter[5,6]

[1]Cognitive Neuroimaging Unit, Psychiatry Service, Veterans Affairs Medical Center, Minneapolis, USA. [2]Division of Neuroscience Research, Department of Psychiatry, University of Minnesota, Minneapolis, USA. [3]Department of Psychology, University of Minnesota, Minneapolis, USA. [4]Psychology Service, Veterans Affairs Medical Center, Minneapolis, MN USA. [5]Department of Clinical Science, Division of Ophthalmology, Karolinska Institute, Huddinge Hospital, Huddinge, Sweden. [6]PET Centre, Uppsala University, Uppsala, Sweden.

Summary. The neural circuits associated with the human monocular near/far response (NFR) were studied in thirteen normal volunteers by changes in relative regional cerebral blood flow (rCBF) measured with $H_2^{15}O$ infusion and positron emission tomography (PET). Behavioral tasks were selected to differentially recruit relevant circuits in three states of activity: resting, fixating, and near/far accommodating. Neural networks selectively activated for eliciting the NFR to blur cues during constant visual fixation occupy posterior structures which include occipital visual regions, temporal cortex, cerebellar hemispheres and vermis. Of note, no significant activation occurred in prefrontal regions. The NFR recruited many regions already active during constant visual fixation, and it engaged circuits partly dissociable from those mediating other types of eye movements as well as those participating in visual fixation.

Introduction

The present experiments sought to identify neural circuits involved in the human near/far response through measurement of neural activity in normal volunteers as indexed by changes in regional cerebral blood flow (rCBF) with positron emission tomography (PET) and the flow radiotracer, $H_2^{15}O$. Monocular accommodative/vergence responses were measured to step changes in target position along the main axis of the dominant eye. The cortical processes that transform the blurred representation of a visual stimulus into an appropriately scaled motor-command signal to be executed by the ciliary muscles are poorly understood (Fig. 1).

80

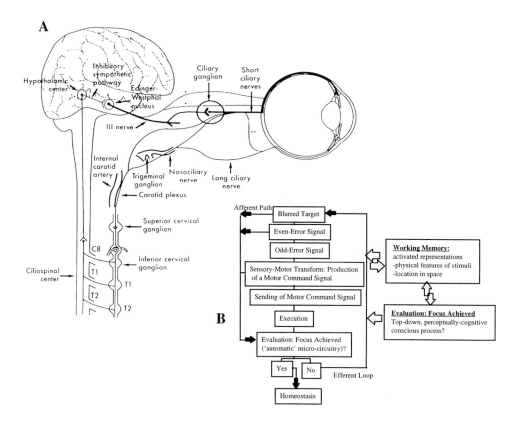

Figure 1. **(A) Human dioptric focusing system and efferent pathways from autonomic nervous system (ANS) to the ciliary muscle. A** The major innervation to the ciliary muscle is parasympathetic and follows the pathway shown by the thick solid lines. The parasympathetic pathway originates in the Edinger-Westphal nucleus and courses with the third nerve, where the fibers travel to and synapse in the ciliary ganglion. The majority of the postganglionic parasympathetic fibers travel to the ciliary muscle via the short ciliary nerves, but some of them (double asterisk) also travel with the long ciliary nerves. There is also evidence for a direct pathway of uncertain functional significance (single asterisk) to the internal eye structures from the Edinger-Westphal nucleus. The sympathetic supply to the ciliary muscle (thin solid lines) originates in the diencephalon and travels down the spinal cord to the lower cervical and upper thoracic segments, to synapse in the spinociliary center of Budge in the intermediolateral tract of the cord. From there, second-order nerves leave the cord by the last cervical and first thoracic ventral roots; these preganglionic fibers run up the cervical chain to synapse in the cervical ganglion. The third-order fibers continue up the sympathetic carotid plexus and enter the orbit, either with the first division of the trigeminal nerve (following the nasociliary division) or independently, where they join the long and short ciliary nerves, in the latter instance passing through the ciliary ganglion without synapsing. CB, cervical vertebra 8; T1 thoracic vertebra 1; T2, thoracic vertebra 2. (with permission from Kaufman, 1992).

(B) Schematic model of accommodative processing during monocular viewing. When an image of a visual target on the retina is blurred, approximately 500 ms elapse between the detection of the blurred target and the complete AR to the target. During this time, neural circuits compute the location of the intended target, decide whether to focus upon the intended target, and initiate the AR. Iterative computations tune the AR based upon initial visual sensory cues, feedback, and comparison to the internal representation of the target. Vergence in the viewing eye is inhibited. During visual accommodation top-down, cognitive-perceptual processing occurs within the accommodative system.

Methods

Thirteen right-handed healthy subjects with normal or normal-corrected visual acuity and right-eye dominance participated in this study after providing written informed consent according to the guidelines of the Veterans Affairs Medical Center (Minneapolis, MN, USA) Human Studies Committee and the Radioactive Drug Research Committee. Three subjects were female, and ten were male. The subjects' mean age was 25 years (range 18-46).

The non-dominant left eye was patched. Space average luminance of the surround was 1.0 cd/m^2 (Quantum Instrument Inc., Photometer LX, Garden City, NJ). Light emitting diodes (LED) were located sagittally at 0.24 m (4.16 D), 1.2 m (0.83 D), and 3.0 (0.33 D) approximately colinearly with the visual axis of the right eye. Three tasks, adapted from Han et al. (1995), were presented from one to three times in counter-balanced order: 1) resting with eyes closed (ECR); 2) continuous fixating upon the active LED at 1.2 m (0.83 D) (FIX); and 3) fixating alternately between the near (0.24 M, 4.16 D) and far (3.0 m, 0.33 D) LED each activated during sequential 2 sec epochs (NEAR/FAR).

Electroencephalographic recordings (EEG and EOG) of eye movements were recorded in each subject and PET scanning session using standardized procedure (ECI, Eaton, OH, USA). The EOG derived from frontal EEG leads demonstrated successful performance of NFR because the occluded eye showed periodic vergence movements in synchrony with the sequential 2 sec stimulation epochs, as expected from the tight coupling of vergence to accommodation.

The camera, ECAT 953B (Siemens, Knoxville, TN, USA), were used with septae retracted and with correction for electronic deadtime and randoms. No decay or scatter correction was used. Attenuation correction used a two-dimensional transmission scan. A slow bolus (30s) intravenous injection of H$_2^{15}$O (0.25 mCi [or 9.2 MBcq] per kg) was initiated upon starting the task (Silbersweig et al., 1993). The activity was integrated from the time the true count rate began to rise until 90 s later. The images were recontructed using a Hanning filter (0.5 cycles/pixel) with filtered-backprojection including non-orthogonal angles (Kinahan and Rogers, 1989). The final image resolution after gaussian blurring was 12 mm full-width at half-maximum. ANALYZE (BIR, Mayo, Rochester, MN, USA) was used for image display.

Results and discussion

The subjects found the task easy to perform. EOG documented the absence of saccades during constant fixation and verified task compliance.

The comparison of interest involved the contrast (paired image comparison) between the conditions of near/far accommodation and that of visual fixation (i.e., NEAR/FAR minus FIX, see Table 1 and Figure 3). This analysis shows that the NFR task recruits circuits beyond those mediating visual fixation; a network comprised of cerebellar, occipital, and temporal regions. The cerebellum was the structure in which most of the activation arose: Bilateral cerebellar hemispheres (foci 3, 4, 7, 8, 9) and vermis (focus 2, 6). The left visual cortex (BA 17/18, focus 1) and the area at right fusiform/inferior temporal (BA 20/37; focus 11) gyrus were activated. In addition, foci of activation occurred in the right and left middle temporal gyri (BA 21; foci 5, 10). Of note, no significant activation was observed in the prefrontal cortex. Activations in FIX vs. ECR conditions and in NEAR/FAR vs. ECR conditions will be reported elsewhere.

Table 1. Brain regions activated by the near/far response.

Foci	Talairach Coordinates			Magnitude of peak activation
(a) NEAR/FAR minus FIX.				
	x	y	z	Z-Score
1. Left visual cortex (BA 17/18)	-8	-80	14	4.17
2. Cerebellar vermis (L)*	-3	-67	-2	3.92
3. Left cerebellar cortex	-28	-69	-2	3.86
4. Right cerebellar cortex	33	-55	-20	3.86
5. Right middle temporal gyrus (BA 21)	51	-46	9	3.81
6. Cerebellar vermis (R)	8	-67	-9	3.65
7. Left cerebellar cortex	-39	-64	-18	3.56
8. Right cerebellar cortex	10	-85	-22	3.37
9. Left cerebellar cortex	-17	-64	-16	3.30
10. Left middle temporal gyrus (BA 21)	-53	8	-16	3.30
11. Right fusiform/inf. gyrus (BA 20/37)	44	-35	-14	3.22

Response magnitudes and locations in coordinates of the atlas of Talairach and Tournoux (1988). x, y, and z (in mm) correspond to right-left, anterior-posterior, and superior-inferior dimensions of the brain. Right, anterior, and superior sides have positive coordinates. BA refers to approximate Brodmann areas as listed in the atlas. Left-most numbers on each line of the table correspond to the numbered major foci in Figure 3. Foci just outside the brain are labeled by an asterisk.

Cerebellar involvement in NFR

The pattern of activation of the cerebellum during NFR, as contrasted with constant fixation, is in agreement with human lesion data (Hain and Luebke, 1990; Kawasaki et al., 1993), and with neurophysiological studies in animals (e.g., Milder and Reinecke, 1983). The research team of Tsukahara and colleagues (Hultborn et al., 1973) were first to propose the critical role of the cerebellum in the near response by elucidating components of the circuitry in the cat (for recent review, see Bando and Toda, 1991). Hultborn et al. (1973) demonstrated that stimulation of the brachium conjunctivum of the cat cerebellum elicits short latency responses in the short ciliary nerve innervating both the pupillary and ciliary muscles. Ijichi et al. (1977) showed that the cat cerebellum contributed to the pupillary light reflex since lesions of the fastigial nucleus produced large changes in the frequency response characteristics of the pupil to changing light levels, and stimulation of the fastigial nucleus produced pupillary dilatation. They hypothesized that the cerebellum plays a regulatory role in the pupillary light reflex of the brainstem and serves to protect retinal pigments from bleaching. Hosoba et al. (1978) stimulated the contralateral interpositius and fasitigial nuclei or ipsilateral interpositus nucleus and detected temporally correlated changes in eye accommodation. They also found that stimulation of cerebellar cortex (lobulus simplex, folium vermis, tuber vermis, and posterior and medial aspect of the lobulus paramedianus), in the very regions reported to show correlated activity with the NFR (Buchtel, 1973) and responses to optic nerve stimulation (Buchtel et al., 1972), produced accommodation. Furthermore, interpositus neurons fire in synchrony with changes in lens accommodation (Bando et al., 1979). Also, single units in the medial midbrain, adjacent to the Edinger-Westphal nucleus that are related to lens accommodation, can be driven by stimulation of the interpositus nucleus (Bando et al., 1984b). Indirect evidence for a bidirectional cerebellar-mesencephalic-cortical network relevant to NFR arises from stimulation studies of the deep cerebellar nuclei and the suprasylvian cortex of the cat, where units associated with lens accommodation reside (Bando et al., 1984a). Stimulation of the deep cerebellar nuclei evokes field potentials in the suprasylvian sulcus (Kato et al., 1987) and stimulation of lateral suprasylvian cortex produces responses in mossy and climbing fibers of the contralateral cerebellar cortex (Kato et al., 1988).

Progress in the neurophysiology of the cerebellar contribution to NFR in non-human primates is more limited. A study of cerebellar stimulation showed distinct regions producing saccades, smooth pursuit, and nystagmus, but did not report upon induced near/far responses (Ron and Robinson, 1973). Total cerebellectomy induced a transient paralysis of vergence while preserving saccadic movements (Westheimer and Blair, 1973). Unilateral excision of a cerebellar hemisphere did not produce abnormal accommodation or pupillary responses (Westheimer and Blair, 1974). Normal monkeys wearing periscopic spectacles show an increase in AC/A ratio. Lesions of the flocculus and ventral paraflocculus do not prevent this plasticity to occur (Judge, 1987). However, Purkinje cells in the flocculus fire in relation to the vergence angle (Miles et al., 1980). Additional evidence from non-human primates for cerebellar participation in NFR comes from studies of single unit activity in the medial part of the nucleus reticularis tegmenti pontis (NRTP). Here, single unit activity can increase with the amplitude of the NFR without involving saccadic or smooth pursuit movements (Gamlin and Clarke, 1995). The NRTP receives afferents from frontal eye fields, superior colliculus, and pretectum, and it is reciprocally connected with the cerebellum. Based upon such data, Gamlin and Clark (1995) postulated that the NRTP is a key intermediary in a cerebropontocerebellar pathway modulating vergence and ocular accommodation. Zhang and Gamlin (1998) recently also demonstrated the involvement of the posterior interposed nucleus of the cerebelium (IP) in vergence and accommodation. Microstimulation of the IP elicited divergence and accommodation for far. The activity of these neurons increased with decreases in vergence angle and accommodation.

A confound to the finding of cerebellar activation during NFR arises from the presence of microsaccades (Fig. 2). These small vertical eye movements, having the same frequency as NFR, could activate oculomotor circuitry: frontal eye fields (FEF), supplementary eye fields (SEF), posterior parietal lobe, vermis and/or cerebelar hemispheres (Fox et al., 1985; Ron and Robinson, 1973; Petit et al., 1993; Paus et al., 1995; Paus, 1996; Petit et al., 1997). In particular, some of the cerebellar hemisphere activation seen here might arise from the microsaccades because these increases were more dominant in the present protocol while the vermal activity replicated across both accommodation studies (See: Richter, Lee and Pardo, this volume).

Frontal Lobe

No rCBF changes occurred in the frontal lobule. If a sustained AR to a blurred target initiates an ocular stabilization processes in frontal areas, to ensure that the target is held in the retinal area of highest visual acuity by inhibiting shifts of fixation to peripheral stimuli as suggested by Richter, Lee and Pardo (this volume), we would expect frontal rCBF decreases to occur (e.g. in FEF). The absence of a predicted decrease in activity may be due to a compatible amount of rCBF increase in the same circuitry caused by the vertical mircosaccades (Fig. 2).

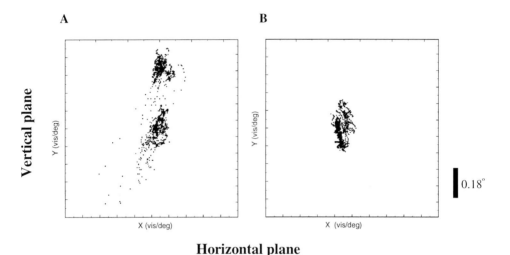

A

B

Vertical plane

Y (vis/deg)

X (vis/deg)

Y (vis/deg)

X (vis/deg)

0.18°

Horizontal plane

Figure 2. Eye movements during accommodation. A Eye position for right fixating eye (left eye occluded) sampled at 50 Hz during monocular accommodation at near (4.16 D) and far (0.33 D) elicited by a square-vawe LED target. **B** Monocular voluntary positive accommodation (left eye occluded) in response to a -5.0 D blurred high-contrast checkerboard pattern (Richter, Lee and Pardo, this volume).

Vertical small-amplitude microsaccades can be observed during accommodation to the NEAR/FAR targets **(A)** although the LEDs were never separated vertically by more than ~0.35°. Earlier studies have shown that even if targets are correctly aligned to elicit only vergence or accommodation in depth, saccades still occur frequently (Enright, 1986). The speeding up of horizontal vergence by vertical small-amplitude microsaccades, is also a well-known phenomena (Enright, 1986). Hence, although the frequency and amplitude of the AR (evoked to the near/far target) and synergistic vergence movements were similar across the two studies, the ocular responses to the near far target were in all likelihood facilitated (Panels show data from a representative subject recorded with the Ober-2 system).

Temporal Lobe

Among the earliest evidence for a contribution of the temporal lobe to NFR, Jampel (1960) showed that faradic stimulation of the area around the superior temporal sulcus (STS), at approximately BA 22 in monkeys, could elicit reproducible accommodation, vergence, and pupilloconstriction. Stimulation of prefrontal, occipital, and other temporal regions failed to elicit this reflex triad. Subsequent research on accommodation has focused upon a feline structure comparable to the STS in monkeys: The lateral suprasylvian area (LSA) in the rostral occipital cortex (Westheimer and Blair,1974; Bando and Toda, 1991). Many neurons in this region respond to target motion, binocular disparity, and target size (Westheimer and Blair, 1974; Toyama et al., 1986). Further research indicates that multiple specialized areas within LSA contain rostral and caudal regions with accommodation units and adjacent pupillocontriction units (reviewed in Bando and Toda, 1991; Sawa et al., 1992). Therefore, these regions appear to contain components of NFR circuitry.

In non-human primates, the STS region contains multiple higher-order visual processing areas (e.g., middle temporal area, MT, medial superior temporal area, MST). These regions having neurons with large receptive fields are specialized for motion analysis (e.g., Albright, 1984). Data from these regions during NFR are lacking. However, Gnadt & Mays (1989) found single units in the medial bank of the monkey STS with activity reflecting the velocity of either accommodation or vergence.

Human MT (V5) is activated during motion stimulation, but this activity is localized to more posterior regions than monkey STS. It is situated just posterior to the junction of the inferior temporal sulcus with the lateral occipital sulcus (Flechsig's Feld 16; Brodmann BA 19/37; Watson et al., 1993) although its precise location with respect to the sulcal anatomy can sometimes be difficult to discern (Tootell et al., 1995). In the present study, both inferior temporal (BA 37) and middle temporal (BA 21) responses appeared during NFR. However, only the BA 37 region is consistent with the location reported for human V5 (Watson et al., 1993; Tootell et al., 1995). Since some subjects reported various degrees of perceived motion, the interpretation of the temporal cortical responses remains ambiguous. The more superior temporal regions may reflect illusory motion processing with potential input into NFR, while the more ventral temporal activity may concern the NFR circuitry related to the reflex triad (accommodation, vergence, pupilloconstriction) as observed in studies of the feline LSA.

Occipital cortex

The activation of left BA 17 and 18 in the paired image comparison between the conditions of NEAR/FAR vs. FIX suggests that these regions may participate in computations relevant to the estimation of blur for use in modulating NFR. Alternately, they may pass information to temporal regions for motion processing. The occipital asymmetry (i.e., left-hemisphere dominance) converges with a qualitative SPECT study of fixation to near and far targets which found greater activation in the hemisphere contralateral to the fixating eye (Hung et al., 1990). Occipital asymmetry (i.e., contralateral rCBF greater than ipsilateral) is also observed during monocular viewing of a checkerboard pattern with either the normal eye or with the amblyopic eye of patients with strabismic amblyopia (Imamura et al., 1997). This asymmetry is also evident during monocular voluntary accommodation (Richter, Lee and Pardo, this volyme).

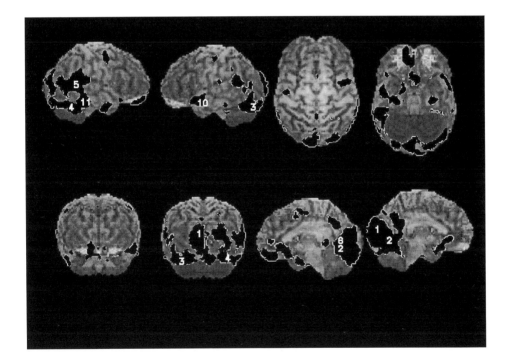

Figure 3. Cerebral activation during visual accommodation (NFR). Contrast between the conditions of near/far accommodation and visual fixation (i.e., NEAR/FAR minus FIX) and surface projections of increased rCBF passing a criterion of P < 0.05 (uncorrected for multiple comparisons). Regions passing this criterion are shown in black and numbered to correspond to the major responses in Table 1 (See discussion).

Conclusion

Posterior brain systems mediate the NFR evoked by blur cues. When networks involved in foveal fixation and visuospatial working memory (Jonides et al., 1993) are dissected from neural circuits mediating NFR, three principal structures emerge: 1) cerebellar vermis and hemispheres; 2) cortices surrounding the superior temporal sulcus and inferior temporal gyrus; and 3) striate and extrastriate cortex. The cerebellar contribution converges with an extensive body of lesion and neurophysiological studies supporting a key role for the cerebellum in the modulation of accommodation and vergence. The activation in temporal regions may reflect human homologues of the feline lateral sylvian area (LSA), which contains accommodation units and interacts with the cerebellum. The temporal activity, particularly in the inferior temporal cortex (V5 complex), may also relate to motion processing, since illusory motion was perceived by some subjects during near/far fixation. The extrastriate cortex may contribute to motion processing as well as to computation of blur to cue the NFR.

No parietal changes in rCBF became evident in the NEAR/FAR minus FIX comparison. IPL and posterior parietal neurons in the macaque are known to carry premotor spatial signals relevant to changes in gaze within 3D space relative to the plane of fixation (Gnadt and Mays, 1995). However, the inferior parietal lobule (IPL, BA 40) were activated in both the NEAR/FAR minus ECR (right BA 40) and the FIX minus ECR comparisons (BA 40 bilaterally). Probably these two oculomotor tasks recruit similar parietal neuronal pools.

The present near/far-task evoked multifocal bilateral cerebellar hemisphere activity unlike the voluntary positive accommodation task (Richter, Lee and Pardo, this volume). Both protocols activated the vermis at similar coordinates. We attribute therefore part of the cerebellar cortex activity as due to an engagement of the saccade system. This system, with largely separate anatomical and physiological substrates, was at least momentarily engaged under the present experimental conditions, unlike in the voluntary positive accommodation responses. The functional role of the vermal activation may relate uniquely to the processing of accommodation. The vermal visual area (lobule VII) receives information from visually responsive neurones including NRTP (Schmahmann and Pandya, 1997). In the present protocol the PET camera was not aligned to detect activity in the deep nuclei.

Acknowledgements

This work was in part supported by the Department of Veterans Affairs, USA, HFSR-F746/95 (GL), University of Minnesota/Karolinska Institute Exchange Program. Dr. Richter was a Postdoctoral Fellow supported in part by the Fulbright Commission, the Swedish Insitute, the Sweden-America Foundation and the Swedish Crown. We thank Dr. Satoshi Minoshima (University of Michigan, Ann Arbor, MI, USA) for providing software for data analysis and Dr. Richard Robb (BIR, Mayo Foundation, Rochester, MN, USA) for providing ANALYZE™ software.

References

Albright, T.D. (1984) Direction and orientation selectivity of neurons in visual area MT of the macaque. *Journal of Neurophysiology* 52:1106-1130.

Bando, T., Ishihara, A., and Tsukahara, N. (1979) Interpositus neurons controlling lens accommodation. *Proceedings of Japanese Academy of Science* 55:153-156.

Bando, T., Yamamoto, N., and Tsukahara, N. (1984a) Cortical neurons related to lens accommodation in posterior lateral suprasylvian area in cats. *Journal of Neurophysiology* 52:879-891.

Bando, T., Tsukuda, K., Yamamoto, N, Maeda, J., Tsukahara, N. (1984b) Physiological identification of midbrain neurons related to lens accommodation in the cat. *Journal of Neurophysiology* 52:870-878.

Bando, T., and Toda, H. (1991) Cerebral cortical and brainstem areas related to the central control of lens accommodation in cat and monkey. *Comparative Biochemistry and Physiology* 98:229-237.

Buchtel, H.A., Iosif, G., Marchesi, G.F., Provini, L., and Strata, P. (1972) Analysis of the activity evoked in the cerebellar cortex by stimulation of the visual pathways. *Experimental Brain Research* 15:278-288.

Buchtel, H.A., Rubia, F.J., and Strata, P. (1973) Cerebellar unitary responses to moving visual stimuli. *Brain Research* 50:463-466.

Enright,J.T. (1986) Facilitation of vergence changes by saccades: Influences of misfocused images and of disparity stimuli in man. *Journal of Physiology* 371:69-87.

Fox, P.T., Fox, J.M., Raichle, M.E., Burde, R.M. (1985) The role of cerebral cortex in the generation of voluntary saccades: A positron emission tomographic study. *Journal of Neurophysiology* 54:348-369.

Gamlin, P.D. and Clarke, R.J. (1995) Single-unit activity in the primate nucleus reticularis tegmenti pontis related to vergence and ocular accommodation. *Journal of Neurophysiology* 73:2115-2119.

Gnadt, J.W. and Mays, L.E. (1989) Posterior parietal cortex, the oculomotor near response and spatial coding in 3-D space. *Society for Neuroscience Abstracts* 15:786.

Gnadt, J. W. and Mays, L.E. (1995) Neurons in monkey parietal area LIP are tuned for eye-movement parameters in 3D space. *Journal of Neurophysiology* 73:280-297.

Hain, T.C., and Luebke, A.E. (1990) Phoria adaptation in patients with cerebellar dysfunction. *Investigative Ophthalmology and Visual Science* 31:1394-1397.

Han, Y., Seideman, M., and Lennerstrand, G. (1995) Dynamics of accommodative vergence movements controlled by the dominant and non-dominant eye. *Acta Ophthalmologica Scandinavica* 73:319-324.

Hosoba, M., Bando, T., and Tsukahara, N. (1978) The cerebellar control of accommodation of the eye in the cat. *Brain Research* 153:495-505.

Hultborn, H., Mori, K., and Tsukahara, N. (1973) The neuronal pathway subserving the pupillary light reflex and its facilitation from cerebellar nuclei. *Brain Research* 63:357-361.

Hung, G.K., Stahl, T., and Powell, A. (1990) Neural activity following monocular accommodation using SPECT. *Medical Science Research* 18:865-868.

Ijichi, Y., Kiyohara, T., Hosoba, M., and Tsukahara, N. (1977) The cerebellar control of the pupillary light reflex in the cat. *Brain Research* 128:69-79.

Imamura, K., Richter, H., Fischer, H., Lennerstrand, G., Franzén, O., Rydberg, A., Andersson, J., Schneider, H., Onoe, H., Watanabe, Y., and Långström, B. (1997) Reduced activity in the extrastriate visual cortex of individuals with strabismic amblyopia. *Neuroscience Letters* 225:173-176.

Jampel, R.S. (1960) Convergence, divergence, pupillary reactions and accommodation of the eyes from faradic simulaion of the macaque brain. *Journal of Comparative Neurology* 115:371-399.

Judge, S.J. (1987) Optically-induced changes in tonic vergence and AC/A ratio in normal monkeys and monkeys with lesions of the flocculus and ventral paraflocculus. *Experimental Brain Research* 66:1-9.

Jonides J., Smith, E.E., Koeppe, R.A., Awh, E., Minoshima, S., and Mintun, M.A. (1993) Spatial working memory in humans as revealed by PET. *Nature* 363:623-625.

Kato, N., Kawaguchi, S., and Miyata, H. (1987) Postnatal development of the retinal and cerebellar projections onto the lateral suprasylvian area in the cat. *Journal of Physiology* 383:729-743.

Kato, N., Kawaguchi, S., and Miyata, H. (1988) Cerebro-cerebellar projections from the lateral suprasylvian visual area in the cat. *Journal of Physiology* 395:473-485.

Kaufman,P. (1992) Accommodation and Presbyopia: Neuromuscular and Biophysical Aspects. In: *Adler's 1992 Physiology of the Eye, Clinical Application 9th Edition, Mosby-Year Book*. (Ed. W.Hurt)

Kawasaki, T., Kiyosawa, M., Fujino, T., and Tokoro, T. (1993) Slow accommodation release with a cerebellar lesion. *British Journal of Ophthalmology* 77:678

Kinahan, P.E. and Rogers, J.G. (1989) Analytic three-dimensional image reconstruction using all detected events. *I E E E Transactions on Nuclear Science* 36:964-968.

Milder, D.G., and Reinecke, R.D. (1983) Phoria adaptation to prisms. A cerebellar-dependent response. *Archives of Neurology* 40:339-342.

Miles, F.A., Braitman, D.J., and Dow, B.M. (1980) Long-term adaptive changes in primate vestibuloocular reflex. IV. Electrophysiological observations in flocculus of adapted monkeys. *Journal of Neurophysiology* 43:1477-1493.

Paus, T., Marrett, S., Worsley, K.J., and Evans, A.C. (1995) Extraretinal modulation of cerebral blood flow in the human visual cortex: implications for saccadic suppression. *Journal of Neurophysiology* 74:2179-2183.

Paus, T. (1996) Location and function of the human frontal eye field: A selective review. *Neuropsychologia* 34:475-483.

Petit, L., Clark, V.P., Ingeholm, J., and Haxby, J.V. (1997) Dissociation of saccade-related and pursuit-related activation in human prefrontal eye fields as revealed by fMRI. *Journal of Neurophysiology* 77:3386-3390.

Petit, L., Orssaud, C., Tzourio, N., Salamon, G., Mazoyer, B., and Berthoz, A. (1993) PET study of voluntary saccadic eye movements in humans: Basal ganglia-thalmocortical system and cingulate cortex involvement. *Journal of Neurophysiology* 69:1009-1017.

Ron, S. and Robinson, D.A. (1973) Eye movements evoked by cerebellar stimulation in the alert monkey. *Journal of Neurophysiology* 36:1004-1022.

Sawa, M., Maekawa, H., and Ohtsuka, K. (1992) Cortical area related to lens accommodation in cat. *Japanese Journal of Ophthalmology* 36:371-379.

Schmahmann, J.D., and Pandya, D.N. (1997) The cerebrocerebellar system. *International Review of Neurobiology* 41:31-60.

Silbersweig, D.A., Stern, E., Frith, C.D., Cahill, C., Schnorr, L., Grootoonk, S., Spinks, T., Clark, J., Frackowiak, R., and Jones, T. (1993) Detection of thirty-second cognitive activation in single subjects with positron emission tomography: A new low-dose $H_2^{15}O$ regional cerebral blood flow three-dimensional imaging technique. *Journal of Cerebral Blood Flow and Metabolism* 13:617-629.

Talairach, J., and Tournoux, P. (1988) *Co-Planar Stereotaxic Atlas of the Human Brain*. New York: Theime Medical Publishers.

Tootell, R.B., Reppas, J.B., Kwong, K.K., Malach, R., Born, R.T., Brady, T.J., Rosen, B.R., Belliveau, J.W. (1995) Functional analysis of human MT and related visual cortical areas using magnetic resonance imaging. *Journal of Neuroscience* 15:3215-3230.

Toyama, K., Komatsu, Y., and Kozasa, T. (1986) The responsiveness of Clare-Bishop neurons to mothion cues for motion stereopsis. *Neuroscience Research* 4:83-109.

Watson, J.D., Myers, R., Frackowiak, R.S., Hajnal, J.V., Woods, R.P., Mazziotta, J.C., Shipp, S., Zeki, S. (1993) Area V5 of the human brain: Evidence from a combined study using positron emission tomography and magnetic resonance imaging. *Cerebral Cortex* 3:79-94.

Westheimer, G. and Blair, S.M. (1973) Oculomotor defects in cerebellectomized monkeys. *Investigative Ophthalmology* 12:618-621.

Westheimer, G. and Blair, S.M. (1974) Functional organization of primate oculomotor system revealed by cerebellectomy. *Experimental Brain Research* 21:463-472.

Zhang H., and Gamlin, P.D. (1998) Neurons in the posterior interposed nucleus of the cerebellum related to vergence and accommodation. I. Steady-state characteristics. *Journal of Neurophysiology* 79:1255-1269.

Spatial contrast sensitivity and visual accommodation studied with VEP (Visual Evoked Potential), PET (Positron Emission Tomography) and psychophysical techniques.

Ove Franzén[1,2,5], Gunnar Lennerstrand[1], José Pardo[3,4] and Hans Richter[1]

[1]*Department of Clinical Science, Karolinska Institute and Huddinge University Hospital, Huddinge, Sweden,* [2]*Mid Sweden University, Sundsvall, Sweden,* [3]*Cognitive Neuroimaging Unit, Veterans Affairs Medical Center,* [4]*Department of Psychiatry, University of Minnesota, Minneapolis, USA, and* [5]*Department of Optometry and Vision Science, University of Latvia, Riga, Latvia*

SUMMARY

Psychophysical scales of a 3,3 c/deg monochromatic checkerboard of variable contrast were compared with steady state visually evoked potentials (VEP) recorded by gross electrodes on the scalp. These estimates of neuronal population responses grew as a power function of physical contrast having an exponent of approximately the same magnitude as the corresponding psychophysical function which gives credence to the validity of the procedures employed. The functional neuroimaging technique of positron emission tomography (PET), based on radioactive decay of a labelled tracer occurring inside the brain, was applied in normal subjects to quantitatively explore the influence of voluntary positive accommodation and also to examine the effect of reduced contrast sensitivity in human strabismic amblyopia. A great asymmetry in metabolic activity was observed in the striate cortex, that is, the Brodmann area 17 (BA 17) activation was strongest contralateral to the dominant viewing eye. The PET scans revealed, however, a high correlation between blood flow increases in the right striate cortex (BA 17) and the left extrastriate cortex (BA 18) during voluntary accommodation, possibly reflecting top-down modulation and reentrant processes. The poor contrast sensitivity in strabismic amblyopia could essentially be explained by deactivation of the ipsilateral extrastriate cortical areas BA 18 and BA 19.

INTRODUCTION

To navigate successfully in three-dimensional space, it is necessary to recognize objects and to know their spatial relationship relative to each other and to oneself. In man and many other animals, spatial pattern vision provides most of the necessary information. Loss of this capacity has a devastating effect on mobility, e.g. food-seeking, avoiding obstacles, etc., and is nearly as incapacitating as complete darkness. Loss of other dimensions of vision has a lesser effect on behavior, in that it is still possible to move about with, for example, impaired color vision, reduced visual fields, at reduced light intensities or with impaired depth perception. The importance of spatial vision to man has always been recognized and has been the subject of considerable study over the years. In the past, the dimensions studied, e.g., visual acuity, depth perception, etc. were chosen without knowledge of the properties of the neural substrate that might mediate these abilities. Recent studies of the mammalian visual system, however, have discovered neural processes that could underlie these capacities and have suggested new dimensions for psychophysical study (Hubel, 1988; Zeki, 1993).

The response of the visual system to contrast can be evaluated in a number of ways. One method borrowed from physical optics and electronics, called linear systems analysis, measures the contrast increment threshold while varying the spatial distribution and gradient of light (Schadé, 1956). Of particular relevance is the fact that this analysis has been applied not only to the human visual system using psychophysical methods but to the single elements of the mammalian visual system employing electrophysiological methods as well. Numerous electrophysiological studies have investigated the response properties of single cells in the cat and monkey retina, lateral geniculate nucleus, and visual cortex and, in general, it has been shown that these cell properties are also found in neurons located in comparable structures in the monkey visual system (e.g. Hubel and Wiesel, 1968; Hubel, 1988; Maffei and Fiorentini, 1973; 1977; De Valois et al., 1979; De Valois and De Valois, 1980).

By means of psychophysical ratio-scaling procedures psychological variables such as brightness, loudness, pitch and pain can be systematically measured which allows the construction of scales of subjective magnitude (Stevens, 1957; Ekman, 1958). In his key paper Stevens (1957) argued that the relationship between sensation magnitude and stimulus intensity was best described by a power function. In accordance with Stevens' dictum that equal stimulus ratios produce equal response ratios, this description of the functional relationship seems to hold for nearly all sense modalities (Gescheider, 1997; Stevens, 1975).

The direct methods of scaling are based on the possibly strong assumption that the subjects are able to make supraliminal judgments of the magnitude of their sensations on a ratio scale level. An evident approach to the problem of testing the sensory process hypothesis of psychophysical scales would be to use neurophysiological and neuroimaging measures from the same organism as an independent criterion in order to get experimental insight into the neural mechanisms underlying sensory perception (Ahlquist and Franzén, 1994; Franzén and Ahlquist, 1996).

Attempts began in the 1960s to make direct correlations between brain events evoked by adequate physical stimuli and psychophysical measures of the perceptual capacity of primates and man that yielded information about the initial representation of sensory events in the cerebral cortex (Mountcastle et al., 1969; Franzén and Offenloch, 1969). Relatively isomorphic representations have been described for several sensory systems (See, e.g. Franzén and Westman, 1991; Franzén et al., 1996).

The aim has never been to study single neurons in isolation, but rather, to reconstruct population events that are going on in the nervous system. It is therefore obvious that the most fruitful path to future research would be to study significant samples of those neuron populations that are assumed to be essential for the perceptual processes (Franzén and Westman, 1991; Wu, 1997).

Environmental signals received by the organism must necessarily produce changes in large ensembles of nerve cells, why the magnitude of an intensive percept may in general be encoded by a population code. Reports pertinent to the major issue of transformation of electrical nervous activity into perception are found in studies carried out at different levels of the nervous system in man and monkey. A neural population model based on the number of activated low-threshold mechanoreceptors (Franzén et al., 1991) predicts satisfactorily the magnitude of tactually evoked responses in the somatic receiving area of the human brain. These evoked responses, mainly composed of postsynaptic potentials in single cortical cells, were in their turn highly correlated with magnitude estimates of touch sensation (Franzén and Offenloch, 1969). In related experiments on the feline somatosensory system, Franzén et al.

(1985) found that the cord dorsum potentials and the cortical evoked potentials grew as power functions of tactile intensity with almost the same exponent, which implies direct proportionality between the population responses at these two levels of the central nervous system. In a study of the neural basis for the sensation of tactile roughness Johnson et al. (1996) concluded that linearity was the best and the most parsimonious description of the coding relationship between spatial variations in the neurophysiological population responses at peripheral / cortical level and the roughness magnitude judgements. Evidence suggests that the magnitude of experimentally induced sharp dental pain is a linear function of the integrated neural A-delta activity in man (Franzén and Ahlquist, 1989; Ahlquist and Franzén, 1994). A vectorial summation model of neuronal population coding was successfully applied (Georgopoulos et al. 1986; Georgopoulos, 1991) to describe and predict the relation of primate cortical motor activity to the direction of upper-limb movements. Each neuron could actually be considered as a multichannel detector, since it can be a member of many cell ensembles of parallel pathways to specify information about different aspects of an event. Georgopoulos (1995) demonstrated that a single cell in primate motor cortex (M1), densely connected to many parts of the cortex and subcortical areas, is broadly tuned and that the calculated population vector of the graded activity (each neural vector is weighted by its activity) pointed in or near the direction of the movement. In this way, the neural image of a sensory or motor event is always embedded in a dynamic, distributed system. Psychophysics is indeed becoming deeply rooted in neuroscience (Bunge and Ardila, 1987).

Progress in molecular and cellular neurobiology has been nothing less than spectacular in the last two decades. The molecular basis of excitation and conduction in neurons, and of synaptic transmission between neurons is well understood. The introduction of novel methods into system neurobiology provides new insights into how we receive and process information from the external world, and act upon it (Kandel et al., 1991; Zigmond et al., 1998).

New, non-invasive methods have opened the way to direct observation of the human brain in action. Positron emission tomography (PET) is a functional brain imaging technique that is a very powerful tool available to the neuroscientist and that offers a new window on the mind / brain problem (Posner and Raichle, 1994). It is based on radioactive decay of a labelled tracer occurring inside the brain. The PET camera can show blood flow to different areas of the brain, and increased blood flow indicates increased brain activity, - a physical process intimately involved with the functional activity of nervous tissue. Since cerebral oxygen consumption and glucose utilization represent synaptic and neural activity, PET scans are able to assess the integrated neuronal function and to reveal what specific brain regions participate and how active they are in processing, e.g., visual information. Recently, the central correlates of voluntary positive visual accommodation (VPA) were measured using PET and $H_2^{15}O$ bolus technique in healthy volunteers with normal vision (Richter et al.,1995).

In the present investigation the following issues will be addressed: 1. neuronal population coding using VEP (visual evoked potentials) and psychophysical techniques 2. relationships among spatially distributed systems in the brain explored by PET under the condition of voluntary accommodation and 3. central processing of contrast information in strabismic amblyopia studied by PET.

METHODS

I. Visual evoked responses (VEP) to a monochromatic checkerboard

VEPs were recorded bipolarly from the scalp with gross silver-silver chloride electrodes. The active electrode was positioned in the midline 2.5 cm above the inion and the indifferent electrode 3 or 7 cm temporally of the active electrode. One ear lobe was grounded. The signals were amplified by a Medelec PA63 preamplifier and a Medelec AA6 Mk III main amplifier. The time-locked, stimulus-evoked potential was obtained by averaging 128 or 256 sweeps in a Medelec AV6 to enhance the signal-to-noise ratio and then recorded on UV sensitive paper (Medelec recording unit). The low and high frequency cut-offs of the amplifiers were set at 1.6 and 80 Hz, respectively.

The extracranial cerebral responses were elicited by a reversing checkerboard pattern with a fundamental spatial frequency of 3.3 c / deg, which is within the intermediate, optimal range of spatial frequencies for driving visual accommodation and for ensuring focusing accuracy

(Owens, 1980). The contrast of the checkerboard (having a two-dimensional Fourier spectrum) was sinusoidally reversed at a frequency of 7 Hz (Kelly 1976; De Valois et al., 1979). The screen had a mean luminance of 40 cd/m^2, and the contrasts of the spatial pattern were 0.10, 0.15, 0.26 and 0.56, which corresponds to intensities between 28-43 dB relative to sensation level. The pattern produced by a Medelec generator and displayed on a TV screen subtending 19 degrees of visual angle was presented in a half-field stimulation mode. The subject fixated the pattern for stimulation of the central retina of the right dominant eye. The left eye was occluded.

II. Psychophysics of spatial contrast

Six subjects were asked, using the direct method of free magnitude estimation, to assess the perceived contrast of the checkerboard by assigning numbers to the stimuli so that the ratio of the psychophysical estimates (numerical assignments to sensory magnitudes) corresponded to the ratio of the contrasts. The stimuli were presented in random sequence. Data analysis allowed for the event that the subjects used different moduli thereby eliminating possible interobserver variability. Geometric means of the magnitude estimates were calculated and normalized before curve-fitting procedures were applied (Kalikow, 1967; Gescheider, 1997). The subjects were also required to provide verbal reports of perceived contrast using an intensity descriptor scale (Borg, 1994). For this purpose the subjects were presented a rating scale consisting of nine categories identified by sensory verbal descriptors. This category scale with ratio-properties consists of nine descriptors (very very weak, very weak, weak, neither weak nor strong, slightly strong, strong, very strong, very very strong and maximal perceived contrast).

As stated above, sensation magnitude (R) is a power function of stimulus intensity (S), $R = k \cdot S^n$ where n is the exponent and k is a constant of proportionality. The psychophysical and VEP data will be displayed in log-log coordinates, because a power function in these coordinates follows a straight line.

III. Neuroimaging

A monochromatic checkerboard used for monocular stimulation was viewed at a distance of 40 cm, and it contained a fixational cross to reduce or eliminate saccadic eye-movements or drifts in smooth pursuit. The PET measurements were performed on an ECAT 953 B camera (Siemens, Knoxville, TN, USA). The subjects were positioned in the scanner and immobilized by means of an individually molded vacuum-operated head-holder.

Neuronal activity was estimated by measuring changes in cerebral blood flow (rCBF) in thirteen normal volunteers who received intravenous injections of O^{15} labeled water into a forearm vein before scans. The Talairach stereotactic coordinates permitted cerebral localization in 3-D space (Talairach and Tournoux, 1988).

Under the first condition the subject was instructed to passively look with the dominant eye (the other eye was occluded) at the checkerboard and to make no effort at accommodating whenever a -5 D lens intermittently interrupted the line of sight. Under the second condition the subject was asked to overcome a -5 D lens alternately placed in front of the eye by exerting voluntary positive accommodation (VPA) to obtain a sharp and clear image of the checkerboard (Stark and Atchinson, 1994; Ciuffreda and Hokoda, 1985).

RESULTS

VEP and psychophysical measures related to physical contrast

For comparison and reference purposes we show in Figure 1 the functional relationship between contrast and contrast-dependent VEPs in response to a phase-reversing vertical sinusoidal grating of 3.5 c/deg spatial frequency (Franzén et al., 1994). The averaged cortical responses to a 3.5 c/deg grating published by Campbell and Maffei (1970) were re-analyzed and mathematically transformed leaving the curve form invariant. These data are also plotted in the same graph. Using the method of least squares a straight line was drawn indicating that

98

a power function with a slope (n) of 0.45 best described the relationship. The binocular VEP is about $\sqrt{2}$ times larger than that measured monocularly (Franzén, 1978; Jakobsson, 1985).

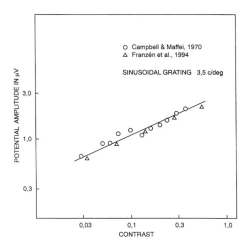

Fig. 1. VEP amplitude as a function of contrast obtained with a sinusoidal grating at a spatial frequency of 3.5 c/deg. (Adapted from Campbell and Maffei, 1970; Franzén et al., 1994).

Steady state occipital responses to contrast-reversal of the 3.3 c / deg checkerboard whose first positive wave has a latency of about 100 msec are displayed for two subjects in Figure 2. These potentials contain the major component at the second harmonic frequency, that is, twice the fundamental temporal frequency of the contrast reversal (Zemon et al., 1986).

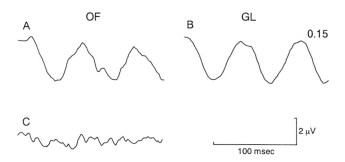

Fig. 2. A, B. Evoked responses from two subjects to a reversing checkerboard with a spatial frequency of 3.3 c/deg. C. Control with a blank screen.

The geometric means of the peak-to-peak amplitude for 12 measurements per level in three subjects are plotted in Fig. 3. The amplitude of these potentials was on the average larger than that evoked by a sinusoidal grating (Fig. 1) and has a somewhat slower growth rate (slope, n = 0.35). The difference in response amplitude remains, even if, as a rough approximation, the checkerboard VEP is divided by $4/\pi$. This is the ratio of the fundamental Fourier components for square and sine gratings of the same spatial frequency, *i.e.*, it represents the contrast energy of the sine wave fundamental in a square wave grating (Campbell and Robson, 1968).

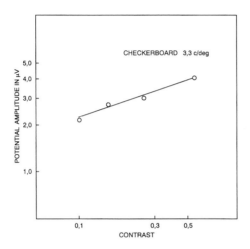

Fig. 3. Plot of VEP amplitude versus checkerboard contrast.

Since the striate cortical cells acting as spatial frequency analyzers are capable of mediating the full range of the subjective contrast function, this neuronal population response may constitute a candidate code for contrast and can therefore be used as a predictor of the psychophysical magnitude function for checkerboard contrast.

The input-output relation of the normalized numerical magnitude estimates of the sensation to the stimulus intensity is a power function with an exponent (n) of 0.48 (Fig. 4). The contrast stimuli fell within the subjective intensity range of weak to strong on the sensory verbal descriptor scale. The sensation curve grew more rapidly than its corresponding neural (VEP) function, which is a common finding in correlative studies of this kind (Franzén and Offenloch, 1969; Franzén and Berkley, 1975; Franzén et al., 1994), - a discrepancy that may be explained by our arbitrary choice of neural response index (Whitsel and Franzén, 1989; Ahlquist and Franzén, 1994).

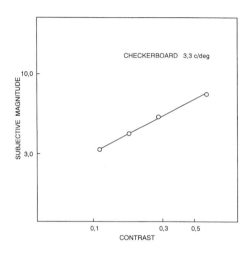

Fig. 4. Normalized free magnitude estimates plotted against checkerboard contrast.

PET measurements

Figure 5 shows the two hemispheres from behind. In this posterior view of the brain, it is evident that voluntary adjustments of the curvature of the crystalline lens increased the cerebral blood flow (voluntary accommodation minus passive viewing) and did so more on the contra lateral than on the ipsilateral side of the primary visual cortex (BA 17). This observation reminds us of the fact that the afferent pathway decussates in the optic chiasm and that more nerve fibers (coming from the nasal hemiretina) cross to the optic tract of the opposite side than those passing on to the ipsilateral optic tract (Weale, 1982).

However, the significant increase in CBF measurements summarized in Figure 6 reveals a high correlation between blood flow changes in the right striate cortex (BA 17) and the left extrastriate cortex (BA 18) during positive accommodation (VPA) which is at odds with the

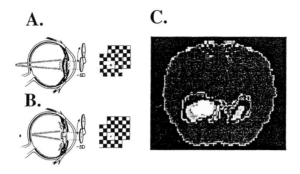

Fig. 5. A. Blurred viewing condition. B. Voluntary accommodation exerted to obtain a clear image of the checkerboard. C. Posterior view of the human striate cortex (BA 17). This positron emission tomographic investigation shows an increased activity under the condition of voluntary accommodation and an obvious asymmetrical activation of the two hemispheres. BA 17 activation was strongest contralateral to the dominant eye.

asymmetrical activation pattern displayed in Figure 5. The two neopallial cortical areas, BA17 and BA 19, are reciprocally connected by way of fibres in the corpus callosum serving the function of midline fusion (Geschwind, 1965; Zeki, 1993).

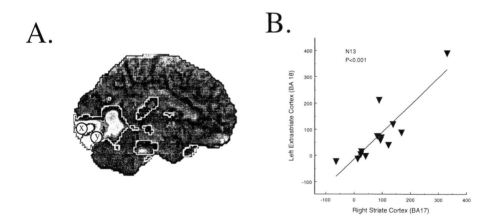

Fig. 6. A. Subtraction CBF images of activated areas BA 17 (x) and BA 18 (y) in thirteen subjects.
B. Plot of increases in the right striate cortex (BA 17) against increases in the left extrastriate cortex (BA 18) under the condition of voluntary accommodatoin.

DISCUSSION

I. Central processing of spatial contrast in normal vision and amblyopia

Visual stimuli such as spatial contrast are transduced into electrical signals by the summation pool of the retinal contrast generating mechanism (Enroth-Cugell and Robson, 1984; Stone, 1965), and ideally, this inflow should be conveyed to the striate cortex of the brain in an invariant form which is an important function of perception (Yilmaz, 1967; Mountcastle, 1975).

The close agreement between the perceptual and cortical responses to spatial contrast of the checkerboard suggests that the transmission of the specific features of the pooled contrast information along dedicated perceptual pathways is approximately linear. Support for such a notion was lent by an investigation on the macaque demonstrating that the activity of cells in the lateral geniculate nucleus in response to visual stimuli was proportional to their prepotentials (Lee, Virsu and Creutzfeldt, 1983), and by another study showing that responses of feline lateral geniculate neurons reflected the response characteristics of the retinal cells (Stone, 1965; Enroth-Cugell and Robson, 1984; Sclar, 1987; Troy and Enroth-Cugell, 1993). In addition, simple cells of layer 4 of the visual cortex combine their synaptic input linearly (Ferster, 1991).

In an elegant study of tactile form perception, Johnson et al. (1991) moved embossed, upper-case letters across the receptive fields of mechanoreceptive fibers innervating a monkey finger. They showed that the isomorphic representation of the stimulus, produced by the peripheral afferent axons, was maintained in BA 3b of postcentral gyrus, where the initial pre-perceptual stages of central processing may occur. Moreover, the spatio-temporal neural population vector was, as noted above, isomorphic with preferred limb movement direction (Georgopoulos, 1991).

Sinusoidal contrast reversal of a 3.7 c/deg grating evoked an apparently normal visual cerebral potential in a child who was diagnosed as functionally blind (Bodis-Wollner et al., 1977). CT scans (computerized tomography), providing images of changes in anatomical structures, revealed that part of BA 17 was preserved but that BA 18 and BA 19 were destroyed thereby impeding the striate visual cortex from communicating with the second and the third visual areas (prestriate cortices) which, in the normal case, project forward in parallell to higher, more specialized areas such as V 4 (color and form) situated in the fusiform gyrus and V 5 (MT, coherent motion) situated in the inferior temporal gyrus (Zeki, 1983; Livingstone and Hubel, 1988; Zeki and Shipp, 1988; Kaas, 1992).

Stimulus-driven electrical activity such as VEPs is, in all likelihood, dominated by components from an early stage of visual cortical processing that may eventually lead to perception. A faithful neural representation of the visual stimulus may be, *in the feedforward process*, a necessary but not sufficient condition for our sensory experience (Franzén and Offenloch, 1969; Franzén, 1976; Franzén et al., 1994).

This indicates, once again, that perception does not take place in the primary visual area of the brain and that the pathway from the striate cortex to, *e.g.*, the inferotemporal cortex may be crucial for the ability to discriminate visual patterns and to recognize different aspects and features of an object. A cortical pathway involving the prestriate and temporal lobe cortices has been postulated to be part of an object channel. A prestriate-parietal pathway is, on the other hand, assumed to be part of a spatial channel. These two parallel pathways, each of which is multichannelled and multimapped, are crosslinked in a complex combinatorial way (Ungerleider and Mishkin, 1982; Spitzer and Richmond, 1991; van Essen et al., 1992; Goodale and Milner, 1992; Gross, 1994; Ungerleider and Haxby, 1994; Bullier and Nowak, 1995). Interestingly, multidimensional scaling analysis of the cortico-cortical connectivity and topological organization of the visual system revealed that the visual areas are subdivided into dorsal and ventral streams (Young, 1992; Young et al., 1994).

Further support for our research philosophy and working hypothesis may be obtained from an investigation on amblyopia carried out by our research group (Imamura et al., 1997). Amblyopia is defined as reduced vision of one eye where no apparent defect or organic lesion can be observed in the eye or the visual pathways (Demer et al., 1988; von Noorden, 1990; Ciuffreda et al., 1991; Demer, 1993; Campos, 1995). It is characterized by impaired form vision, usually reduced visual acuity and contrast sensitivity (Lundh, 1983; Koskela, 1986), and it is due to retarded visual development of the visual system most commonly caused by strabismus (squint) in childhood or anisometropia (differences in refraction between the two eyes). It is very difficult to find pure anisometropic amblyopia, since there is nearly always a part of abnormal binocular interaction involved (von Noorden, 1995).

It is generally thought that amblyopia can only be treated during childhood when the nervous system still shows plasticity, and that the best time is during the sensitive period in visual development (Daw, 1991). Although there is a short time scale to development and to the activity-dependent wiring of the neuronal connections of the cerebral circuits and to shaping sensory pathways (Shatz, 1990; Meister et al., 1991), evidence is accumulating that the period of plasticity may not be restricted to the first years, but it may persist to some degree in the adult brain (Campos, 1995).

From animal experiments it is obvious that adequate sensory experience is an important builder of the architecture of the brain, for artificial strabismus in kittens produces significant alterations in visual cortex and leads to ocular dominance shifts and to the formation of two separate populations of monocular cortical neurons and accordingly to an almost complete loss of binocularity (Hubel and Wiesel, 1965). The same effect of an abnormal visual experience has been observed in amblyopic monkeys whose striate cortical neurons exhibited deficits in contrast sensitivity (Kiorpes, 1996; Movshon et al., 1987; Eggers et al., 1984) and stereopsis (Crawford et al., 1984).

To be able to quantitatively examine the relationship between reduced visual contrast sensitivity in human strabismic amblyopia and its concomitant activation pattern, we investigated, with use of PET and the $H_2^{15}O$ bolus technique, changes in the regional cerebral blood flow (rCBF) induced by monocular visual stimulation (monochromatic black and white checkerboard) of eight individuals (two of them had also a marked anisometropia) with this disorder, who had all undergone strabismus operation (Imamura et al., 1997). The physical contrast and spatial frequency of the checkerboard were set just above the individual amblyopic detection threshold and then used for stimulating the amblyopic as well as the healthy, dominant eye in order to enhance the differential effect on grating visibility (Franzén et al., 1994).

The Vistech CS System (VCTS 6500) consisting of gratings of different spatial frequencies and contrasts was utilized to determine the contrast sensitivity function (reciprocal of detection contrast threshold). Figure 7A shows that human vision is more sensitive to intermediate spatial frequencies than to low and high spatial frequencies and that perceptibility of the same contrast stimuli was severely affected in the amblyopic eye.

Statistical analysis of subtracted images showed significant increases in rCBF by the stimulation of the sound eye localized bilaterally in BA17, 18 and 19. The striate cortex could readily be activated by the amblyopic eye, although its response was reduced. The results suggested that the poor contrast sensitivity (Fig. 7 A) in amblyopia is essentially associated with a deactivation in identifiable regions of the ipsilateral extrastriatal areas 18 and 19 (Fig. 7 B-D).

The most important stimulus for accommodation seems to be a blurred image which triggers the accommodative system to adjust the curvature of the crystalline lens thereby changing its refractive power. Phillips and Stark (1977) performed experiments with target blur in a closed feedback loop with accommodation which allowed the subject to alter the amount of blur by accommodating. The authors concluded that defocus blur, the improper focus of the eye, is a sufficient stimulus to accommodate.

The focusing process leads evidently to increases of the cerebral blood flow in the striate cortex (> 10%) in comparison to the condition under which the eye is exposed to defocus blur without accommodating (Fig. 6). The great asymmetry in metabolic activity induced by the input from the dominant eye is particularly worthy paying attention to, since, in Rhesus monkeys with normal binocular vision, about 50 % of the axons cross to the contralateral side of the visual pathway. The same holds true for cats whose cortical cells receive approximately equal input from the two eyes. This means that the two retinae of monkey and cat project about equally to the brain hemispheres (Hubel and Wiesel, 1972; Wiesel et al., 1974; Kennedy et al., 1976; Dowling, 1992).

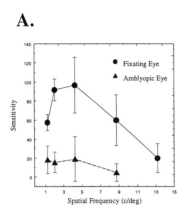

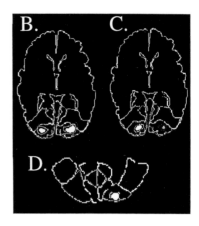

Fig. 7. A. Contrast sensitivity functions for sine-wave gratings for the healthy fixating eye and the amblyopic eye of eight adults with strabismic amblyopia. B. PET images after contrast stimulation of the sound fixating eye. C. Stimulation of the amblyopic eye. D. Image subtraction of the normal eye contrast from the amblyopic eye contrast revealing a deactivation of the ipsilateral BA 18 and BA 19.

The asymmetry observed in humans is therefore of great theoretical and clinical interest because the deactivation of BA 18 och BA 19 in strabismic amblyopia was localized to the ipsilateral side of the amblyopic eye (Imamura et al., 1997). A relatively weak activation of the amblyopic eye in combination with a strong contralateral input from the healthy eye exerting a powerful inhibition may explain the suppression of the visual neuronal activity in the ipsilateral hemisphere (Fig. 7 B-D) and, consequently, the functional impairment of spatial contrast vision (Fig. 7 A). The notion of an inhibitory action of the sound eye is supported by the experimental finding that a deprived eye (unilateral eyelid suture in cat) is capable of more vigorously triggering cortical cells when the intracortical GABA-mediated suppression of the viewing eye is interrupted (Duffy et al., 1976; Sur, 1993) following administration of bicuculline (this drug blocks the inhibition-generating transmitter gamma-amino-butyric acid, GABA). Furthermore, the cortical neural territory (and its dendritic arbor structure) of the occluded eye is not diminished by binocular competition, if the cortical postsynaptic cell activity is blocked by GABA (Hata and Stryker, 1994) showing that a cortical territory is generally activity-dependent and dynamically maintained (Whitsel and Franzén, 1989; Merzenich et al., 1991; 1996). A dramatic illustration of the dynamics of adaptive cortical plasticity was found in a PET study of blind Braille reading subjects whose primary visual cortical area was activated during this tactile task (Sadato et al., 1996).

Distributed representation, correlate of binding, and reentrance

A fundamental question of the neural population encoding is how parallel inputs/outputs converge to form a global, unified percept from a pointillist retinal pattern and to extract an invariant structure of the external world. According to a hypothesis advanced by Singer et al. (1995) a percept or perceptual grouping could be generated by a binding process ascribed to a coherent 40 Hz firing in ensembles of adjacent cortical cells, that is, binding occurs by temporal signal correlations in interconnected neuronal populations. The precise timing of the cell firing relative to the whole cell group, occurring in the millisecond range, may be controlled by inhibitory interneurons (Buzsaki and Chrobak, 1995). A neuroanatomical substrate for this synchronized activity in spatially separated cells or modules may be found in long-range horizontal axons that extend for several mm along the cortical surface (Ts'o and Gilbert, 1988; Gilbert and Wiesel, 1989).

This binding model for coding at the population level (Singer and Gray, 1995) has recently been tested on strabismic cats which developed amblyopia in the squinting eye (Roelfsma et al., 1994). The authors found in area 17 a reduced synchronization of responses evoked by moving gratings that were presented to the amblyopic eye. They suggested that decreased synchronization of neural discharges could also bring about an inefficient drive of neuron assemblies in other visual areas, and that it could thus constitute a correlate of strabismic amblyopia. The model does not predict, however, deactivation of the ipsilateral extrastriate areas as shown in our human PET study.

Practically every part of the cerebral cortex is connected with subcortical structures by afferent and efferent projecting axons. The parvocellular (P) and the magnocellular (M) layers of lateral geniculate nucleus (LGN) of the thalamus send strong parallel projections to V 1 (BA 17) but not to V 2 (BA 18) why the high correlation between the PET measurements, obtained in the reciprocally connected areas BA 17 and BA 18 (Fig 6), cannot be accounted for by a common input from LGN (Friston, 1994), but rather be understood by a cognitive top-down facilitation of the visual process under the condition of voluntary accommodation. Cooling deactivation of V1 in the monkey depresses the activity or rather puts to silence the nerve cells in V2, V3 and V4 but not in V5, because its subcortical input bypasses V1 (Schiller and Malpeli, 1977; Payne et al., 1996).

Feedback projections coming from the primary visual cortex are approximately 80% of the total input to LGN (Schiller, 1986). This is just one instance of a top-down modulation of the ascending visual stream which demonstrates that the retinal images are not passively received and analyzed by the striate cortex (Zeki, 1993).

The number of extrastriate feedback connections from primate and human area V 2 to V1 is much larger than the influence from V 1 on V 2 neurons (Friston, 1994; Budd, 1998). Furthermore, this feedback is estimated to be approximately ten times the geniculocortical afference to area V 1 (Peters et al., 1994), which implies that a major portion of the input to the neocortical neural machinery is of internal rather than of external origin (Douglas and Martin, 1991).

Dynamic self-regulatory feedback mechanisms of the accommodative system provides the individual with a clear image of a target or an object (Stark et al., 1965; Richter and Franzén, 1994). When the subject exerts voluntary positive or negative accommodation to obtain a sharp retinal image, attention is evidently directed to this location of the scene which enhances perceptual sensitivity (Luck et al., 1994).

In a recent PET study activation of the posterior parietal lobes was observed when the subjects were asked to attend to targets placed to the right or to the left of the visual field (Corbetta et al., 1993). For good reasons, we assume that the mechanisms of the widely distributed neuromodulatory network of the posterior parietal lobe is critically involved in focusing visual attention (Lynch et al., 1977; Mesulam, 1981; Wurtz et al., 1982; Driver and Mattingley, 1995) and also implicated in voluntary accommodation (Franzén et al., 1987; Gnadt and Mays, 1989; Bando and Toda, 1991; Richter and Franzén, 1994; Richter et al., 1998; Sakata et al., 1997). Visual attention and voluntary accommodation may enhance the selectivity in V 1 / V 2 (Payne et al., 1996) and may, by reentrant signals (Tononi et al., 1998), contribute to the synchronization of the neural temporal patterning. This process may lead to the high transcallosal correlation between blood flow measures recorded in the right BA 17 and the left BA 18 of the two hemispheres (Fig. 6) and, possibly, to the generation of conscious awareness of the visual input.

All these investigations taken together provide support for the proposal (Edelman, 1993) that higher brain functions such as visual perception and accommodation depend on parallel and / or sequential action of large ensembles of complex neural networks that constitute reentrant, distributed systems subserving distributive functions.

ACKNOWLEDGEMENTS

This work was supported by grants from the Swedish Medical Research Council, the Swedish Council for Research in Humanities and Social Sciences, Work Environment Fund, Mid Sweden University and the Sweden-America Foundation.

REFERENCES

Ahlquist, M., and Franzén, O. (1994) Encoding of the subjective intensity of sharp dental pain. Endodontics and Dental Traumatology 10: 153-166.

Bando, T. and Toda, H. (1991) Cerebral cortical and brainstem areas related to the central control of lens accommodation in cat and monkey. Comparative Biochemistry and Physiology 98: 229-237.

Bodis-Wollner, I., Atkin, A., Edward, R. and Wolkstein, M. (1977) Visual association cortex and vision in man: Pattern-evoked occipital potentials in a blind boy. Science 198; 629-63.

Bolz, J. and Gilbert, C.D. (1989) The role of horizontal connections in generating long receptive fields in the cat visual cortex. European Journal of Neuroscience 1: 263-268.

Borg, G. (1994) Psychophysical Scaling: An overview. In: J. Boivie, P. Hansson and U. Lindblom. (eds): *Touch, Temperature, and Pain in Health and Disease: Mechanisms and Assessments.* Progress in Pain Research and Management, Vol. 3. IASP Press, pp. 27-38.

Budd, J. (1998) Extrastriate feedback to primary visual cortex in primates: A quantitative analysis of connectivity. Proceedings of the Royal Society, London, 265: 1037-1044.

Bunge, M. and Ardila, R. (1987) *Philosophy of Psychology.* Springer Verlag, Berlin.

Buzsáki, G. and Chrobak, J.J. (1995) Temporal structure in spatially organized neuronal ensembles: A role for interneuronal networks. Current Opinion in Neurobiology 5: 504-510.

Campbell, F.W. and Maffei, L. (1970) Electrophysiological evidence for the existence of orientation and size detectors in the human visual system. Journal of Physiology 207: 635-652.

Campbell, F.W. and Robson, J.G. (1968) Application of Fourier analysis to the visibility of gratings. Journal of Physiology 197; 551-566.

Campos, E. (1995) Amblyopia. Survey of Ophthalmology 40: 23-39.

Ciuffreda, K.J. and Hokoda, S.C. (1985) Effect of instruction and higher level control on the accommodative response spatial frequency profile. Ophthalmic and Physiological Optics 5: 221-23.

Ciuffreda, K.J., Levi, D.M. and Selenow, A. (1991) *Amblyopia. Basic and clinical aspects.* Butterworth-Heineman, Boston.

Corbetta, M., Miezin, F.M., Shulman, G.L. and Petersen, S.E. (1993) A PET study of visuospatial attention. Journal of Neuroscience 93: 1202-1226.

Crawford, M.L.J., Smith III, E.L., Harwerth, R.S. and von Noorden, G.K. (1984) Stereoblind monkeys have few binocular neurons. Investigative Ophthalmology and Visual Science 25; 779-781.

Daw, N.W. (1991). Mechanisms of plasticity in the visual cortex. Investigative Ophthalmology and Visual Science 25: 4168-4179.

Demer, J.L. (1993) Positron emission tomographic studies of cortical function in human amblyopia. Neuroscience and Behavioral Reviews 17: 469-476.

Demer J.L., von Noorden, G.K., Volkow, N.D. ad Gould, K.L. (1988) Imaging of cerebral blood flow and metabolism in amblyopia by positron emission tomography. American Journal of Ophthalmology 105: 337-347.

De Valois, K.K., De Valois, L. and Yund, E.W. (1979) Responses of striate cortex cells to grating and checkerboard patterns. Journal of Physiology 291; 483-505.

De Valois, R.L. and De Valois, K.K. (1980) Spatial Vision. Annual Review of Psychology 31: 309-341.

Douglas, R.J. and Martin, K.A.G. (1991) A functional microcircuit for cat visual cortex. Journal of Physiology 440: 735-769.

Dowling, J. E. (1992) *Neurons and Networks.* An Introduction to Neuroscience. Harvard University Press, Cambridge, Massachusetts.

Driver, J. and Mattingley, J.B. (1995) Selective attention in humans: Normality and pathology. Current Opinion in Neurobiology 5: 191-197.

Duffy, F.H., Burchfiel, J.L. and Conway, J.L. (1976) Bicuculline reversal of deprivation amblyopia in the cat. Nature 260: 256-257.

Edelman, G.M. (1993) Neural Darwinism: Selection and reentrant signaling in higher brain function. Neuron 10: 115-125.

Eggers, H.M., Grizzi, M.S. and Movshon, J.A. (1984) Spatial properties of striate cortical neurons in esotropic macaques. Investigative Ophthalmology and Visual Science 25 (suppl.), 278.

Ekman, G. (1958) Two generalized ratio scaling methods. Journal of Psychology 45: 287-295.

Engel, A.K., König, P., Kreiter, A.K., Schillen, T.B. and Singer, W. (1992) Temporal coding in the visual cortex: New vistas on integration in the nervous system. Trends in Neurosciences 15: 218-226.

Enroth-Cugell, C. and Robson, J.G. (1984) Functional characteristics and diversity of cat retinal ganglion cells. Investigative Ophthalmology and Visual Science 3:250-267.

Ferster, D. (1991) Linearity of synaptic interactions in the assembly of receptive fields in cat visual cortex. Current Opinion in Neurobiology 4: 563-568.

Franzén, O. (1976) Somatosensory potentials from the exposed cortex in monkey and from the scalp in man related to the sensory magnitude of tactual stimulation. In. Y. Zotterman (ed.) *Sensory Functions of the Skin.* Pergamon Press, Oxford, pp. 119-127.

Franzén, O. (1978) On binocular vision. Scandinavian Journal of Psychology 19: 223-229.

Franzén, O. and Ahlquist, M. (1989) The intensive aspect of information processing in the intradental A-delta system in man – A psychophysiological analysis of dental sharp pain. Behavioural Brain Research 33: 1-13.

Franzén, O. and Ahlquist, M. (1996) Central representation and coding of sharp dental pain in man. *8th World Congress of Psychophysiology.* Tampere, Finland. A 169.

Franzén, O. and Berkley, M. (1975) Apparent contrast as a function of modulation depth and spatial frequency: A comparison between perceptual and electrophysiological measures. Vision Research 15: 655-660.

Franzén, O., Johansson, R. and Terenius, L. (1996) *Somesthesis and the neurobiology of the somatosensory cortex.* Birkhäuser Verlag, Basel.

Franzén, O., Kenshalo, Sr. D., Essick, G. (1991) Neural population encoding of touch intensity. In: Franzén, O. and Westman, J. (eds): *Information Processing in the Somatosensory System.* Macmillan Press, London, pp. 71-80.

Franzén, O., Lennerstrand, G. and Richter, H. (1994) Brain potential correlates of supraliminal contrast functions and defocus. International Journal of Human-Computer Interaction 6: 155-176.

Franzén, O. and Offenloch, K. (1969) Evoked response correlates of psychophysical magnitude estimates for tactile stimulation in man. Experimental Brain Research 8: 1-18.

Franzén, O., Richter, H. and von Sandor, R. (1987) Vision monitoring of VDU operators and relaxation of visual stress by means of a laser speckle system. In: B. Knave and P.-G. Widebäck (eds.), *Work with Display Units 86.* Elsevier Science Publishers (North Holland), Amsterdam, pp. 539-551.

Franzén, O. and Westman, J. (eds): *Information Processing in the Somatosensory System.* Macmillan Press, London.

Franzén, O., Thompson, F. and Davenport, P. (1985) The role of recruitment of cutaneous mechanoreceptors in spinal and cortical intensity functions. Acta Physiologica Scandinavica 124 (suppl. 542), 431.

Friston, K.J. (1994) Functional and effective connectivity in neuroimaging: A synthesis. Human Brain Mapping 2:56-78.

Georgopoulos, A.P. (1991) Higher order motor control. Annual Review of Neuroscience 14: 361-377.

Georgeopoulos, A.P. (1995) Current issues in directional motor control. Trends in Neurosciences 18: 506-510.

Georgopoulos, A.P., Schwartz, A.B. and Kettner, R.E. (1986) Neuronal population coding of movement direction. Science 233: 1416-1419.

Gescheider, G. (1997) *Psychophysics. The Fundamentals.* Lawrence Erlbaum Associates, Publishers, London.

Geschwind, N. (1965) Disconnexion syndromes in animals and man. Brain 88: 237-294, 585-644.

Gilbert, C.D. and Wiesel, T.N. (1989) Columnar specificity of intrinsic horizontal and corticocortical connections in cat visual cortex. Journal of Neuroscience 97: 2432-2442.

Gnadt, J.W. and Mays, L.E. (1989) Posterior parietal cortex, the oculomotor near response and spatial coding in 3-D space. Neuroscience Abstracts 15: 786.

Goodale, M.A. and Milner, A.D. (1992) Separate visual pathways for perception and action. Trends in Neurosciences 15: 20-25.

Gross, C.G. (1994) How inferior temporal cortex became a visual area. Cerebral Cortex 5: 455-469.

Hata, Y. and Stryker, M.P. (1994) Control of thalamocortical afferent rearrangement by postsynaptic activity in developing visual cortex. Science, 265, 1732-1735.

Hubel, D. (1988) *Eye, Brain and Vision.* Scientific American Science Libraries, 22, New York.

Hubel, D. and Wiesel, T.N. (1965) Binocular interaction in striate cortex of kittens reared with artificial squint. Journal of Neurophysiology 28: 1041-1059.

Hubel, D. and Wiesel, T.N. (1968) Receptive fields and functional architecture of monkey striate cortex. Journal of Physiology 195: 215-243.

Hubel , D. and Wiesel, T.N. (1972) Laminar and columnar distribution of geniculo-cortical fibres in the Macaque monkey. Journal of Comparative Neurology 146: 421-450.

Imamura, K., Richter, H., Fischer, H., Lennerstrand, G., Franzén, O., Rydberg, A., Andersson, J., Schneider, H., Onoe, H., Watanabe, Y. and Långström, B. (1997) Reduced activity in the extrastriate visual cortex of individuals with strabismic amblyopia. Neuroscience Letters, 225: 173-176.

Jakobsson, P. (1985) Binocular interaction in the human visual evoked potential. Linköping University Medical Dissertations No. 192, Linköping, Sweden.

Johnson, K.O., Phillips, J.R., Hsiao, S.S. and Bankman, I.N. (1991) Tactile pattern recognition. In: O. Franzén and J. Westman (eds.) *Information processing in the somatosensory system.* Macmillan Press, London, pp. 305 – 318.

Johnson, K.O., Hsiao, S.S. and Blake, D.T. (1996) Linearity as the basic law of psychophysics: Evidence from studies of the neural mechanisms of roughness magnitude estimation. In: O. Franzén, R. Johansson and L. Terenius (eds.) *Somesthesis and the Neurobiology of the Somatosensory Cortex.* Birkhäuser Verlag, Basel, pp. 213-228.

Kaas, J.H. (1992) Do humans see what monkeys see? Trends in Neurosciences 15: 1–3.

Kalikow, D.N. (1967) Psychofit. Unpublished computer program (Fortran) for the analysis of magnitude estimates. Brown University, RI, USA.

Kandel, E.R., Schwartz, J.H. and Jessell, T.M. (1991) *Principles of Neural Science.* Edward Arnold, London.

Kelly, D.H. (1976) Pattern detection and the two-dimensional Fourier transform: Flickering checkerboards and chromatic mechanisms. Vision Research 16: 277-287.

Kennedy, C., Des Rosiers, M.H., Sakurada, O., Shinohara, M., Reich, M., Jehle, J.W. and Sokoloff, L. (1976) Metabolic mapping of the primary visual system of the monkey by means of the autoradiographic [^{14}C]deoxyglucose technique. Proceedings of the National Academy of Sciences 73: 4230-4234.

Kiorpes, L. (1996) Development of contrast sensitivity in normal and amblyopic monkeys. In: F. Vital-Durand, J. Atkinson and O.J. Braddick (eds.) *Infant vision.* Oxford University Press, Oxford, pp. 3-15.

112

Koskela, P.U. (1986) Contrast sensitivity in amblyopia. II. Changes during pleoptic treatment. Acta Ophthalmologica 64: 563-569.

Lee, B.B., Virsu, V. and Creutzfeldt, O.D. (1983) Linear signal transmission from prepotentials to cells in the macaque lateral geniculate nucleus. Experimental Brain Research 52: 50-56.

Livingstone, M. and Hubel, D. (1988) Segregation of form, color, movement and depth: Anatomy, physiology, and perception. Science 240: 740-749.

Luck, S.J. Girelli,, M., McDermott, M.T. and Ford, M.A. (1997) Bridging the gap between monkey neurophysiology and human perception: An ambiguity resolution theory of visual selective attention. Cognitive Psychology 33: 64-87.

Luck, S., Hillyard, S.A., Moulona, M., Woldorff, M.G., Clark, V.P. and Hawkins, H.L. (1994) Effects of spatial cueing on luminance detectability: Psychophysical and electrophysiological evidence for early selection. Journal of Experimental Psychology and Human Perceptual Performance 20: 87-94.

Lundh, B.L. (1983) Clinical contrast sensitivity. Linköping University Medical Dissertations, No. 144, Linköping, Sweden.

Lynch, J.C., Mountcastle, V.B., Talbot, W.H. and Yin, T.C.T. (1977) Parietal lobe mechanisms for directed visual attention. Journal of Neurophysiology 2: 362-389.

Maffei, L. and Fiorentini, A. (1973) The visual cortex as a spatial frequency analyser. Vision Research 13: 1255-1267.

Maffei, L. and Fiorentini, A. (1977) Spatial frequency rows in the striate visual cortex. Vision Reserach 17: 257-264.

Meister, M., Wong, R.O.L., Baylor, D.A. Shatz, C.J. (1991) Synchronous bursts of action potentials in ganglion cells of the developing mammalian retina. Science 252: 939-943.

Mesulam, M.-M. (1981) A cortical network for directed attention and unilateral neglect. Annals of Neurology 10: 309-325.

Merzenich, M.M., Allard, T.T. and Jenkins, W.M. (1991) Neural ontogeny of higher brain function. Implications of some recent neurophysiological findings. In: O. Franzén and Jan Westman (eds.) *Information processing in the somatosensory system*. Macmillan Press Ltd., London, pp. 193-209.

Merzenich, M.M., Wang, X., Xerri, C. and Nudo, R. (1996) Functional plasticity of cortical representations of the hand. In: O. Franzén, R. Johansson and L. Terenius (eds.) *Somesthesis and the Neurobiology of the Somatosensory Cortex*. Birkhäuser Verlag, Basel, pp. 249-269.

Mountcastle, V.B., Talbot, W.H., Sakata, H. and Hyvärinen, J. (1969) Cortical neuronal mechanisms in flutter vibration studied in unanesthetized monkeys. Neuronal periodicity and frequency discrimination. Journal of Neurophysiology 32: 452-484.

Mountcastle, V.B. (1975) The view from within: Pathways to the study of perception. The Johns Hopkins Medical Journal 136: 109-131.

Movshon, J.A., Eggers, H.M., Gizzi, M.S., Hendrickson, A.E., Kiorpes, L. and Boothe, R.G. (1987) Effects of early unilateral blur on the macaque's visual system: II. Physiological observations. Journal of Neuroscience 7: 1340-1351.

Owens, D.A. (1980) A comparison of accommodative responsiveness and contrast sensitivity for sinusoidal gratings. Vision Research 20: 159-167.

Payne, B.R., Lomber, S.G., Alessandro, E., Bullier, V. and Bullier, J. (1996) Reversible deactivation of cerebral network components. Trends in Neurosciences 19: 535-542.

Phillips, S. and Stark, L. (1977) Blur: A sufficient accommodative stimulus. Documenta Ophthalmologica 43: 65-89.

Posner, M.I. and Raichle, M.E. (1994) *Images of Mind*. Scientific American Library, New York.

Peters, A., Payne, B.R. and Budd, J. (1994) A numerical analysis of the geniculocortical input to striate cortex in the monkey. Cerebral Cortex 4: 215-229.

Reiter, H.O. and Stryker, M.P. (1988) Neural plasticity without postsynaptic action potentials: Less active inputs become dominant when kitten visual cortical cells are pharmacologicaly inhibited. Proceedings from the National Academy of Sciences, U.S.A. 85:3623-3627.

Richter, H., Abdi, S., Han, Y., Lennerstrand, G., Franzén, O., Andersson, J., Pardo, J.V., Schneider, H. and Långström, B. (1998) Higher order control processes and plasticity in neural CNS circuitry subserving negative voluntary accommodation. Investigative Ophthalmology and Visual Science 29: 1047.

Richter, H. and Franzén, O. (1994) Velocity percepts of apparent laser speckle motion modulated by voluntary changes of visual accommodation: Real-time, in-vivo measurements of the accommodative response. Behavioural Brain Research 62: 93-102.

Richter, H., Lee. J.E. and Pardo, J. (1995) Central correlates of voluntary visual accommodation in humans measured with $H_2^{15}O$ Water and PET. Presented at the ARVO Annual Meeting, Fort Lauderdale, Florida, May 14 – May 19, 1995. Abstracts.

Roelfsema, P.R., König, P., Engel, A.K., Sireteanu, R. and Singer, W. (1994) Reduced synchronization in the visual cortex of cats with strabismic amblyopia. European Journal of Neuroscience 6: 1645-1655.

Sadato, N., Pascual-Leone, A., Grafman, J., Ibanez, V., Deiber, M.P., Dold, G. and Hallett, M. (1996) Activation of the primary visual cortex by Braille reading in blind subjects. Nature 380: 526-528.

Sakata, H., Taira, M., Kusunoki, M., Murata, A. and Tanaka, Y. (1997) The parietal association cortex in depth perception and visual control of hand action. Trends in Neurosciences 20: 350-357.

Schadé, O.H. Sr. (1956) Optical and photoelectric analogue of the eye. Journal of Optical Society of America 46: 721-739.

Schiller, P.H. and Malpeli, J.G. (1977) The effect of striate cortex cooling on area 18 cells in the monkey. Brain Research 126: 366-369.

Schiller, P.H. (1986) The central visual system. Vision Research 26, 1351-1386.

Sclar, G. (1987) Expression of "retinal" contrast gain control by neurons of the cat's lateral geniculate nucleus. Experimental Brain Research 66: 589-596.

Shatz, C.J. (1990) Impulse activity and the patterning of connections during CNS development. Neuron 5: 745 – 756.

Singer, W. and Gray, C.M. (1995) Visual feature integration and the temporal correlation hypothesis. Annual Review of Neuroscience 18: 555-586.

Singer, W., Engel, A.K., Kreiter, A.K., Munk, M.H.J., Neuenschwander, S. and Roelfsema, P.R. (1997) Neuronal assemblies: Necessity, signature and detectability. Trends in Cognitive Sciences 1: 252-260.

Spitzer, H. and Richmond, B.J. (1991) Task difficulty: Ignoring, attending to, and discriminating a visual stimulus yield progressively more activity in inferior temporal neurons. Experimental Brain Research 83: 340-348.

Stark, L.R. and Atchinson, D.A. (1994) Subject instructions and methods of target presentation in accommodation research. Investigative Ophthalmology and Visual Science 35: 528-537.

Stark, L.W., Takahashi, Y. and Zames, G. (1965) Absence of an odd-error signal mechanism in human lens accommodation. IEEE Transactions on Systems Science and Cybernetics SSC-1: 75-83

Stevens, S.S. (1957) On the psychohysical law. Psychological Review 64: 153-181.

Stevens, S.S. (1975) *Psychophysics: Introduction to its perceptual, neural and social prospects.* Wiley, New York.

Stone, J. (1965) A quantitative analysis of the distribution of ganglion cells in the cat's retina. Journal of comparative Neurology 124: 333-352.

Sur, M. (1993) Cortical specification: Microcircuits, perceptual identity and an overall perspective. Perspectives on Developmental Neurobiology 1: 109-113.

Talairach, J. and Tournoux, P. (1988) Coplanar stereotactic atlas of the human brain. Thieme, Stuttgart.

Tononi, G., Edelman, G.M. and Sporns, O. (1998) Complexity and coherency: Integrating information in the brain. Trends in Cognitive Sciences 2: 474-483.

Troy, J.B. and Enroth-Cugell, C. (1993) X and Y ganglion cells inform the cat's brain about contrast in the retinal image. Experimental Brain Research 93: 383-390.

Ts'o, D. and Gilbert, C. (1988) The organization of chromatic and spatial interactions in the primate striate cortex. Journal of Neuroscience 8: 1712-1727.

Ungerleider, L.G. and Haxby, J.V. (1994) "What" and "where" in the human brain. Current Opinion in Neurobiology 4: 157-165.

Ungerleider, L.G. and Mishkin, M. (1982) Two cortical visual systems. In: D.J. Ingle, M.A. Goodale and R.J.W. Mansfield (eds.) *Analysis of Visual Behaviour.* MIT Press, Cambridge, Massachusetts, pp. 549-586.

Van Essen, D.C., Anderson, C.H. and Felleman, D.J. (1992) Information-processing in the primate visual system: An integrated systems perspective. Science 255: 419-423.

Weale, R.A. (1982) *Focus on Vision.* Hodder and Stoughton, London.

Whitsel, B.L. and Franzén, O. (1989) Dynamics of information processing in the somatosensory cortex. In: C. von Euler, I. Lundberg and G Lennerstrand (eds.). *Brain and Reading.* Macmillan Press, London, pp. 129-137.

Wiesel, T. and Gilbert, C.D. (1991) Neural mechanisms of visual perception. In: D. Man-Kit Lam and C.D. Gilbert (eds) *Neuronal Mechanisms of Visual Perception.* Gulf Publishing Company, Houston, pp. 7-33.

Wiesel, T.N., Hubel, D.H. and Lam, D.M.K. (1974) Autoradiographic demonstration of ocular-dominance columns in the monkey striate cortex by means of transneuronal transport. Brain Research 79: 272-279.

von Noorden, G.K. (1990) *Binocular vision and ocular motility.* Theory and management of strabismus. C.V. Mosby, St Louis, MO.

Wu, G. (1997) Functional organisation and population behaviour of human peripheral nerve fibres. A microneurography study. Doctoral Dissertation, Karolinska Institute, Huddinge University Hospital, Huddinge, Sweden.

Wurtz, R.H., Goldberg, M.E. and Robinson, D.L. (1982) Brain mechanisms of visual attention. Scientific American 246: 124-135.

Yilmaz, H. (1967) Perceptual invariance and the psychophysial law. Perception and Psychophysics 2: 533-538.

Young, M.P. (1992) Objective analysis of the topological organization of the primate cortical visual system. nature 358: 152-155.

Young, M.P., Scannel, J.W., Burns, G. and Blakemore, C. (1994) Analysis of connectivity: Systems in the cerebral cortex. Review of Neuroscience 5: 227-250.

Zeki, S. and Shipp, S. (1988) The functional logic of cortical connections. Nature 335: 311-317.

Zeki, S. (1993) *A vision of the brain.* Blackwell, Oxford, England.

Zemon, V., Victor, J.D. and Ratliff, F. (1986) Functional subsystems in the visual pathways of humans characterized using evoked potentials. In: R.Q. Cracco and I. Bodis-Wollner (eds.) *Frontiers of Neuroscience. Evoked Potentials.* Alan R. Liss, New York, pp. 203-210.

Zigmond, M.J., Bloom, F.E., Landis, S.C., Roberts, J.L. and Squire, L.R. (1998). Eds. *Fundamental Neuroscience.* Academic Press, London.

Accommodation and the through-focus changes of the retinal image

W.N.Charman

Department of Optometry and Vision Sciences, UMIST, PO Box 88, Manchester M60 1QD, U.K.

Summary: The optical changes in the retinal image as a function of focus are reviewed in relation to the initiation, direction and maintenance of the accommodation response. For any spatial frequency component in the object, the deterioration of ocular modulation transfer with defocus is more rapid when the spatial frequency is high or when the pupil is larger. Optical aberrations, particularly longitudinal chromatic aberration, appear to be of some value in guiding dynamic responses when initial errors of focus are modest (<2 D). Fluctuations in accommodation at low temporal frequencies (<0.6 Hz) may help to maintain the mean level of the steady-state response to any stimulus. Under conditions where only low spatial frequency information is available to the control system, due to the characteristics of the object (spatial frequency spectrum), observing conditions (e.g.low luminance) or observer (e.g.amblyopia), the steady-state response is less accurate.

Introduction

Under most circumstances, the accommodation system of the young human eye is remarkably effective in changing the eye's focus to produce a retinal image whose clarity meets the visual system's needs under its current observing conditions. Most of us take this effectiveness for granted until, in middle age, it declines due to the advance of presbyopia.

Although a wide variety of stimulus factors, such as binocular disparity, proximity and size change, normally contribute to the efficiency of the accommodative mechanism, it is reasonable that any attempt to understand its function should start with the way in which the retinal image changes as a function of the accuracy of focus. Three related aspects of these changes will be considered here:

(i) The effect of focus on the spatial characteristics of the retinal image. It would be expected that the detection of a deterioration in the image quality caused by change in the dioptric distance of the object of regard could serve to initiate a response, i.e that optical blur could serve as a stimulus to accommodation.

(ii) The extent to which differences produced by optical aberration in the retinal images at equal dioptric distances on either side of optimal focus could provide "odd error" cues to ensure that ocular focus change is in the direction which improves, rather than worsens, the image quality.

(iii) The possible role of small, time-varying changes in retinal image quality produced by fluctuations in accommodation in guiding the initial response and maintaining the "steady state" accommodation to a fixed target.

The basic response

Before discussing these aspects in more detail, it will be helpful to remind ourselves of the typical characteristics of the accommodation response. When observing a target at a fixed dioptric distance, the focus of the eye shows small temporal fluctuations with an amplitude of about 0.2 D and a frequency spectrum extending to a few Hz, and often having a strong component at 1-2 Hz (see, e.g., Charman and Heron, 1988; Winn and Gilmartin, 1992; Denieul and Corno-Martin, 1994; Winn, 1996, this meeting, for reviews). A change in target distance results in accommodation response lasting about 0.5-1.0 sec after a latency or reaction time of about 0.3 sec (fig.1a). If the mean or "steady state" level of response is plotted as a function of the stimulus level, "leads" of accommodation occur at low dioptric stimulus levels and "lags" at higher levels (fig.1b) (see, e.g. Ciuffreda, 1991 for review). The slope of the linear portion of the response/stimulus curve varies with many factors. In particular, it reduces if the target is of low luminance or contains predominantly low spatial frequencies (e.g. Heath, 1956; Nadell and Knoll, 1956; Johnson, 1976; Tucker and Charman, 1986; Charman, 1986), although it is relatively robust against changes in object contrast, provided contrast does not approach threshold levels (e.g. Ciuffreda and Rumpf, 1985; Tucker et al., 1986; Ward, 1987).

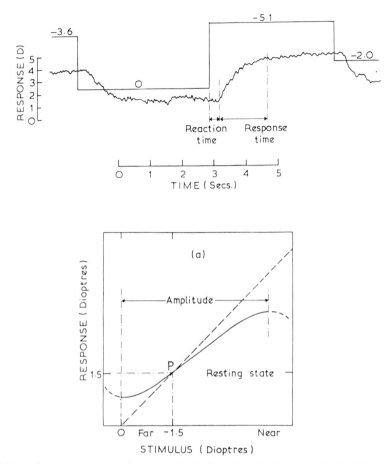

Fig.1. (top) Dynamic response to a step changes in stimulus level. Note the fluctuations in the initial and final levels. (bottom) Schematic steady state stimulus/response curve. Stimulus and response are equal at the resting state or tonic accommodation level for the individual eye, typically equal to about 1.5 D.

Image quality as a function of focus

For an aberration-free eye with a circular pupil, geometrical optics predicts that defocus of the image of a monochromatic point object will produce a circular retinal blur patch whose diameter increases linearly with the magnitude of the dioptric defocus and the pupil diameter (e.g. Bennett and Rabbetts, 1974; Smith, 1982). The blur patches for positive and negative errors of focus will be identical. The images of other objects will be given by the convolution of these blur patches with the object's luminance distribution. The gradient of retinal illuminance for the image of an edge, for example, will steadily reduce as the defocus increases.

Diffraction modifies these predictions, particularly when the images are close to optimal focus, but does not change their essential nature. Fig.2, for example, shows for two pupil diameters the change in modulation transfer (i.e. effectively the modulation in the retinal image of a sinusoidal object grating of 100% contrast) as focus is changed. It will be noted that low spatial frequency components in any image are relatively robust against defocus, whereas at high spatial frequencies image contrast falls rapidly. Thus if the observed object is initially well out of focus, the optical information available to guide the initial response is likely to be at low spatial frequencies (e.g. Bour, 1981), whereas for small changes in focus higher spatial frequencies can play a role. This makes studies of the dependence of response dynamics on the magnitude of stimulus change of particular interest: to date there few such studies have been carried out (e.g. Tucker and Charman, 1979). Fig.2 also illustrates that the effects at any spatial frequency scale inversely with pupil diameter, with sensitivity to defocus reducing as the pupil diameter is decreased.

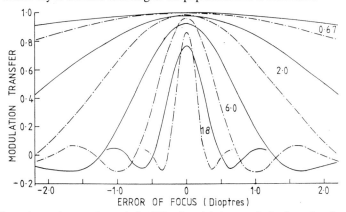

Fig.2. Through- focus changes in monochromatic (l= 535 nm) modulation transfer for 3 mm (continuous curves) and 5 mm (chain-dotted curves) pupil diameters at the spatial frequencies (c/deg) indicated.

Ocular aberration will, in general, tend to reduce modulation transfer and make its changes about optimal focus less symmetrical, particularly for larger diameters of pupil: modulation transfer will also be dependent on grating orientation and shifts in spatial phase may be introduced (e.g. Walsh and Charman, 1989).

Not surprisingly, these modulation transfer results are paralleled by typical results for through-focus changes in visual acuity when the eye is under cycloplegia and unable to accommodate; losses in modulation transfer at higher spatial frequencies result in diminished acuity (fig.3). Note the fall in acuity (or increase in minimum angle of resolution) with defocus is much slower with small pupils, this being exploited practically in the optometric pinhole test. In fact the pupil dependence of the type shown in fig.3 is influenced not only by diffraction and aberration but also by the Stiles-Crawford effect, which can be considered as an apodisation of the pupil and tends to increase depth-of-focus for the larger pupils (e.g. Campbell, 1957; Metcalf, 1965).

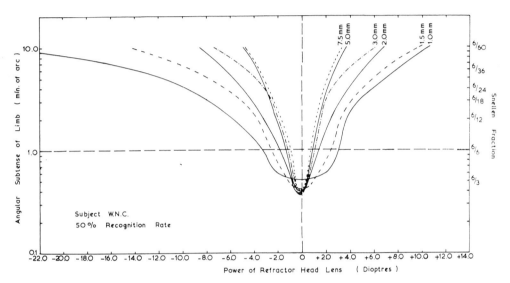

Fig.3. Visual acuity as a function of applied spectacle lens power for an eye under cycloplegia for the pupil diameters indicated (after Tucker and Charman, 1975).

It follows from curves of the type shown in figure 2 and 3 that, if the focus of the retinal image is changed by a small amount, this has little effect on the image characteristics if the retinal image is initially at optimal focus, but a much greater effect if the image is initially slightly defocused, since the changes depend on the gradients of the curves. This increase in sensitivity to change in focus for slightly defocused images was originally noted by Campbell and Westheimer (1958), see fig.4, and elaborated upon by Walsh and Charman (1988). Minimal focus changes of about 0.1 to 0.2 D can be detected and it is of interest that accommodation responses can be triggered by stimulus changes of similar (Ludlam et al., 1968) or even smaller magnitude (Kotulak and Schor, 1986a). The possible relationship between lags and leads in steady-state accommodation and characteristics of the type shown in fig.4 will be discussed further below.

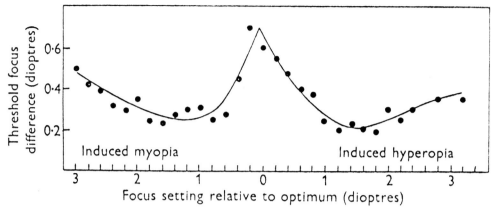

Fig.4. Sensitivity to periodic (2 Hz) focus change of a small, high contrast object lit by white light as a function of the initial level of defocus for an eye under cycloplegia with a 3 mm artificial pupil. Note that the smallest changes in focus can be detected when the eye is initially slightly out of focus (after Campbell and Westheimer, 1958) .

Optical cues to correct direction of accommodation change

Although authors such as Fincham (1951, 1953) had attempted to study the factors influencing the direction of the initial focusing response, it was the advent of automatic recording optometers that opened up this question to more thorough study. It is again important to remember that, in addition to the optical cues to be discussed here, there are numerous non-optical cues, such as binocular disparity (e.g. McLin et al., 1988), proximity (e.g. Rosenfield and Gilmartin, 1990; North et al., 1993), size change (e.g. Kruger and Pola, 1985) and perceived depth in 2-dimensional scenes (e.g. Takeda et al., 1990) that can potentially initiate and guide a response.

Errors in response direction

In fact, in most experimental studies a significant proportion of responses are found to be in the wrong direction, irregular or absent for some subjects, even when no particular effort is made to minimise potential cues (e.g. Campbell and Westheimer, 1959; Stark and Takahashi, 1965; Smithline, 1974; Charman and Heron, 1979; Bour, 1981). It is, perhaps, unfortunate that many authors tend either to omit such data as unreliable, only presenting the data of "trained" subjects who give consistent responses, or to conceal the effect selecting the "best" responses or by averaging many responses, thereby giving a distorted impression of typical performance under the experimental conditions in use.

The observation of 50% error rates for the initial response in some subjects led Stark and his colleagues (Troelstra et al, 1964; Stark and Takahashi, 1965; Stark, 1968) to suggest that under monocular conditions there was no odd-error optical cue and that accommodative tracking was therefore a trial and error function. This would, however, now seem to be an extreme view and it is worth re-examining the main possible optical cues that are normally available to guide accommodation. These are illustrated in fig. 5.

Longitudinal chromatic aberration

Fincham (1951, 1953) suggested that one important cue to the direction of correct focus was provided by the longitudinal chromatic aberration (LCA), the eye having some 2 D of relative myopia for the extreme blue end of the spectrum as compared with the extreme red. Fincham hypothesised that the eye could make use of the differences in the colour fringes on either side of correct focus to guide the initial response in the correct direction (see also Aggarwala et al., 1995). In support of this suggestion, he found that in monochromatic light 60% of his subjects experienced difficulty in accommodating, whereas they had no problems in white light. Some further support for this view was provided by Campbell and Westheimer (1959), although their subject who initially found difficulty in monochromatic illumination learnt relatively quickly to make use of other cues to produce correct responses (unlike those of Kruger et al., 1993).

Later authors were more sceptical of the value of cues provided by LCA (see, e.g. Kruger et al., 1993 for review) but there is now strong evidence, based on the effects of neutralising, reversing or otherwise manipulating the LCA, that normal LCA favourably influences dynamic accommodation responses (see, e.g. Kruger et al., 1993; Stone et al., 1993; Aggarwala et al., 1995; Kruger, 1996, this meeting). It is, however, less important for the static response (e.g. Bobier et al., 1992).

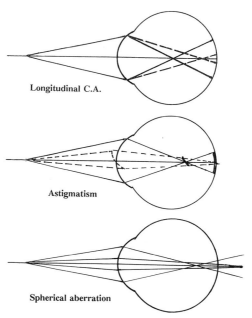

Fig.5. Possible "odd error" optical cues to guide the initial response direction. All these aberrations produce asymmetry between the images on either side of optimal focus. In the case of longitudinal chromatic aberration, the blue rays are shown as continuous, the red as dashed.

Astigmatism

Campbell and Westheimer (1959) found that relatively small amounts of astigmatism provided by a cylindrical lens could help improve the directional accuracy of the initial response. There has been no subsequent attempt to follow up this suggestion, e.g. by exploring whether the correction of small amounts of natural ocular astigmatism impairs the response in any way.

Spherical and other aberrations

Campbell and Westheimer (1959) thought that these could be useful. Since responses in the correct direction can still be obtained with pupil diameters <3 mm, even though monochromatic aberrations are small under these conditions, and contact lenses introducing different amounts of aberration have little effect upon the response, it seems unlikely that these aberrations usually play any major role. Asymmetric wave aberration could, however, by inducing focus-dependent lateral shifts in the image provide useful cues: such aberrations are known to exist in many eyes (e.g. Walsh and Charman, 1989).

Summary of optical cues

Although individuals may make use of a wide variety of other cues to reduce the possibility of error in the initial direction of the response, it now appears that the "odd error" cue provided by LCA can make a useful contribution under some circumstances. In general the nature of the optical cues means that they are likely to be most effective where only modest (<2 D) changes in object vergence are involved.

Role of fluctuations in accommodation

Following the clear demonstration of the existence of fluctuations in accommodation, a variety of authors speculated that they might be used in the feedback control system for accommodation (e.g. Alpern, 1958; Fender, 1958; Crane, 1966; Charman and Tucker, 1978; Hung et.al., 1982; Kotulak and Schor, 1986b; Walsh and Charman, 1988). The basic idea as proposed by the early workers was that the fluctuations play a "hunting" role (e.g. Alpern, 1958). In simple terms, guidance for the initiating and maintaining the response would be achieved by noting whether a fluctuation which increased accommodation led to an increase in image contrast and clarity, in which case this would be the appropriate direction for the main response, or decreased it, in which case the preferred response would be a decrease in accommodation. Almost all these models were based on the idea that it was the fluctuation peak at 1-2 Hz that was particularly important.

The relatively recent finding that the low frequency fluctuations (<0.6 Hz) increase with decrease in pupil diameter and luminance level, when ocular depth-of-focus is larger, and high frequency fluctuations do not (Gray et al., 1993a,b; Winn, 1996, this meeting), suggests strongly that it is these slower fluctuations which could play the "hunting" role, rather than those at higher temporal frequencies, which may simply be noise introduced by the ocular pulse (Winn et al., 1990).

With accommodation latencies of a third of a second or less, even 2 Hz oscillations would be at the limit of effectiveness for initiating a correct response. If, then, the significant frequency components are of lower frequency, their contribution to control must lie primarily in the maintenance of the steady-state response rather than in response to stimulus change.

How might the fluctuations act? A critical observation here is that the non-unit slope of the response/stimulus curve (fig.1a) implies that the errors in response increase with the difference between the stimulus and the tonic accommodation level of the individual. Evidently, any given amplitude and frequency of fluctuation will produce temporal changes in the contrast of each component spatial frequency in the retinal image, the amplitude of these temporal changes depending upon the mean position of focus (see fig.5, where the fluctuations are, for simplicity shown as being sinusoidal with a single frequency). Fluctuation about optimal focus produces very small changes in contrast, whereas fluctuation about a focus where modulation transfer changes steeply with focus produces large contrast changes.

From the point of view of accommodation control, it would seem reasonable that what would be important to the system at each spatial frequency would be the ratio of the modulation change in the image, DM, to the mean modulation M, where $DM/M = (dT/dF)_F \cdot DF/T(F)$, where $T(F)$ is the modulation transfer and $(dT/dF)_F$ is the gradient of the through-focus change in modulation transfer at the mean level of defocus, F, and DF is the amplitude of the focus fluctuations. Very similar arguments have been used by Kotulak and Schor (1986b).

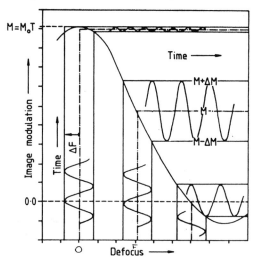

Fig.6. Schematic illustration of the changes in the modulation of the retinal image of a grating that are produced by small, equal, sinusoidal oscillations of focus at three different mean positions of focus (after Walsh and Charman, 1988).

On this basis, Charman and Tucker (1978) and Walsh and Charman (1988) argued that as the accommodation demand increased, larger values of DM/M would be required to maintain the system at the higher level of response. This could only be achieved through larger errors, or lags, in mean accommodation which would move the responses onto the steeper parts of the through-focus modulation transfer curve to generate larger values of DM/M. Thus it would be expected that the lags in response would increase with stimulus level, as observed. Note too that this process would be independent of the contrast of the original grating component, as is required by the experimental observation that the response/stimulus curve for any target is relatively robust against changes in object contrast. It is, of course, necessary that the retinal modulation should always exceed threshold.

With targets of broad spatial bandwidth, the spatial frequency of greatest importance would depend upon the observing conditions. For example, in low luminances where high spatial frequencies were below threshold and the response depended on low spatial frequencies with their relative insensitivity to defocus, larger lags would be expected, as observed. Reductions in pupil diameter would make modulation transfer at any spatial frequency less sensitive to change in focus (see fig.1). Thus one would expect that increases in both the steady-state error and the amplitude of the fluctuations would be required to maintain the necessary level of DM/M. With continued reduction in pupil diameter, response would remain at the tonic level, since the changes in modulation produced by fluctuation would be insufficient to provide adequate feedback.

Discussion

Although the broad outlines of the role of optical factors seem clear, much needs to be done to determine how they are integrated with the numerous other factors contributing to the accommodation response. It is already apparent that under many circumstances non-optical cues may dominate over cues given by the retinal image quality (e.g. Kruger and Pola, 1985; 1987), so that any model based on optical effects alone cannot be adequate. Although this paper has been concerned with retinal blur produced by dioptric defocus of a "sharp" stimulus, it is of interest to note that under some circumstances an initial accommodation response can also be induced by blurring the object while maintaining its distance constant (Smithline, 1974; Phillips and Stark, 1977).

References

Aggarwala, K.R., Nowbotsing, S. and Kruger, P.B. (1995) Accommodation to monochromatic and white-light targets. *Investigative Ophthalmology and Visual Science,.* **36**; 2695-2705.

Alpern, M. (1958) Variability of accommodation during steady fixation at various levels of illumination. *Journal of the Optical Society of America.* **48**; 193-197.

Bennett, A.G. and Rabbetts, R.B. (1984) *Clinical Visual Optics* First edition, Butterworths, London, pp.67-68.

Bobier, W.R., Campbell, M.C.W. and Hinch, M. (1992) The influence of chromatic aberration on the static accommodation response. *Vision Research* **32**; 823-832.

Bour, L.J. (1981) The influence of the spatial distribution of a target on the dynamic response and fluctuations of the accommodation of the human eye. *Vision Research* **21**; 1287-1296.

Campbell, F.W. (1957). The depth-of-field of the human eye. *Optica Acta* **4**; 157-164.

Campbell, F.W. and Westheimer, G. (1958) Sensitivity of the eye to differences in focus. *Journal of Physiology.* **143**; 18P.

Campbell, F.W. and Westheimer, G. (1959) Factors influencing the accommodation responses of the human eye. *Journal of the Optical Society of America* **49**; 568-571.

Charman, W.N. (1986) Static accommodation and the minimum angle of resolution. *American Journal of Optometry & Physiological Optics* **63**, 915-921.

Charman, W.N. and Heron, G. (1979) Spatial frequency and the dynamics of the accommodation response. *Optica Acta* **26**; 217-228.

Charman, W.N. and Heron, G. (1988) Fluctuations in accommodation: a review. *Ophthalmic and Physiological Optics* .**8**; 153-164.

Charman, W.N. and Tucker, J. (1978) Accommodation as a function of object form. *American Journal of Optometry & Physiological Optics* **55**; 84-92.

Ciuffreda, K.J. (1991) Accommodation and its anomalies. *In*: W.N.Charman (ed.) *Vision and Visual Dysfunction: Vol.1. Visual Optics and Instrumentation*, Macmillan, London, Chapter 11, pp. 231-279.

Ciuffreda, K.J. and Rumpf, D. (1985) Contrast and accommodation in amblyopia. *Vision Research.* **25**; 1445-1457.

Crane, H.D. (1966) *A theoretical analysis of the visual accommodation system in humans.* Report NASA CR-606, NASA, Washington, D.C.

Denieul and Corno-Martin (1994) Mean response and oscillations of accommodation with colour and contrast. *Ophthalmic and Physiological Optics,* **14**; 184-192.

Fender, D.H. (1964) Control mechanisms of the eye. *Scientific American* **211**; 24-33.

Fincham, E.F. (1951) The accommodation reflex and its stimulus. *British Journal of Ophthalmology* **35**; 381-393.

Fincham, E.F. (1953) Factors controlling ocular accommodation. *British Medical Journal* **9**; 18-21.

Gray, L.S., Winn, B. and Gilmartin, B. (1993a) Accommodative microfluctuations and pupil diameter. *Vision Research.* **33**; 2083-2090.

Gray, L.S., Winn, B. and Gilmartin, B. (1993b) Effect of target luminance on microfluctuations in accommodation. *Ophthalmic and Physiological Optics* **13**; 258-265.

Heath, G.G. (1956) The influence of visual acuity on the accommodative responses of the eye. *American Journal of Optometry & Archive of American Academy in Optometry.***33**; 513-524.

Hung, G.K., Semmlow, J.L., Ciuffreda, K.J. (1982) Accommodative oscillation can enhance average accommodative response: a simulation study. *IEEE Transactions on Systems Science and Cybernetics* **SMC-12**; 594-598.

Johnson, C.A. (1976) Effects of luminance and stimulus distance on accommodation and visual resolution. *Journal of the Optical Society of America* **66**; 138-142.

Kotulak, J.C. and Schor, C.M. (1986a) The accommodative response to subthreshold blur and to perceptual fading during the Troxler phenomenon. *Perception* **15**; 7-15.

Kotulak, J.C. and Schor, C.M. (1986b) A computational model of the error detector of human visual accommodation. *Biological Cybernetics* **54**; 189-194.

Kruger, P.B. (1996) The chromatic stimulus for accommodation to stationary and moving targets. 1st International Symposium on "Accommodation/vergence mechanisms in the visual system", Stockholm, September 1996.

Kruger, P.B. and Pola, J. (1985) Changing target size as a stimulus for accommodation. *Journal of the Optical Society of America. A*, **2**; 1832-1835.

Kruger, P.B., Mathews, S., Aggarwala, K.R. and Sanchez, N. (1993) Chromatic aberration and ocular focus: Fincham revisited. *Vision Research* **33**; 1397-1411.

Ludlam, W.M., Wittenberg,S., Giglio, E.J. and Rosenberg, R. (1968) Accommodative responses to small changes in dioptric stimulus. *American Journal of Optometry & Archive of American Academy in Optometry* **45**; 483-506.

McLin, L.N., Schor, C.M. and Kruger, P.B. (1988) Changing target size (looming) as a stimulus to accommodation and vergence. *Vision Research* **28**; 883-898.

Metcalf, H. (1965) Stiles-Crawford apodisation. *Journal of the Optical Society of America* **55**; 72-74.

Nadell, M.C. and Knoll, H.A. (1956) The effect of luminance, target configuration and lenses upon the refractive state of the eye. Parts I and II. *American Journal of Optometry & Archive of American Academy in Optometry* **33**; 24-42 and 86-95.

North, R.V., Henson, D.B. and Smith, T.J. (1993) Influence of proximal, accommodative and disparity stimuli upon vergence system. *Ophthalmic and Physiological Optics* **13**; 239-243.

Rosenfield, M. and Gilmartin, B. (1990) Effect of target proximity on the open-loop accommodative response. *Optometry and Vision Science* **67**; 74-79.

Smith, G. (1982) Ocular defocus, spurious resolution and contrast reversal. *Ophthalmic and Physiological Optics* **2**; 5-23.

Smithline, L.M. (1974) Accommodative response to blur. *Journal of the Optical Society of America* **64**; 1512-1516.

Stark, L. (1968) *Neurological Control Systems: Studies in Bioengineering.* Plenum Press, New York, Chapters 2 and 3, pp.205-229.

Stark, L. and Takahashi, Y. (1965) Absence of an odd-error signal mechanisms in human accommodation. *IEEE Transaction on Biomedical Engineering* **BME-12**; 138-146.

Stone, D., Mathews, S. and Kruger, P.B. (1993) Accommodation and chromatic aberration: effect of spatial frequency. *Ophthalmic and Physiological Optics* **13**; 244-252.

Takeda, T., Iida, T. and Fukui, Y. (1990)n Dynamic eye accommodation evoked by apparent distances. *Optometry and Vision Science* **67**; 450-455.

Troelstra, A., Zuber, B.L., Miller, D. and Stark, L. (1964) Accommodative tracking: a trial and error function. *Vision Reserarch* **4**; 585-595.

Tucker, J. and Charman, W.N. (1975) The depth-of-focus of the human eye for Snellen letters. *American Journal of Optometry & Physiological Optics* **52**; 3-21.

Tucker, J. and Charman, W.N. (1979) Reaction and response times for accommodation. *American Journal of Optometry & Physiological Optics* **56**; 490-503.

Tucker, J. and Charman, W.N. (1986) Depth of focus and accommodation for sinusoidal gratings as a function of luminance. *American Journal of Optometry & Physiological Optics* **63**; 58-70.

Tucker, J., Charman, W.N. and Ward, P.A. (1986) Modulation dependence of the accommodation response to sinusoidal gratings. *Vision Research* **26**; 1693-1707.

Walsh, G. and Charman, W.N. (1988) Visual sensitivity to temporal change in focus and its relevance to the accommodation response. *Vision Research* **28**; 1207-1221.

Walsh, G. and Charman, W.N. (1989) The effect of defocus on the contrast and phase of the retinal image of a sinusoidal grating. *Ophthalmic and Physiological Optics* **9**; 398-404.

Ward, P.A. (1987) The effect of stimulus contrast on the accommodation response. *Ophthalmic and Physiological Optics* **7**; 9-15.

Winn, B. (1996) The dynamics of the steady state accommodation response. 1st International Symposium on "Accommodation/vergence mechanisms in the visual system", Stockholm, September 1996.

Winn, B. and Gilmartin, B. (1992) Current perspective on microfluctuations of accommodation. *Ophthalmic and Physiological Optics* **12**; 252-256.

Winn, B., Pugh, J.R., Gilmartin, B. and Owens, H. (1990) Arterial pulse modulates steady-state ocular accommodation. *Current Eye Research* **9**, 971-975.

Accommodative microfluctuations: a mechanism for steady-state control of accommodation

Barry Winn

Department of Optometry, University of Bradford, Richmond Road, Bradford, West Yorkshire, England.

Summary: Rapid and continuous fluctuations in ocular focus are known to occur when the eye views a stationary stimulus. The advent of high-speed infra-red optometers has established that these microfluctuations of ocular accommodation have two dominant components: a low frequency component (LFC) of <0.6Hz and a high frequency component (HFC) between 1.0 and 2.3Hz. Although the retinal image blur associated with microfluctuations has the potential to guide and maintain optimum accommodation levels, there is no consensus with regard to the respective contribution of each of the dominant frequency components. In an attempt to clarify the role of the accommodation microfluctuations we have conducted a series of experiments to investigate their nature and aetiology. Manipulation of stimulus parameters and viewing conditions along with the use of topical drugs has allowed induced changes in the response waveform to be investigated.

A significant between-subject variation in the HFC was found and led us to consider the relationship between this component and other physiological mechanisms which provide intraocular rhythmic variation. Simultaneous measurements of ocular accommodation and systemic arterial pulse frequency demonstrate that the location of the HFC is significantly correlated with arterial pulse frequency. We have also identified a relationship between the power in the LFC and the quality of the retinal image. However, the failure of the HFC to vary with stimulus conditions in a consistent manner supports the result that it is associated with accommodative plant noise rather than being under neurological control. There is an increased requirement for low-frequency modulation of retinal image contrast when the feedback system is placed under duress suggesting that the LFCs play a role in the control of sustained accommodation. The studies to date suggest that the complex waveform of the accommodative microfluctuations is a consequence of the combination of neurological control and physiological rhythmic variation: the former attributable to the LFCs; the latter to arterial pulse.

Introduction

The temporal instability in ocular accommodation has attracted the interest of numerous investigators over the last 50 years. The pioneering work of Collins (1937) using an ingenious electronic refractometer was followed by the more systematic investigations of Arnulf and his colleagues in the mid 1950's using an elaborate double-pass ophthalmoscopic method to observe directly, or photographically, the temporal variations in retinal image quality of gratings (Arnulf, Dupuy & Flamant, 1951a,b). The technique was later refined using an image intensifier but the results of these early investigations suggested that, rather than being a spurious characteristic of the accommodation response, microfluctuations could provide a mechanism for maintaining optimum mean response levels by monitoring small changes in contrast of the retinal image.

It was, however, the collaboration of Campbell, Westheimer and Robson (1959a,b) that produced special insight into the nature of accommodative microfluctuations. Their objective continuously-recording infra-red optometer (Campbell & Robson, 1959) had sufficient time resolution to demonstrate not only the magnitude of microfluctuations but also their temporal characteristics and importantly their frequency spectra. A number of research laboratories (Denieul, 1982; Kotulak & Schor, 1986a; Winn, Pugh, Gilmartin and Owens, 1990a) have now consolidated their observation that the nominally steady-state accommodation response exhibits temporal variations which can be characterised by two dominant regions of activity: a low frequency component (LFC<0.6Hz) and a high frequency component (HFC between 1.0-2.3Hz). These microfluctuations occur with a root-mean-square (r.m.s.) value of between approximately 0.02D and 0.2D at temporal frequencies of up to 5Hz and are positively correlated with increases in stimulus vergence (Denieul, 1982; Kotulak & Schor, 1986a).

A functional role has been attributed to these microfluctuations as they offer a means by which a directional cue could be derived from an even-error stimulus (Alpern, 1958; Crane, 1966; Kotulak & Schor, 1986b; Charman & Heron, 1988) However, there is no consensus regarding the respective contribution made by each of the dominant components in the accommodation control process. In an attempt to clarify the role of the accommodation microfluctuations we have conducted a series of experiments to investigate their nature and aetiology.

Measurement of accommodation

For the studies described below the following instrumentation and methods were employed. Accommodation microfluctuations were measured with a modified Canon Autoref R1 (IR) objective optometer (Pugh & Winn, 1988). Pupil sizes of greater than 3.8mm were used to ensure that the optometer output was independent of pupil size (Winn, Pugh, Gilmartin & Owens, 1989).

Eye movements were monitored with a Hamamastu C3160 Percept Scope. Eye movement artefacts could be eliminated from the accommodation records if fixation was maintained within 1 degree.

Frequency characteristics of central and peripheral zones of the lens

The diminished fluctuations in HFCs evident with small pupils has led to the proposal that they may not contribute directly to accommodative control but may be a form of "plant noise" derived from the mechanical and elastic properties of exposed parts of the peripheral lens and its support structure (Charman, 1983). The proposal was based on observations of Foucalt knife-edge shadow patterns (Berny, 1969; Berny & Slansky, 1970) from which it was deduced that greater wavefront aberration occurs where the zonular fibres attach to the peripheral lens.

To test this proposal we have measured the magnitude and frequency of the microfluctuations in central and peripheral regions of the lens on 5 young emmetropic subjects while they viewed a near target (Winn, Pugh, Gilmartin & Owens, 1990a). The pupil of each subject was dilated with the mydriatic 10% phenylepherine hydrochloride to allow recording in each lens zone. This procedure causes a slight reduction in accommodative amplitude (0.8±0.3D) but allows sufficient reserve for young subjects to comfortably fixate a target placed at a vergence of -4D. The mean pupillary dilation of 8.5mm was insufficient to allow measurement through entirely discrete lens zones as the optometer utilises a circular region of approximately 3.5mm diameter in continuously recording mode. An overlap of up to 10% in zone area was tolerated throughout the study.

Comparison of the frequency response from each lens zone was achieved using power spectrum analysis. The form of the power spectra was found to be similar for central and peripheral zones although an overall reduction in magnitude was observed in the periphery (figure 1).

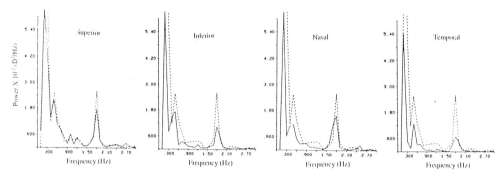

Figure 1. Power spectra of the accommodation microfluctuations are presented for central and peripheral zones of the lens (subject - NF). The dashed line is the power spectrum for the central region and the solid line for the peripheral lens zone indicated. Note the characteristic LFC (0.5Hz) and the HFC (1.8Hz) of the steady-state response. (from Winn, Pugh, Gilmartin & Owens, 1990a).

The HFCs are thus a consistent feature of accommodative microfluctuations in the central area of the lens and not simply a spurious feature of the periphery. Although the magnitude of response varies across the lens the frequency characteristics remain consistent. This is not surprising as it is difficult to conceive of a mechanism that would produce regions within the lens which have different frequency characteristics. The results add support to recent mathematical models which suggest that the lens capsule acts as a distributor of force across the lens surface.

Arterial pulse

The source of the HFCs is of interest as they are not derived from differences across the lens and do not appear to be related to changes in stimulus parameters (Denieul & Corno, 1986). A significant between-subject variation in the location of the HFC was found in the above

133

experiment which led us to consider the relationship between HFCs and other physiological systems which provide intraocular rhythmic variation (Winn, Pugh Gilmartin & Owens, 1990b). Simultaneous measurements of ocular accommodation and systemic arterial pulse made on 20 visually normal subjects demonstrated that the location of the HFC is significantly correlated ($r=0.99$, $P<0.001$) with arterial pulse frequency (figure 2a). Significant inter-subject variation in the location of the HFC has been reported by previous studies but has not been related to arterial pulse. This may be due to the large variation in arterial pulse frequency and the high resting levels of some subjects in the sample which may have caused previous workers to overlook this relationship.

Exercise-induced changes in pulse rate were investigated on 3 young subjects and demonstrated that the correlation between the HFC and arterial pulse frequency is retained (figure 2b).

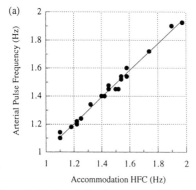
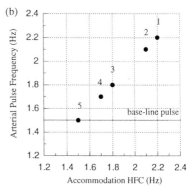

Figure 2. (a) Correlation between arterial pulse frequency and the HFC of accommodation for group data (n=20; y=0.06+0.95x). (b) Correlation between arterial pulse frequency and the HFC of accommodation for the recovery phase (sequence 1-5) of exercise-induced changes in pulse. The subject was a 21 year old female whose base-line pulse was 1.5Hz under the experimental conditions and was typical of the sample. (from Winn, Pugh, Gilmartin & Owen, 1990b).

Two unilateral aphakics aged 21 years and 40 years were recruited to conduct a control experiment in an attempt to identify any potential artefact. These subjects provide the ideal control as the crystalline lens has been removed ensuring any change in response is not the result of ocular accommodation. The relationship between arterial pulse and the HFC was demonstrable on the sound eye but absent from the aphakic eye (figure 3).

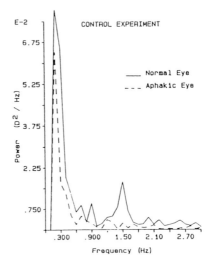

Figure 3. Power spectra from each eye of a 40 year old unilateral aphakic. The absence of the HFC in the aphakic eye suggests the activity is principally derived from the accommodative plant. (from Winn, Pugh, Gilmartin & Owen, 1990b).

A recent study (Collins, Davis & Wood, 1995) has confirmed the presence of this relationship and suggested that the LFC is attributable to respiration. The frequency resolution for reliable identification of low frequency signals requires extended recording periods (frequency resolution = 1/time) which could introduce additional artefacts using the Canon R1. However, the notion that LFCs may be linked to respiratory mechanisms is of interest and is worthy of further investigation using different instrumentation.

The intraocular mechanisms by which arterial pulse modulates the HFC are likely to involve rhythmic changes in both choroidal blood flow and IOP. In addition to reducing IOP the magnitude of IOP pulse is also reduced significantly by beta-receptor antagonists possibly due to their vasoconstrictive action or the choroidal vasculature. Modulation of intraocular pressure (IOP) pulse using topical beta-blockers indicate that factors relating to IOP and ocular vasculature affect the magnitude and frequency composition of the accommodative fluctuations (Owens, Winn, Gilmartin & Pugh, 1991). Following a double-blind protocol against saline, timolol was shown to reduce the r.m.s. value of the fluctuations (figure 4). A reduction in the peak location of the HFC was observed in all subjects following treatment with timolol and was correlated with the 4-6 beats min⁻¹ reduction in arterial pulse known to result from systemic absorption of timolol via ocular routes.

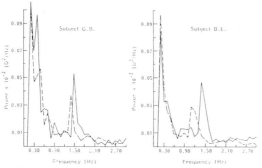

Figure 4. Power spectra of the accommodation microfluctuations for 2 subjects whose eyes have been treated with timolol maleate 0.5%. Timolol induces an overall reduction in power and a frequency shift in the HFC which is consistent with a reduction in pulse rate of 6 beats min⁻¹. The solid line is the pre-treatment response and the dashed line the post-treatment response. (from Owens, Winn, Gilmartin & Pugh, 1991).

Stimulus conditions

The above studies suggest that the complex waveform of the accommodative microfluctuations is the result of both neurological control and physiological rhythmic variation. The LFC seems to offer a mechanism for neurological control although the HFC may be utilised by the contrast detection mechanism as a component of the aggregate accommodation response. The microfluctuations will modulate retinal image contrast which induces a change in the steady-state response to maintain a constant level of retinal image contrast and ensures that the object of regard is clearly focused. Degradation of the retinal image (Heath, 1956) or imposition of open-loop conditions (Westheimer, 1957) will require an increase in the magnitude of the microfluctuations if the same information is to be provided to the contrast detection mechanism. If the LFC alone offers control of steady-state level then a change in this component of the response would be expected when appropriate changes in stimulus conditions are introduced. This hypothesis has been tested by modulating depth-of-focus and luminance.

Depth-of-focus

The accommodation response was recorded on 3 visually normal subjects while they viewed a stationary high contrast (90%) photopic (100 cdm⁻²) target through a series of artificial pupils (0.5, 1, 2, 3, 4 and 5 mm diameter) (Gray, Winn & Gilmartin, 1993a). The stimulus was placed at a vergence equal to the individual subjects's open-loop level of accommodation to ensure that no change in mean accommodation level occurred when pupil size was changed.

Variation in the response profile was observed for the 3 smallest pupils (0.5, 1, and 2mm) used in the study and can be characterised by an increase in low frequency drift (figure 5).

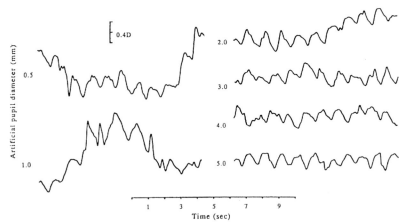

Figure 5. Accommodation responses for each pupil condition over a 10 second recording period for a typical subject. (from Gray, Winn & Gilmartin, 1993a).

Analysis of the data in terms of r.m.s. values and the power in each of the principal components (HFC and LFC) demonstrates an increase in the magnitude of response for pupils smaller than 3mm which is attributable to changes in the LFC (figure 6).

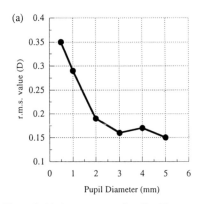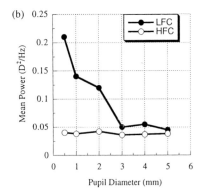

Figure 6. (a) Average r.m.s. for all subjects as a function of pupil size. One-way ANOVA for the factor of artificial pupil diameter revealed a significant variation in r.m.s. values with changing pupil diameter (P=0.0001). (b) Mean power in the LFC and HFC for the microfluctuations as a function pupil diameter for all subjects. (from Gray, Winn & Gilmartin, 1993a).

Luminance

Low frequency drifts in the accommodation response have been reported when the system is placed under open-loop conditions (Westheimer, 1957). It has also been reported that the slope of the stimulus/response curve decreases when the luminance of the target is reduced (Johnson, 1976). To investigate this issue further accommodation microfluctuations were recorded on 3 visually normal subjects as the luminance of the stimulus was varied from 0.002 to 11.63 cd m^{-2} in nine equal logarithmic steps (Gray, Winn & Gilmartin, 1993b). The target was again placed at a vergence equal to the individual subjects's open-loop level of accommodation to ensure the static level did not change during the experimental trails.

Accommodation responses were constant for target luminances greater than 0.010 cd m^{-2} but increased and became more variable for luminances equal to or below this value (0.002, 0.004 and 0.010 cd m^{-2}). Analysis of the data in terms of r.m.s. values and the power in each of the principal components demonstrates an increase in the magnitude of response for luminance levels < 0.010 cd m^{-2} which is attributable to changes in the LFC (figure 7).

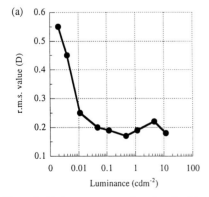 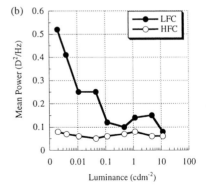

Figure 7. (a) Average r.m.s. values for all subjects as a function of target luminance. ANOVA revealed the variation in r.m.s. value with changing luminance level to be significant (P=0.0001). (b) Mean power in the LFC and HFC as a function of target luminance for all subjects. (from Gray, Winn & Gilmartin, 1993b).

The results from the above studies are consistent with the working hypothesis that an increase in the accommodative microfluctuations will occur when the stimulus is degraded or the accommodation loop is opened which allows the model of steady-state control to be developed further

Control of steady-state accommodation

The complex waveform of the accommodative microfluctuations is a consequence of the combination of neurological control and physiological noise: the former is attributable to the LFCs; the latter to the HFCs. The HFC does not vary with changes in stimulus conditions and supports the view that this component is modulated by arterial pulse. The increase in the magnitude of the accommodation microfluctuations when stimulus conditions are degraded or when the system is placed under open-loop conditions is due to changes in the LFC. A functional role for the microfluctuations as an error-detector in the accommodation system is therefore probably related to the maintenance of focus on a stationary target.

An elegant model of accommodation control has been developed (Kotulak & Schor, 1986b) whereby the accommodative error detector can extract relevant information to convert the even-error blur signal into an odd-error signal to ensure maintenance of the steady-state response. The model is primarily based on the fast "2Hz" component of the response but could equally apply to the LFC. The results of the above studies indicate that the LFC is the more likely component of control which can easily be incorporated into the model. It is proposed that magnitude and directional information can be extracted from the visual signal using the first derivatives of the two time functions; temporal changes in retinal image contrast and temporal changes in crystalline lens power. Comparing the signs of the first derivatives of these two time function provides information relating to the direction of the accommodative error. If the eye is overaccommodated the lens power function will be out of phase with the retinal image contrast function. If the eye is underaccommodated the two functions will be in phase. Retinal image contrast is directly related to the focus error at any instant, thus magnitude information can be extracted by a direct comparison of the two first derivatives.

In conclusion, it is likely that the accommodation microfluctuations have an important role in control of accommodation through modulation of retinal image contrast. Their presence enhances the detection of defocus by providing instantaneous feedback to the contrast detection mechanism which ensures an appropriate response to the prevailing stimulus conditions.

Acknowledgements
I would like to thank the following individuals for their contributions to the studies outlined in this chapter: John R. Pugh, Bernard Gilmartin, Helen Owens, Lyle S. Gray, Niall C. Strang and Helena Culhane. Their expertise made it possible to perform the studies and their enthusiasm made it a joy to conduct. The studies were supported in part by the British College of Optometrists and the Visual Research Trust which is gratefully acknowledged.

References

Alpern, M. (1958) Variability of accommodation during steady fixation at various levels of illuminance. *Journal of the Optical Society of America* 48: 193-197.

Arnulf, A., Dupuy, O. and Flamant, F. (1951a) Les microfluctuations d'accommodation de l'oeil et l'acuité visuelle pour les diamétres pupillaires naturelles. *Comptes Rendus de la Academie des Science, Paris*, 232: 339-341.

Arnulf, A., Dupuy, O. and Flamant, F. (1951b) Les microfluctuations de l'oeil et leur influence sur l'image rétiniene. *Comptes Rendus Hebdomadaire de l'Académie des Sciences, Paris*, 232: 438-450.

Berny, F. (1969) Etude de la formation des images rétiniennes et determination de l'aberration de sphericité de l'oeil human. *Vision Research*, 9: 977-990.

Berny, F. and Slansky, S. (1970) Wavefront determination resulting from Foucault test as applied to the human eye and visual instruments. *In* J. Home (ed.): *Optical instruments and techniques,* Oriel, Newcastle-upon-Tyne, pp. 375-386.

Campbell, F.W. and Robson, J.G. (1959) High-speed infra-red optometer. *Journal of the Optical Society of America*, 49: 268-272.

Campbell, F.W., Robson, J.G. and Westheimer, G. (1959) Fluctuations of accommodation under steady viewing conditions. *Journal of Physiology*, 145: 579-594.

Campbell, F.W., Westheimer, G. and Robson, J.G. (1959) Significance of fluctuations of accommodation *Journal of the Optical Society of America*, 48: 669.

Charman, W.N. (1983) The retinal image in the human eye. *In* N. Osborne and G. Chader (eds.): *Progress in Retinal Research*, Vol. 2, Pergamon, Oxford pp. 1-50.

Charman, W.N. and Heron, G. (1988) Fluctuations in accommodation: a review. *Ophthalmic and Physiological Optics*, 8: 153-164.

Collins, G. (1937) The electronic refractometer. *British Journal of Physiological Optics*, 1: 30-40.

Collins, M., Davis, B. and Wood, J. (1995) Microfluctuations of steady-state accommodation and the cardiopulmonary system. *Vision Research*, 35: 2491-2502.

Crane, H.D. (1966) A theoretical analysis of the visual accommodation system in humans. *Report NASA CR-606,* NASA, Washington.

Denieul, P. (1982) Effects of stimulus vergence on near accommodation response, microfluctuations and accommodation and optical quality of the human eye. *Vision Research*, 23: 561-569.

Denieul, P. and Corno, F. (1986) Accommodation et contraste. *L'optometrie*, 32: 4-8.

Gray, L.S., Winn, B. and Gilmartin, B. (1993a) Accommodative microfluctuations and pupil diameter. *Vision Research*, 33: 2083-2090.

Gray, L.S., Winn, B. and Gilmartin, B. (1993b) Effect of target luminance on microfluctuations of accommodation. *Ophthalmic and Physiologica Optics*, 13: 258-265.

Heath, G.G. (1956) The influence of visual acuity on accommodative responses of the eye. *American Journal of Optometry Archives of American Academy of Optometry*, 33: 513-524.

Johnson, C.A. (1976) Effects of luminance and stimulus distance on accommodation and visual resolution. *Journal of the Optical Society of America*, 66: 138-142.

Kotulak, J.C. and Schor, C.M. (1986a) Temporal variations in accommodation during steady-state conditions. *Journal of the Optical Society of America. A*, 3: 223-227.

Kotulak, J.C. and Schor, C.M. (1986b) A computational model of the error detector of human visual accommodation. *Biological Cybernetics*, 54: 189-194.

Owens, H., Winn, B., Gilmartin, B. and Pugh, J.R. (1991) Effect of a topical beta-adrenergic receptor antagonist on the dynamics of steady-state accommodation. *Ophthalmic and Physiological Optics*, 11: 99-104.

Pugh, J.R. and Winn, B. (1988) Modification of the Canon Autoref R1 for use as a continuously recording optometer. *Ophthalmic and Physiological Optics*, 9: 451-454.

Westheimer, G. (1957) Accommodation measurements in empty visual fields. *Journal of the Optical Society of America.*, 47: 714-718.

Winn, B., Pugh, J.R., Gilmartin, B. and Owens, H. (1989) The effect of pupil size on static and dynamic measurements of accommodation using an infra-red optometer. *Ophthalmic and Physiological Optics*, 9: 277-283.

Winn, B., Pugh, J.R., Gilmartin, B. and Owens, H. (1990a) The frequency characteristics of accommodative microfluctuations for central and peripheral zones of the human crystalline lens. *Vision Research*, 30: 1093-1099.

Winn, B., Pugh, J.R., Gilmartin, B. and Owens, H. (1990b) Arterial pulse modulates steady-state ocular accommodation. *Current Eye Research*, 9: 971-975.

Pharmacology of accommodative adaptation

Bernard Gilmartin

Aston University , Dept. of Vision Sciences, Aston Triangle, Birmingham B4 7ET, UK

Summary. Whilst it is incontrovertible that innervation of ciliary smooth muscle is mediated principally by the parasympathetic nervous system there is evidence for a supplementary inhibitory system which is mediated by the sympathetic nervous system. Sympathetic inhibition is relatively small (around 2D), slow (time courses are of the order of 20 to 40s) and augmented by substantial concurrent levels of background parasympathetic activity. Reference is made to *in vivo* studies in humans which have intervened with peripheral autonomic innervation at the neuroeffector junction by antagonist or agonist activity at autonomic receptors [notably muscarinic receptors and alpha-1 and beta (-1 and -2) adrenoceptors] following topical instillation of various ophthalmic drugs.
The inhibitory nature and magnitude of sympathetic input was evident from early studies on tonic accommodation but only when the method of measurement involved a level of cognitive demand sufficient to trigger significant background parasympathetic activity. The effect of non-selective (e.g. timolol) and selective (e.g. betaxolol) beta adrenoceptor antagonists on microfluctuations of accommodation reflect principally the systemic cardioselective properties of these agents in that a reduction in heart rate and hence arterial pulse (via systemic absorption) reduces correspondingly the higher frequency components of the microfluctuations.
The effects of timolol and betaxolol on within-task measures of accommodative response have shown effects on AC/A ratios and dynamic responses to sinusoidal target movement which are consistent with the inhibitory and temporal response properties of sympathetic innervation mentioned earlier. The results on dynamic responses have, however, been equivocal, and serve to demonstrate that significant inter-subject variation in access to a sympathetic facility can occur.
Recent work has considered whether inter-subject variation is a function of refractive error. Of particular interest was the notion that a deficit in sympathetic inhibition may be a precursor to the onset and development of late-onset myopia. Post-task accommodative hysteresis, a possible consequence of such a deficit, was assessed by measuring the time-course of regression of accommodation when open-loop (darkness) conditions were immediately imposed following far and near tasks. For the proposal to be feasible only late-onset myopes should exhibit post-task responses which fail to differentiate between selective and non-selective beta - antagonist agents. As the overall profile of responses to both classes of beta - adrenoceptor antagonism was equivalent for each of three different refractive groups used (emmetropes N=6; early-onset myopes N=6; late-onset myopes; N=6) it was concluded that a propensity to late-onset myopia is not associated with a specific deficit in sympathetic inhibition.
An outline is given of putative mechanisms for the generalised myopic shifts that have been shown to occur across refractive groups following substantial nearwork. Special reference is made to the myopic shifts that are evident in a proportion of individuals with incipient presbyopia as they highlight the potential role of the posterior ciliary anchoring ligaments in modulating the biomechanics of linkage between anterior and posterior segments of the eye.

Introduction

The aim of this chapter is to outline how, in pharmacological terms, the net effect of central control processes in accommodation and vergence are translated into peripheral activity; in particular the accommodative response that follows contraction of ciliary smooth muscle. The content will be restricted to autonomic nervous system (ANS) control of accommodation which is, of course, the principal control system, and to *in vivo* work on accommodation responses on humans during sustained near vision.

Peripheral neuroeffector activity is mediated predominantly by the parasympathetic system via the action of the transmitter acetylcholine on muscarinic receptors (Kaufman, 1992). Most structures innervated by the ANS have, however, dual innervation and there is a range of *in vivo* and *in vitro* pharmacological and physiological evidence for a sympathetic presence in terms of the action of noradrenaline on two sub-classes of post-synaptic receptors: beta receptors and alpha adrenoceptors (Gilmartin, 1986). In recent years specific subtypes of autonomic receptors have beem identified: muscarinic receptor subtype M3 is concerned with contraction of ciliary smooth muscle (Pang, *et al.,* 1994) and two sub-divisions (-1 and -2) of alpha and beta adrenoceptors have been linked to sympathetic inhibition of accommodation (i.e. negative as opposed to positive accommodation). The cross-linkage that can occur between cholinergic and adrenergic transmitters and receptors at pre- and post-synaptic sites provides a complex mechanism for modulation of transmitter release which is, to date, not fully understood (Starke, Gothert and Kilbinger, 1989). Furthermore these synaptic interactions make the functional consequences of dual innervation difficult to unravel in terms of accommodative adaptation to sustained visual tasks.

The approach adopted by our group has been to intervene with peripheral autonomic innervation at the ciliary muscle neuroeffector junction by blocking or stimulating respective cholinoceptors and adrenoceptors using topical instillation of various autonomic drugs (Gilmartin *et al.,* 1992). Whereas it would be valuable to intervene with the system at both pre- and post-synaptic sites *in vivo* work on humans imposes ethical contraints on agents and procedures that can be used which means that normally only conventional topical agents may be employed. Unfortunately these agents are often less inherently specific in their receptor activity which diminishes their value in experimental terms. Receptor specificity is also related to dose level at the receptor site which is subject to the vagaries of pharmacokinetics of drug absorption across the cornea. In addition the ocular effects can be confounded with central nervous system effects as inevitably there will be some systemic absorption of topically instilled drugs.

The results of various drug studies will be reviewed with reference to the three phases of accommodative response that might be considered when investigating responses to sustained viewing of a target: the pre-task (or pre-adaptive) tonic level; within-task closed-loop responses; post-task regression of within-task responses towards the pre-task tonic level. The locus of these post-task regressions are stable indexes (Strang, Winn and Gilmartin, 1994) and can be used to deduce the nature of within-task accommodative adaptation. Type A, the most common, gives a very rapid regression to base-line levels, within 10s, and occasionally a counter adaptation, indicated by an initial negative component, may also be present. Type B gives a more monotonic regression and has a time-course of around 20s to 40s. Type C, only occurs occasionally and exhibits an retarded regression which can extend over several minutes.

Tonic accommodation

A distinguishing feature of tonic accommodation is its significant inter-subject variation and, given that parasympathetic innervation predominates in accommodation, it was clear from our initial studies that the variation is due principally to variations in parasympathetic innervation of the ciliary muscle (Gilmartin and Hogan, 1985). Using a laser optometer a variation of 2.5D is illustrated in Fig. 1A for a group of ten young subjects. Topical instillation of the non-selective muscarinic receptor antagonist tropicamide (0.5%) caused a hyperopic shift in tonic accommodation which was directly proportional to its pre-drug level so that the distribution of tonic accommodation collapsed completely after 24min with a mean hyperopic shift of 1.24D.

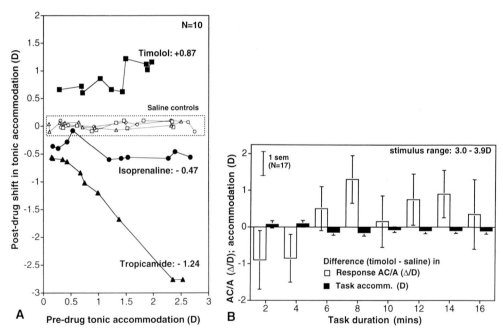

Figure 1. A: the effect of topical instillation of autonomic drugs on tonic accommodation (TA) measures with a He-Ne laser optometer in darkness. Data points represent the mean of five TA measurements (typical sd = 0.2) for each individual subject for each respective experimental condition (redrawn from Gilmartin and Hogan, 1985). B: the effect of timolol maleate (0.5%) on response AC/A (after Rosenfield and Gilmartin, 1987).

Fig. 1A also illustrates shifts in tonic levels with drugs that act on the sympathetic system. A significant mean hyperopic shift of 0.47D was found following stimulation with the non-selective beta adrenoceptor agonist isoprenaline sulphate (3%). The hyperopic shift varied little between subjects which is consistent with the agonist action of isoprenaline, that is it will simply add to the prevailing level of sympathetic activity and thus produce a constant shift across tonic levels. In contrast the non-selective beta adrenoceptor antagonist timolol maleate (0.5%) produced a significant mean myopic shift of 0.87D which showed a weak relationship with tonic accommodation. The shift increased from 0.6D to 1.2D as pre-drug tonic levels increased. Unlike isoprenaline the antagonist action of timolol is such that its effect will depend on the prevailing level of sympathetic activity and if this varies then timolol's antagonist effect will vary. Figure 1A gives some indication that this is indeed the case: beta antagonism and hence sympathetic input increases as tonic levels levels increase.

Autonomic components of tonic accommodation

As tonic levels are clearly determined by the prevailing level of parasympathetic activity another way of interpreting the progressive myopic shift with timolol is that background parasympathetic or accommodative activity will augment inhibitory sympathetic activity. This early data on tonic accommodation was in part consistent with some aspects of the functional role of sympathetic inhibition as indicated by the drug and nerve stimulation experiments on monkey by Törnqvist (1967) and Hurwitz *et al.* (1972). We can compile a picture of a predominant parasympathetic system interacting with a supplementary sympathetic system. The principal features of sympathetic input is that it is inhibitory, relatively small, probably no more than 2D, and relatively slow; time courses range between 20 and 40s compared to the 1 or 2s for the parasympathetic system. Importantly it is augmented by concurrent parasympathetic activity. Augmentation occurs on two counts: first, sympathetic inhibition will only become apparent when there is something to inhibit and hence there is a base-line requirement for parasympathetic activity. Second, parasympathetic activity above this level appears to augment sympathetic input directly but not to an extent greater than 2D even for very high parasympathetic levels (Gilmartin, 1986; Gilmartin and Bullimore, 1987).

Therefore the popular notion that tonic accommodation represents an equilibrium or fulcrum for parasympathetic and sympathetic innervation of the ciliary muscle is only likely to be valid when the level of background parasympathetic innervation is relatively high. Sympathetic input requires a threshold level of parasympathetic activity (derived from a variety endogenous and exogenous sources, an example of the latter being the mental effort induced by subjective measures of accommodation) which is triggered and modulated no doubt via the pre- and post-synaptic neuro-modulation systems referred to earlier.

Effects of topical autonomic drugs on intra-ocular vasculature

The likelihood of drugs with muscarinic and beta receptor activity having significant vascular effects is much less than that associated with agents acting on alpha adrenoceptors. These drugs can induce such effects owing to the presence of alpha receptors in the smooth muscle walls of intraocular blood vessels thus making it difficult to isolate genuine smooth muscle receptor effects. There is, however, some potential for specific ciliary smooth muscle effects as pharmacological work on excised human ciliary muscle has identified a small population of alpha-1 inhibitory receptors (Zetterström and Hahnenberger, 1988). We have been unable to demonstrate any effects on tonic accommodation with the alpha-1 agonist phenylephrine HCl (10%) or the alpha-1 antagonist thymoxamine HCl (0.5%). In contrast alpha-1 receptor activity has significant effects on amplitude of accommodation which is likely to be linked to changes in ciliary body volume. For example, a decrease in volume will result from the vasoconstriction effect of the agonist phenylephrine which will increase ciliary collar diameter and hence produce a decrease in amplitude of accommodation of around 2D in young subjects (Rosenfield *et al.* 1990).

Effects of topical autonomic drugs on within-task accommodative response

The accommodative response to a specific stimulus involves microfluctuations of accommodation and, of course, an associated synkinetic response of the vergence system. We have investigated the effect of topical instillation of timolol and betaxolol on microfluctuations while viewing a 4D near stimulus using the well-known Canon R1 open-view IR optometer in continuous recording mode (Owens *et al.*, 1990). Microfluctuations of accommodation will be addressed elsewhere in this publication but, in summary, the results reflect principally the systemic cardioselective characteristics of both betaxolol and timolol (related to their beta-1 adrenoceptor action) in that a reduction in heart rate and hence arterial pulse reduces correspondingly the peak frequency of the higher frequency components of ocular microfluctuations.

Work with Mark Rosenfield considered the effect of timolol on response AC/A (Rosenfield and Gilmartin, 1987). It was hypothesised that blocking the inhibitory input of the sympathetic system with timolol would result, for a given accommodative response, in less net drive from the parasympathetic system and hence a reduction in the response AC/A. Fig. 1B shows this to be the case: a significant drop in AC/A occurred which subsided after the first 4 minutes of the 16 minute near task and then tended to increase over the course of the task. Unfortunately we never pursued this route for investigating autonomic control with other agents nor were able to clarify the nature of these adaptation effects.

Of special interest was the need to differentiate the temporal characteristics of dual innervation and to gain better control over the potential for interaction effects between accommodation level and intra-ocular pressure (IOP) which can be reduced as much as 4 mm Hg in young normal subjects with topical beta adrenoceptor drugs. Because beta-2 sympathetic receptors predominate in excised human ciliary smooth muscle inhibition will occur with the predominantly non-selective beta receptor antagonist timolol but to a much lesser extent if at all with the predominantly beta-1 receptor antagonist, betaxolol. Both agents will, however, reduce IOP by virtue of beta activity elsewhere in the eye and therefore betaxolol serves as a useful control agent. There is no evidence that either agent will affect pupil size.

The effect of beta-adrenoceptor drugs on dynamic measures of accommodation

In recent work direct measures of dynamic accommodation have been made using the Canon R1 optometer (Strang *et al.* 1994). Accommodation responses were recorded continuously to sinusoidal Maltese Cross target movement within a Badal optical system over a 2D range from 2 to 4D. The target frequency was varied from 0.05 to 0.5Hz. It was hypothesised that if rate of change of background parasympathetic activity is a significant factor in dual innervation, then the lower frequencies would induce a combined parasympathetic/sympathetic response as there is sufficient time for the parasympathetic system to trigger the slower inhibitory sympathetic system and thus produce an optimum response. Conversely, for the higher frequencies the faster parasympathetic system would predominate in providing an optimum response owing to insufficient time to induce sympathetic activity. For the hypothesis to be valid only the lower frequency responses should be susceptible to the blocking effects of timolol as it blocks the inhibitory beta-2 sympathetic receptors. Betaxolol responses should match the saline control over the whole frequency range as its selective beta-1 receptor antagonism restricts its principal effects to the lowering of the IOP.

Preliminary data has provided some support for the hypothesis: reduced gain is found at lower frequencies when beta-2 adrenoceptors are blocked with timolol whereas the betaxolol gain matched that of saline. No difference in gain was evident between any of the treatments at the faster frequences as the conditions are not conducive to inhibitory sympathetic activity. Whereas these tracking experiments have provided some insight into the temporal properties of autonomic control of accommodation, the results on both gain and phase measurements (and related step responses) were inconclusive a finding which might reflect the possibility that only certain individuals have access to an inhibitory sympathetic facility.

Post-task measures of accommodative regression

The extent of inter-subject variation in sympathetic facility has become evident from a series of experiments on Phase 3, where reinstatement of open-loop conditions after a period of sustained viewing offers an alternative means of assessing within-task adaptation. The question posed was whether the role of within-task sympathetic inhibition could be to modify the magnitude and duration of post-task shifts in tonic accommodation. For example it may be that sympathetic inhibition could contribute to the rapid Type A response mentioned earlier or that a a deficit in inhibition could lead to the retarded Type C response.

To go some way to answering these questions our most recent published work in this area looked at the effect of topical instillation of timolol, betaxolol and saline on post-task regression patterns following a 3min fixation of a far task set at 0.2D and a near task which was set at 4D above the actual response level found for the far task (Gilmartin and Winfield, 1995). Accommodation responses were measured with the Canon R-1 optometer. The recording system was modified such that accommodation could be sampled in single-shot mode at 1.5s intervals over a 60s post-task period. The targets were single-line high-contrast 6/9 equivalent black-on-white letters on an internally illuminated Snellen Chart. The experiment was double-blind throughout and had a 2-day washout between drug instillations. Eighteen undergraduate subjects were used aged between 19 and 23. The subjects were drawn from three refractive groups: emmetropes, early onset myopes and late-onset myopes (that is myopia onset after 16 years of age; the mean age of onset for the group was just under 19 years).

Raw data for subject EW, a late-onset myope, is illustrated in Fig. 2A and compares post-task regressions for timolol and betaxolol trials. The betaxolol regression matched that of its respective saline control, that is a rapid Type A regression to the mean tonic level following both far and near tasks. Similar results were found for timolol following the far task although an overshoot was found in the later stages of recording which we have found on several occasions but are uncertain as to what it might represent. Of particular note for this subject is the retarded regression to the tonic level following the near task in the timolol trial. To condense the data the far and near regression plots were subtracted and the difference expressed as a percentage of the mean difference between far and near accommodation levels occurring during the final 15s of the task period. This was carried out for each of the 7 mean-sphere readings located symmetrically about each 10s data point.

The positive shift with timolol (relative to betaxolol) found for subject EW indicates a retardation or attenuation of regression from which it is deduced that sympathetic inhibition is present (Fig. 2B). EW was one of 6 subjects making up the late-onset myope group and whilst she was representative of those members of the group showing a timolol effect a range of reponse profile was found.

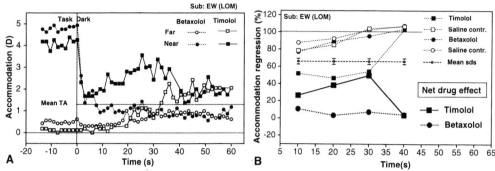

Figure 2. A:Post-task regression patterns for late-onset myope EW for timolol and betaxolol trials. B: Condensed data for subject EW. The top plots show the summarized data for each respective drug and saline trials. Error bars represent, for each time period, the mean standard deviation (derived from each set of 7 data points) for all conditions. In the bottom plots account has been taken of the respective saline controls and the net drug effect is shown as the relative differences in regression between timolol and betaxolol (after Gilmartin and Winfield, 1995).

The overall profile of response to the drug treatments was found not to differ significantly between each of the refractive groups used; examples of clear effects were counter balanced by subjects without effects.

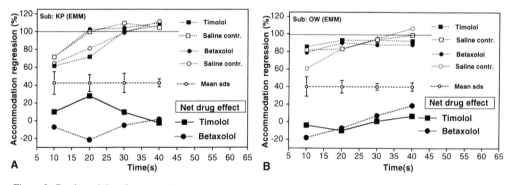

Figure 3: Condensed data for post-task regression patterns of two subjects drawn from the emmetropic group. A= data for subject KP; B= data for subject OW. See Fig 2B legend for details (after Gilmartin and Winfield, 1995).

For example a positive shift with timolol relative to betaxolol was found for emmetrope KP (Fig. 3A) but not for subject OW (Fig. 3B). Similar examples of the diversity of response could be drawn from the early-onset myopia group.

We concluded that only certain individuals have the facility for sympathetic inhibition and that this is independent of refractive status. Our thoughts have been that myopia, and in particular late-onset myopia (as it has a more well-defined link with sustained accommodation), may result from a progressive sequence: first, there is a specific deficit in sympathetic inhibition of ciliary smooth muscle; second, the deficit attenuates post-task regression to the tonic position which may result in a subsequent propensity to accommodative hysteresis following sustained near vision; finally, the hysteresis triggers micro-adaptational processes which accummulate to a critical level (perhaps via an iterative ratchet-type response with regard to accommodative gain) where structural recalibration takes place - for this to result in genuine myopia the structural correlate would have to be elongation of the posterior vitreous chamber. Our work suggests that if there is a link between a deficit in sympathetic inhibition and the onset and development of myopia induced by near-work in young adults then it should occur across refractive groups rather than being restricted to the late-onset myopia group. There is evidence that this is in fact the case (O'Neal & Connon, 1987).

Conclusions

There is good support, in the context of human accommodation studies, for the major characteristics of sympathetic inhibition outlined earlier. Additional factors are the inter-subject variation of sympathetic facility and the possibility of a link between sympathetic inhibition and the development of myopia. In this regard there is a clear need for a longitudinal study on the relationship between profile of autonomic response to accommodative demand and refractive error development.

Although the chapter has centred on events occurring at the neuro-effector junctions of ciliary smooth muscle there are other sites which have the potential for pharmacological intervention, notably where the ciliary muscle is attached anteriorly by means of tendons to the scleral spur and trabecular meshwork and posteriorly to the elastic network of Bruch's membrane of the choroid (Tamm *et al.* 1991). Much attention has been paid to the pharmacology of the anterior segment owing to its relevance to uveal-scleral outflow routes for aqueous humour and associated treatment of chronic open-angle glaucoma. It is likely that pharmacological intervention with the retraining force provided by the posterior anchoring ligaments with respect to the forward movement of the ciliary muscle on contraction will prove to be of special interest in the future. To date the pharmacology of this area is not well defined.

Our laboratory is currently considering aspects of retinal stretch in the posterior segment of the globe in anisomyopia and more accessible ways of determining posterior retinal contour than that obtained with special imaging techniques such as the ocular MRI (Logan, Gilmartin and Dunne, 1995). We suspect that these anchoring sites may be important in relation to the development of myopia which is a feasible proposition as it appears that these posterior attachments are in turn linked to the scleral canal such that the whole accommodative plant is linked to the posterior sections of the globe (Kaufman, 1992). Whether the inherent restraining elasticity of these ligaments contributes to presbyopia is also of interest. We have recently done some preliminary work on the epidemiology of myopia onset and development which occurs during the incipient phase of presbyopia. Indications are that around 15% of individuals show a mean myopic shift of around 0.75D during this time which, if shown to be axial, could expose some intriguing links between presbyopia, myopia and possibly uveal-scleral outflow mechanisms (Gilmartin, 1995).

Acknowlegements: I am especially indebted to the following individuals for their contributions to the studies described in this chapter: Robert Hogan, Mark Bullimore, Mark Rosenfield, Helen Owens, John Pugh, Niall Strang, Lyle Gray, Barry Winn, and Nicola Winfield.

References:

Gilmartin, B. (1986). A review of the role of sympathetic innervation of the ciliary muscle in ocular accommodation. *Ophthalmic and Physiological Optics, 6*, 23-37.
Gilmartin, B. (1995). The aetiology of presbyopia: a summary of the role of lenticular and extralenticular structures. *Ophthalmic and Physiological Optics, 15*, 431-437.
Gilmartin, B, and Bullimore, M.A. (1987). Sustained near-vision augments sympathetic innervation of the ciliary muscle. *Clinical Vision Science, 1*, 197-208.
Gilmartin, B., Bullimore, M. A., Rosenfield, M., Winn, B. and Owens, H. (1992). Pharmacological effects on accommodative adaptation. *Optometry and Visual Science, 69*, 276-282.
Gilmartin, B. and Hogan, R. E. (1985). The relationship between tonic accommodation and ciliary muscle innervation. *Investigative Ophthalmology and Visual Science, 26*, 1024-1029.
Gilmartin, B. and Winfield, N.R. (1995) The effect of topical ß-adrenoceptor antagonists on accommodation in emmetropia and myopia. *Vision Research, 35*, 1305-1312.
Hurwitz, B.S., Davidowitz, J., Chin, N.B. and Breinin, G.B. (1972) The effects of the sympathetic nervous system on accommodation: 1. Beta sympathetic nervous system. *Archives Ophthalmology , 87*, 668-674.
Kaufman, P.L. (1992). Accommodation and presbyopia: neuromuscular and biophysical aspects. In Hart Jr. M.H. (Ed), *Adler's Physiology of the eye* (pp. 406-407). 9th edition, St. Louis: Mosby-Year Book.
O'Neal, M. R. and Connon, T. R. (1987). Refractive error change at the United States Air Force Academy - class of 1985. *American Journal of Optometry and Physiological Optics, 62*, 344-354.
Logan, N.S. Gilmartin, B. and Dunne, M.C.M. (1995). Computation of retinal contour in anisomyopia. *Ophthalmic and Physiological Optics, 15*, 363-366.
Owens, H., Winn, B., Gilmartin, B. and Pugh, J. R. (1991). The effect of topical beta-adrenergic antagonists on the dynamics of steady-state accommodation. *Ophthalmic and Physiological Optics, 11*, 99-104.
Pang, I.H., Matsumoto, S., Tamm, E. and DeSantis, L. (1994). Characterization of muscarinic receptor involvement in human ciliary muscle cell function. *Journal of Ocular Pharmacology,10*, 125-136.
Rosenfield, M. and Gilmartin, B. (1987). Oculomotor consequences of beta-adrenoceptor antagonism during sustained near vision. *Ophthalmic and Physiological Optics, 7*, 127-130.
Rosenfield, M., Gilmartin B., Cunningham, E. and Dattani, N. (1990). The influence of alpha-adrenergic antagonists on tonic accommodation. *Current Eye Research 9*, 267-272.
Starke, K., Gothert, M. and Kilbinger, H. (1989). Modulation of neurotransmitter release by presynaptic autoreceptors. *Physiological Revue, 69*, 864-989.
Strang, N. C., Winn, B. and Gilmartin, B. (1993). Temporal closed-loop measures of accommodation show that sustained accommodation augments sympathetic innervation of ciliary muscle. *Investigative Ophthalmology and Visual Science (suppl.), 34*, 1310.
Strang, N. C., Winn, B. and Gilmartin, B. (1994). Repeatability of post-task regression of accommodation in emmetropia and late-onset myopia. *Ophthalmic and Physiological Optics, 14*, 88-91.
Tamm, E., Lütjen-Drecoll, E., Jungkuntz, W. and Rohen, J. W. (1991). Posterior attachment of the ciliary muscle in young, accommodating old, presbyopic rhesus monkeys. *Investigative Ophthalmology and Visual Science, 32*, 1678-1692.
Törnqvist, G. (1967) The relative importance of the parasympathetic and sympathetic nervous systems for accommodation in monkeys. *Investigative Ophthalmology and Visual Science, 6*, 612-617.
Zetterström, C. and Hahnenberger, R. (1988). Pharmacological characterization of human ciliary muscle adrenoceptors *in vitro*. *Experimental Eye Research, 46*, 421-430.

Behavioral links between the oculomotor and cardiovascular systems

R.A. Tyrrell, M.A. Pearson and J.F. Thayer[1]

Clemson University, Department of Psychology, Clemson, SC 29634-1511, USA

[1]*University of Missouri, Department of Psychology, Columbia, MO 65211, USA*

Summary. Among the more consistent and intriguing findings relating to tonic levels of accommodation and vergence is that individual differences are large. Previous research into the functional consequences of these individual differences has been impressive. Less is known, however, about the physiological source of inter-subject variations in dark focus and dark vergence. Given that the autonomic nervous system appears to be involved in determining both oculomotor and cardiovascular tone, it is possible that shared autonomic control could be reflected by behavioral similarities across these two systems. This chapter reviews evidence that addresses this possibility and highlights three recent studies that included psychophysiological measures of cardiovascular behavior.

In one such study, both dark focus and dark vergence tended to be nearer during sympathetic dominance of the heart than during parasympathetic dominance. In another study, subjects read a story at 15 cm for 20 minutes. While roughly half of the subjects experienced neither oculomotor nor cardiovascular consequences, those subjects who did experience adaptation of the oculomotor resting states (> 0.1 D or > 0.1 MA) also tended to experience an increase in heart rate that appeared to be due primarily to parasympathetic withdrawal. Thus cardiovascular parameters covaried with oculomotor adaptation, and individual differences in the tendency to react to near work are consistent across the cardiovascular and oculomotor systems. A third study tested whether oculomotor responses to near work can be influenced by the demands inherent in the near task. In this study, both emmetropes and late-onset myopes performed a "low stress" task and a "high stress" task for 20 minutes at 15 cm. Measurements of cardiovascular activity assessed before and during the task confirmed that the stress manipulation was effective. Oculomotor adaptation varied across the tasks: In both refractive groups, significant dark focus adaptation followed the "low stress" task but was inhibited in the "high stress" task. Dark vergence adaptation depended on both task and refractive group.

These studies provide support for a behavioral link between the oculomotor and cardiovascular systems and provide new incentive for the exploration of the relevance of autonomic mechanisms in the regulation of accommodation and vergence. It is emphasized that autonomic behavior is complex, and that behavioral manipulations of autonomic activity preserve the natural and complex interactions among autonomic mechanisms.

Introduction

Among the more consistent and intriguing findings relating to tonic levels of accommodation and vergence is that individual differences are large. Previous research into the functional consequences of these individual differences has been impressive (see, for example, Owens &

Andre, this volume). Less is known, however, about the physiological source of variations in dark focus and dark vergence. Given that the autonomic nervous system appears to be involved in determining both oculomotor and cardiovascular tone, it is possible that shared autonomic control could be reflected by behavioral similarities across these two systems. This chapter reviews evidence that addresses this possibility and highlights three recent studies that included psychophysiological measures of cardiovascular behavior.

Theoretical views of autonomic control of accommodation have evolved over the years. Conceptual models have ranged from those that included only a role for parasympathetic forces, to those that involved a simple push-pull relationship between parasympathetic and sympathetic forces, to the model proposed by Gilmartin wherein sympathetic influences are seen as relatively small, slow, and inhibitory (Gilmartin, 1986). While accumulating physiological evidence has begun to reveal the complexity of the relationships, much of the behavioral evidence has involved measuring behavioral responses to pharmacological manipulations of autonomic activity. Given that the function of the autonomic nervous system is to facilitate adaptation to changing environmental and behavioral demands, surprisingly little research has included behavioral manipulations of autonomic activity.

Investigations that have examined accommodative responses to behavioral manipulations of autonomic activity ("arousal") have yielded interesting, although seemingly discrepant findings. While some studies measured outward accommodative reactions to stressors such as mental activity, cold pressor, and viewing gruesome slides (Miller & Takahama, 1987; 1988), other studies reported that inward accommodative reactions were associated with anxiety, anger, hostility, and with recovery from physical exercise (Leibowitz, 1976; Miller, 1978; Miller & LeBeau, 1982; Ritter & Huhn-Beck, 1993). Importantly, however, these studies generally did not include measures of autonomic activity, making interpretation difficult.

Although the question of whether binocular vergence is influenced by autonomic forces is less certain and has received less attention, there is empirical support for such an influence. Both injections of epinephrine and electrical stimulation of the superior cervical ganglion give rise to dose- and voltage-dependent tonic contractions of extraocular muscles in intact cats (Eakins & Katz, 1967; Eakins & Katz, 1971). Also, the oculocardiac reflex (i.e., traction or pressure on the extraocular rectus muscles inducing a drop in heart rate) is mediated by the vagus (parasympathetic) nerve (Alexander, 1975; Blanc, Hardy, Milot, & Jacob, 1983). Further, Miller and Takahama (1988) found that dark vergence was significantly more distant during relaxation than in other conditions, and that subjective arousal ratings were positively correlated with dark vergence.

Assessment of the autonomic consequences of behavioral manipulations of arousal is necessary due to the complexity of the autonomic nervous system. For example, a manipulation that effectively increases subjective reports of arousal may or may not have induced actual autonomic changes. Further, although measurements of changes in heart rate can indicate that an autonomic reaction has taken place, such measurements are not sufficient to untangle the relative contributions of sympathetic and parasympathetic forces. An increase in heart rate, for example, might be due to sympathetic activation, parasympathetic withdrawal, or a combination of these effects. Recent advances in psychophysiology have provided a number of non-invasive techniques to separate sympathetic and parasympathetic influences on the cardiovascular system. One such technique involves the examination of heart rate variability (e.g., Akselrod, et al., 1981; Latson, 1994; Saul, 1990). The spectral decomposition of variations in heart rate, for example, typically yields two primary sources of heart rate variability — a relatively low frequency component (centered around 0.1 Hz) and a relatively high frequency component (centered around the respiratory frequency). The power in the high frequency component is a consequence of respiratory sinus arrhythmia and is a useful index of parasympathetic activity (e.g., Pagani, et al., 1986). The power in the low frequency component is primarily due to sympathetic influences, although sympathetic-parasympathetic interactions may also be relevant (e.g., Pagani, Rimoldi, & Milliani, 1992; Pagani, Lombardi, & Malliani, 1993). Thus, the development of this and other psychophysiological techniques (e.g., impedance cardiography; Sherwood, et al., 1990) allows researchers to assess the autonomic consequences of their behavioral manipulations of arousal. When oculomotor behavior is measured simultaneously, the possibility that shared autonomic control can yield behavioral links between the oculomotor and cardiovascular systems can be tested directly.

Effects of behavioral manipulations of ANS activity on dark focus and dark vergence

One recently reported study measured within-subjects variations in dark focus, dark vergence, and a variety of cardiovascular parameters during conditions that were intended to induce autonomic fluctuations (Tyrrell, Thayer, Friedman, Leibowitz, & Francis, 1995). In this study, three healthy young adults repeatedly experienced a variety of conditions, including two that were intended to induce specific autonomic responses. During the *cold face stress* condition, which had previously been shown to induce parasympathetic activation by mimicking the human dive reflex (Brick, 1966; Kawakami, Natelson, & DuBois, 1967; Thayer & Kohler, 1993), the subject sat still while an experimenter held a plastic bag of cold water (8–10 °C) between the subject's forehead and the autorefractor's forehead support. During the *grip strength / shock-avoidance* condition, the subject squeezed a hand dynamometer with his/her left hand at roughly 25% of their maximum grip strength. Periodically during this task, an experimenter would warn the subject that if he or she did not maintain adequate grip strength an electrical shock would be delivered to the subject's lower leg. Each subject received 3-5 actual shocks and roughly twice as many warnings during the task. (Each subject had earlier determined the intensity of the shock.) During each four-minute condition two measures of dark vergence (using a Vergamatic) and 20-40 measures of dark focus (using a Canon R-1 autorefractor) were made, EKG signals were sampled at 1 kHz for later off-line analysis, and an automated blood pressure monitor assessed systolic and diastolic blood pressure every minute. To help distinguish sympathetic and parasympathetic influences on the heart, subjects were asked to maintain a paced respiration at 15 breaths per minute (0.25 Hz), aided by the sound of a metronome.

Analysis of cardiovascular parameters revealed that, compared to the cold face task, the grip strength / shock avoidance task resulted in a significant decrease in heart period (i.e., increased heart rate) and a cardiovascular pattern suggestive of a shift away from parasympathetic domination of the heart and towards sympathetic domination. Given this, it was interesting to find that both dark focus and dark vergence were nearer during the grip strength / shock avoidance task than during the cold face task. These shifts, of 0.23 D and 0.15 MA, respectively, are depicted in Figure 1. The direction of the task-induced shift in dark focus is not consistent with previous reports that dark focus is nearer during relaxation than during tasks such as mental activity or viewing aversive slides (e.g., Miller & Takahama, 1987; 1988), but is consistent with reports of inward accommodative shifts in anxious, angry, or physically active subjects (Leibowitz, 1976; Miller, 1978; Miller & LeBeau, 1982; Ritter & Huhn-Beck, 1993). The task-induced shift in dark vergence is similar in magnitude and direction to the effect reported by Miller and Takahama (1988) that dark vergence was approximately 0.15 MA farther during relaxation than during their other conditions.

Thus this study documented that behavioral manipulations which result in changes in cardiovascular parameters also result in changes in the oculomotor resting states: Both dark focus and dark vergence tended to be nearer during sympathetic dominance of the heart than during than during parasympathetic dominance.

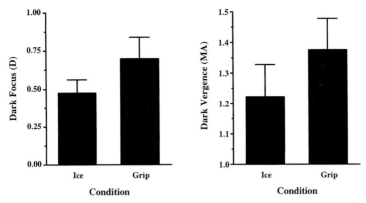

Figure 1. Mean dark focus and mean dark vergence (+1 standard error of the mean) during the cold face task ("Ice") and during the grip strength / shock avoidance task ("Grip"). Both dark focus and dark vergence were nearer during the cold face task, when cardiovascular parameters suggested a relative sympathetic dominance. Data from Tyrrell, et al. (1995).

Oculomotor-cardiovascular covariation during near visual work

While the above data stand in agreement with the notion that behavioral manipulations that are traditionally used to influence autonomic control of the cardiovascular system also influence the oculomotor resting states, the data do not address the possibility that the converse is also true. Another recent study tested whether a task that is typically used to influence oculomotor functioning — near visual work — can also influence cardiovascular functioning (Tyrrell, Thayer, & Leibowitz, 1994).

In that study, 17 young adults each spent 20 minutes reading a Sherlock Holmes story from a VDT positioned at a distance of 15 cm (6.67 D). During the reading period, the subject's electrocardiogram was sampled at 1.0 kHz during five adjoining four-minute intervals, and an automated blood pressure monitor assessed systolic and diastolic blood pressure at one minute increments. Immediately before and after the reading period, dark focus and dark vergence were measured using a Canon R-1 Autorefractor and a Vergamatic, respectively. The subject was again instructed to maintain paced breathing at 15 breaths per minute during both the oculomotor measurements and the reading periods.

As expected, comparisons between pre- and post-task measures revealed that, at a group level, dark focus and dark vergence tended to adapt to nearer post-task positions. However, examination of the individual differences in the magnitude of the oculomotor adaptation revealed that for both resting states, only roughly half of the participants exhibited the expected adaptation. To determine whether cardiovascular changes covaried with oculomotor adaptation, the cardiovascular data were examined separately for those who did experience greater than 0.10 diopter (for accommodation) or 0.10 meter-angle (for vergence) of oculomotor adaptation (*"Adaptors"*) and those who not (*"Non-adaptors"*). Six subjects exhibited adaptation in both resting states, three exhibited only dark vergence adaptation, four exhibited only dark focus adaptation, and four exhibited no adaptation.

To determine if the time spent on the reading task resulted in systematic cardiovascular changes, the significance of linear contrasts was tested within a number of cardiovascular parameters. In both Dark Vergence Adaptors and Dark Focus Adaptors, heart period significantly decreased during the reading period. This is indicative of an acceleration of heart rate during the 20 minute reading period.

To explore the autonomic basis for this cardiac acceleration, an autoregressive spectral decomposition of the variations in heart period was conducted (Colombo, et al., 1990). Inspection of the linear trends evident during the near reading task revealed the autonomic basis for the reduction in heart period: In both groups of Adaptors, high frequency power significantly decreased during the reading period. Together, these results indicate that those who experienced oculomotor adaptation also experienced a parasympathetically mediated acceleration of heart rate. The decrease in heart period and in high frequency power was particularly strong between the first two 4-minute periods, suggesting that the cardiovascular effects were strongest in the early stages of the near reading task.

By comparison, the two groups of Non-adaptors exhibited fewer and smaller cardiovascular changes. With the exception of a trend for the heart period of the Dark Vergence Non-Adaptors to decrease, no significant linear trends were exhibited by either group of Non-adaptors. The fact that both groups of Adaptors experienced cardiovascular effects that the Non-adaptors largely did not suggests that these cardiovascular changes were not simply due to the near work itself. Rather, the cardiovascular parameters covaried with the oculomotor adaptation.

At this point, it is not known why some subjects experienced both oculomotor and cardiovascular effects of near reading while other subjects experienced neither effect. However, it is interesting that the mean pre-task systolic and diastolic blood pressure of the Dark Vergence Adaptors was significantly higher than that of the Dark Vergence Non-Adaptors. That is, prior to the near reading task, those subjects who would later experience dark vergence adaptation had significantly higher blood pressure than those who would later experience no such adaptation. Thus it appears possible either that blood pressure mediates oculomotor adaptation or that both blood pressure and the predisposition to adapt are mediated by a third, as yet unspecified, variable. No other significant differences existed between Adaptors and Non-Adaptors in terms of their pre-task cardiovascular parameters.

Effects of task demands on adaptation of dark focus and dark vergence

Given that behavioral variations of autonomic activity can modify the static oculomotor resting states, and that near work can result in cardiovascular consequences, it seemed possible that the effects of near work on oculomotor adaptation might depend on the demands inherent in the near task. A recent study addressed this possibility by asking subjects to engage in two different near tasks (Pearson, Tyrrell, Thayer, & Jiang, 1996).

In this study, ten college-aged emmetropes and ten college-aged late-onset myopes (mean reported age of myopia onset = 14.8 years) each performed a "low stress" task and a "high stress" task. During the "low stress" task, subjects read a Sherlock Holmes story at their own pace. During the "high stress" task, subjects competed for financial rewards on a demanding visual search / reaction time task. The task required subjects to search a 5 x 3 matrix of two-digit numbers for two target numbers and to respond by indicating whether none, one, or both target numbers were present. The time during which the matrix was presented decreased by 0.12 s with each correct response and increased by 0.36 s with each incorrect response. Both tasks were completed binocularly at 15 cm (6.67 D), both lasted 20 minutes, and all subjects performed the two tasks on different days. Dark focus (Canon R-1) and dark vergence (Vergamatic) were assessed immediately before and after each task.

Cardiovascular activity was assessed before and during the tasks ($n = 12$), and confirmed that the task manipulation was effective: Relative to the "low stress" task, the "high stress" task resulted in a parasympathetically mediated acceleration of heart rate (and higher subjective arousal scores) in both refractive groups. What is more interesting, however, is the oculomotor reactions.

In both refractive groups, significant dark focus adaptation (post-task minus pre-task) followed the "low stress" task. In contrast, neither refractive group experienced significant dark focus adaptation in the "high stress" task. When viewed together with the pattern of cardiovascular changes, these results indicate that in both groups accommodative adaptation was inhibited during parasympathetic withdrawal. Regarding dark vergence adaptation, a significant interaction indicated that dark vergence adaptation varied with both task *and* group: While the emmetropes experienced greater dark vergence adaptation in the "high stress" task than in the "low stress" task, the myopes reacted oppositely. Thus while late-onset myopes' dark vergence adaptation was inhibited during parasympathetic withdrawal, the emmetropes' adaptation was actually increased.

In interpreting these oculomotor effects, it is important to remember that the cardiovascular arousal that was produced by the "high stress" task was due to parasympathetic withdrawal, and that the oculomotor consequences of a near task that behaviorally induces sympathetic activation remain to be tested. In any case, the results suggest that parasympathetic activity is relevant for oculomotor adaptation to near work. When behavioral demands result in a parasympathetically mediated acceleration of heart rate, dark focus adaptation is inhibited in both myopes and in late-onset myopes, and dark vergence is inhibited in late-onset myopes.

Conclusions

Taken together, the studies reviewed above provide support for a behavioral link between the oculomotor and cardiovascular systems. These studies also provide new incentive for the exploration of the relevance of autonomic mechanisms in the control and regulation of accommodation and vergence. Additional investigations into these topics are likely to reveal findings that have important theoretical and practical consequences.

At a theoretical level, further research into autonomic mechanisms may reveal the source of inter-individual and intra-individual differences in the oculomotor resting states. Such findings may in turn prove to be relevant for the understanding of oculomotor contributions to the development and progression of late-onset myopia, particularly if autonomic parameters (tonic or phasic) can be shown to distinguish between late-onset myopes and emmetropes.

On a practical level, given that behavioral and autonomic demands can influence dark focus and dark vergence, it is possible that these demands can influence visual performance in stressful conditions. For example, if autonomic arousal can induce inward shifts in dark focus and dark vergence, this may in turn reduce some individuals' ability to focus and converge on distant stimuli. In addition, further investigations into autonomic behavior as it relates to near work may complement existing research that has linked oculomotor parameters to reports of visual fatigue (e.g., Owens & Wolf-Kelly, 1987; Lie & Watten, 1994; Tyrrell & Leibowitz, 1990), and result in a better understanding of the mechanisms underlying the subjective experiences that follow near work. Regardless, the finding summarized earlier that near visual work can induce cardiovascular changes is, to our knowledge, the first empirical evidence that a visually stressful activity can also be cardiovascularly stressful (Tyrrell, Thayer, & Leibowitz, 1994).

It seems obvious that much work in this domain remains to be done, and it is believed that the research that has been summarized here has methodological implications for future research. Given that behavioral demands can influence autonomic activity which, in turn, can influence oculomotor behavior, researchers should be aware of the possibility that the oculomotor behaviors they are measuring and inducing may have limited generalizability to situations with differing behavioral demands. In addition, care should be taken when designing experiments to minimize the possibility of autonomic arousal being confounded with other variables. Another methodological implication concerns the utility of behavioral manipulations of autonomic activity. While much has been learned about the autonomic control of accommodation from pharmacological studies, it should be remembered that for both accommodative and cardiovascular behavior, sympathetic and parasympathetic forces are known to interact in important and complex ways (e.g., Gilmartin, 1986; Uijtdehaage & Thayer, 1989). Thus, when neural activity in one of the two autonomic branches is pharmacologically blocked, the resulting behavior may not reflect the natural consequences of the remaining branch. Behavioral manipulations of autonomic activity have the advantage of preserving the natural interactions among autonomic mechanisms. This advantage is nicely complemented by the availability of a variety of non-invasive techniques developed by psychophysiologists to assess autonomic activity in the cardiovascular system.

Acknowledgments
The authors thank Bai-Chuan Jiang, Herschel W. Leibowitz, and D. Alfred Owens for their insightful discussions.

References

Akselrod, S., Gordon, D., Ubel, F.A., Shannon, D.C., Barger, A.C., & Cohen, R.J. (1981). Power spectrum analysis of heart rate fluctuation: A quantitative probe of beat-to-beat cardiovascular control. *Science, 213,* 220-222.

Alexander, J.P. (1975). Reflex disturbances of cardiac rhythm during ophthalmic surgery. *British Journal of Ophthalmology, 59,* 518-524.

Blanc, V.F., Hardy, J., Milot, J., & Jacob, J. (1983). The oculocardiac reflex: A graphical and statistical analysis in infants and children. *Canadian Anaesthetists' Society Journal, 30,* 360-369.

Brick, I. (1966). Circulatory responses to immersing the face in water. *Journal of Applied Physiology, 21,* 33-36.

Colombo, R., Mazzuero, G., Soffiatino, F., Ardizzoia, M., & Minuco, G. (1990). A comprehensive PC solution to heart rate variability analysis in mental stress. *1989 IEEE Computers in Cardiology,* 475-478.

Eakins, K.E., & Katz, R. (1971). The pharmacology of extraocular muscle. In P. Bach-y-Rita, C. C. Collins, & J. E. Hyde (Eds.), *The Control of Eye Movements* (pp. 237-258). New York: Academic Press.

Eakins, K.E., & Katz, R.L. (1967). The effects of sympathetic stimulation and epinephrine on the superior rectus muscle of the cat. *Journal of Pharmacology and Experimental Therapeutics, 157*, 524-531.

Gilmartin, B. (1986). A review of the role of sympathetic innervation of the ciliary muscle in ocular accommodation. *Ophthalmic & Physiological Optics, 6*, 23-37.

Kawakami, Y., Natelson, B.H., & DuBois, A.B. (1967). Cardiovascular effects of face immersion and factors affecting diving reflex in man. *Journal of Applied Physiology, 23*, 964-970.

Latson, T.W. (1994). Principles and applications of heart rate variability analysis. In I. C. Lynch (Ed.), *Clinical Cardiac Electrophysiology: Perioperative Considerations* (pp. 307-349). Philadelphia, PA: J.B. Lippincott Co.

Leibowitz, H.W. (1976). Visual perception and stress. In G. Borg (Ed.), *Physical Work and Effort* (pp. 25-37). Oxford: Pergamon.

Lie, I., & Watten, R.G. (1994). VDT work, oculomotor strain, and subjective complaints: An experimental and clinical study. *Ergonomics, 37*, 1419-1433.

Miller, R.J. (1978). Mood changes and the dark focus of accommodation. *Perception & Psychophysics, 24*, 437-443.

Miller, R.J., & LeBeau, R.C. (1982). Induced stress, situationally-specific trait anxiety, and dark focus. *Psychophysiology, 19*, 260-265.

Miller, R.J., & Takahama, M. (1988). Arousal-related changes in dark focus accommodation and dark vergence. *Investigative Ophthalmology & Visual Science, 29*, 1168-1178.

Miller, R.M., & Takahama, M. (1987). Effects of relaxation and aversive visual stimulation on dark focus accommodation. *Ophthalmic & Physiological Optics, 7*, 219-223.

Owens, D.A., & Andre, J.T. (this volume). *Predicting performance in difficult conditions: A behavioral analysis of normal variations of accommodation.*

Owens, D.A., & Wolf-Kelly, K. (1987). Near work, visual fatigue, and variations of oculomotor tonus. *Investigative Ophthalmology & Visual Science, 28*, 743-749.

Pagani, M., et al. (1986). Power spectrum analysis of heart rate and arterial pressure variabilities as a marker of sympatho-vagal interaction in man and conscious dog. *Circulation Research, 59*, 178-193.

Pagani, M., Lombardi, F., & Malliani, A. (1993). Heart rate variability: Disagreement on the markers of sympathetic and parasympathetic activities. *Journal of the American College of Cardiology, 22*, 951.

Pagani, M., Rimoldi, O., & Malliani, A. (1992). Low-frequency components of cardiovascular variabilities as markers of sympathetic modulation. *Trends in Pharmacological Science, 13*, 50-54.

Pearson, M.A., Tyrrell, R.A., Thayer, J.F., & Jiang, B.C. (1996). Adaptation of oculomotor resting states depends on both task and refractive status. *Investigative Ophthalmology & Visual Science, 37*, S166.

Ritter, A.D., & Huhn-Beck, H. (1993). Dark focus of accommodation and nervous system activity. *Optometry and Vision Science, 70*, 532-534.

Saul, J.P. (1990). Beat-to-beat variations of heart rate reflect modulation of cardiac autonomic outflow. *News in Physiological Science, 5*, 32-37.

Sherwood, A., Allen, M.T., Fahrenberg, J., Kelsey, R.M., Lovallo, W.R., & van Doornen, L.J.P. (1990). Methodological guidelines for Impedance Cardiography. *Psychophysiology, 27*, 1-23.

Thayer, J.F., & Kohler, S.S. (1993). Cardiovascular and metabolic adjustments to cold face stress. *Psychophysiology, 30*, S64.

Tyrrell, R.A., & Leibowitz, H.W. (1990). The relation of vergence effort to reports of visual fatigue following prolonged near work. *Human Factors, 32*, 341-357.

Tyrrell, R.A., Thayer, J.F., Friedman, B.H., Leibowitz, H.W., & Francis, E.L. (1995). A behavioral link between the oculomotor and cardiovascular systems. *Integrative Physiological and Behavioral Science, 30*, 46-67.

Tyrrell, R.A., Thayer, J. F., & Leibowitz, H.W. (1994). Cardiovascular changes covary with oculomotor adaptation induced by near work. *Investigative Ophthalmology & Visual Science, 35*(4) , 1282.

Uijtdehaage, S.H.J., & Thayer, J.F. (1989). Sympathetic-parasympathetic interactions in heart rate response to laboratory stressors. *Psychophysiology, 26*, S63.

Development of accommodation and vergence in infancy

Louise Hainline

Infant Study Center, Psychology Dept. Brooklyn College, Brooklyn NY 11210 USA

Summary: Proper visual development depends on the behaviors of accommodation and vergence to create high resolution, registered images for each eye. Accommodation has been reported to be absent or poor in infants under about 2 months. Our research suggests that average behavior at this age will depend on the particular subjects included in the sample, as young infants' performance is variable. Some one-month-old infants vary accommodation very little regardless of target distance, others fail to relax accommodation at far, and others seem to accommodate reasonably well. Vergence to static targets located at difference distances, on the other hand, appears quite good early in life, even for infants who show less than perfect accommodation. These results are attributed to developments in infant spatial vision, with the suggestion that improvements with age in the visibility of higher spatial frequencies play an important role in improving accommodation; vergence appears not to depend as heavily on good visual acuity, possibly being driven more by the visibility of middle and low spatial frequencies which change less with development in the first half year. The discrepancy between accommodation and vergence at the earliest ages could stem from a lack of linkage between the two systems or from a large "dead zone" in the linkages early in life; present data do not allow a choice between these possibilities.

Introduction

It is fairly well established that the primate visual system needs certain inputs to establish its neural connections properly. Images need to be in reasonably good focus -- raising animals with lenses that prevent clear vision results in apparently permanent losses of spatial vision, amblyopia, particularly if there is a difference in image quality between the two eyes. Images also need to be in reasonable registration. Raising animals with induced eye-turns causes apparently permanent ocular misalignments, strabismus. Both manipulations are often associated with a failure to develop normal binocular functions, particularly stereopsis. So, eventually, both accommodation and vergence must reach some minimal level of functionality before the brain is able to complete the task of establishing high quality, binocular vision. This chapter will discuss research on the status of accommodation and vergence in young infants, to evaluate the starting point for these functions, the foundation on which are built the mature abilities described else where in this volume.

Development of Accommodation

Space does not permit a full review of the human developmental work on accommodation and vergence. For up-to-date references, the interested reader is referred to sources such as Daw (1995) and individual chapters in Simons (1993) and Viral-Durand, Atkinson and Braddick (1996). Probably the first systematic study of infant accommodation was done using retinoscopy by Haynes, White and Held in 1965. This study, widely cited, reported that infants younger than 2 months did not vary accommodation at all, instead showing a fixed accommodative plane at about 19 cm (5.25D of demand) regardless of target distance. Target visibility was not kept constant as distance increased, however, leading Banks to repeat the study in 1980; he reported that one- to two-month-old infants did show monotonic changes in accommodation, but that their accommodative responses had too shallow a slope when fit with a linear function, with a slope of about 0.5, rather than 1, as expected from adults. This result was caused by the fact that infants failed to relax accommodation for more distant targets. Banks attributed the poor accommodation at distance to optical factors, particularly to a high depth of focus due to small pupils for the youngest infants, resulting in reduced feedback from changes in accommodation. This coupled with their poor spatial vision resulted in poor accommodative response at distance. The problem with the depth-of-focus explanation is that it fails to take note of the fact that as a result of a corresponding increase in the axial length of the eyeball as pupil size increases, that actually depth-of-focus is relatively constant from infancy to adulthood (Banks and Bennett, 1988; Hainline and Abramov, 1992). Nevertheless, much of the data suggests that infants under 2 months show less appropriate accommodation than older infants, particularly for more distant targets.

Our own work on infant accommodation and vergence (e.g., Hainline, Riddell, Grose-Fifer and Abramov, 1992) has been done with a photorefraction system that allows objective measurement of accommodative and convergence behavior simultaneously from both eyes (Abramov, Hainline and Duckman, 1990). The system is based on measurement of crescents of light in images of the pupils that represent retinal blur circles created as a reference light source passes twice through the eyes' optics; crescent size and position convey the plane of focus of each eye, which can be referred to the demand created by the location of the target (a small lighted doll in an otherwise darkened room) on each trial. Vergence is gauged from the measurement of the relative positions of the camera flash's Purkinje image in the two eyes, relative to pupil center. Our data on infant accommodation agree with the general finding of the earlier studies when one looks at average behavior for infants of different ages. After 2 months or so (see Fig. 1), there is little difference in average accommodation between infants and adults. On average, as Banks reported, younger infants tend to overaccommodate for all but the closest target, with error between target demand and refractive plane increasing as target distance increases.

When one looks at the behavior of individual infants, however, the picture gets more complicated. First of all, for most infants and even naive, uninstructed adults, the accommodative response function for targets such as these are not linear; rather the curves tend to be better fit by

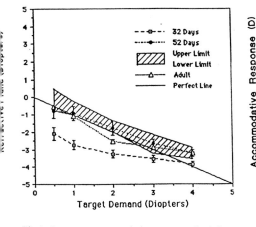

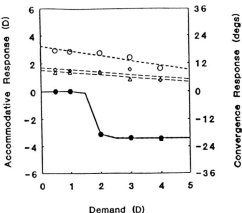

Fig 1: Average accommodative response for infants and adults. Data for the two youngest groups and adults plotted separately. The hatched zone represents the range of behaviors shown by other ages (60-365 days).

Fig. 2: Individual accommodation (solid line) and vergence (binocular response; dotted line, dashed line monocular responses) for one 45 day-old infant. Note the non-linear accommodative response function.

reverse sigmoid functions, with several "zones" of accommodation across targets, rather than a smoothly graded, linear response (Fig 2). This observation, of course, requires data from more than 2 or 3 points which is why it may have escaped notice before; we typically have data from 4-5 target demands for each subject. Looking carefully at the data from the youngest infants, we find that in our sample of 82 infants 2-months old and younger (larger than the 22 in this age range measured by Haynes, et al. (1965) and 12-15 in different groups aged 2-months and younger observed by Banks), a little less than half displayed the flat accommodative behavior reported by Haynes, et al. some 30 years ago. In our study, about a third showed a graded response across targets which if fit with a linear function showed a shallow slope of around a half as observed on average by Banks some 15 years ago. A small group of infants showed a more or less appropriate response, and a few actually underaccommodated for the far targets, looking more like hyperopes than myopes. (Incidentally, under cycloplegia, infants of this age are more likely to be hyperopic than emmetropic or myopic, so we are looking at behavioral accommodation, not the residual optical status of the system.) The average accommodative response continues to improve over the first few months, but the most marked change happens between 1 and 3 months (see Fig. 1).

Development of Vergence

There are fewer studies of the development of vergence than accommodation, and the studies that exist disagree markedly on how well infants verge to different target planes, and more basically, how well they align their eyes in general. Most of the laboratory studies report reasonably good eye alignment and vergence to targets at discretely different target planes (e.g., Slater and Findlay, 1975; Aslin, 1977; Hainline, et al., 1992; Thorn, Gwiazda, Cruz, Bauer, and Held, 1994), although data on infants under 2-3 months have been sparse. These studies have tended to use carefully specified stimuli, fixed viewing distances, and careful subject placements. In the clinical literature, however, there are reports (Archer, 1993) that young infants exhibit a high degree of large eye turns, with 30-60 deg. misalignments reported observationally at a near viewing distance. One study by Archer and colleagues (Sondhi, Archer, and Helveston, 1988) reported that more than 70% of all one-month-olds showed evidence of congenital exotropia, not accounted for by infants' larger angle λ (the angle between the pupillary axis and the visual axis; the pupillary axis is more divergent in most subjects than the optic axis. Angle λ is larger in infants than in adults due to their shorter eyeballs which often results in an appearance of symmetrical exotropia, even when infants are viewing correctly.). The target was generally the observer's face, at an uncontrolled but near distance, so the results do not explicitly relate to changes in vergence per se. However, the poor alignment reported is markedly different from reports of infants' near eye alignment in more well-controlled studies of vergence.

There have been suggestions (e.g., Aslin, 1977; Thorn, et al., 1994) that dynamic vergence, vergence to targets moving in depth, may be poor in young infants, although important variables such as exact target distance and speed of target movement have not been varied systematically. Nor have the actual vergence eye movements been recorded for later viewing, to ask whether dynamic vergence, like accommodation, might be better at near than at far in early infancy. There is also one report by Aslin (1989) that infants under 2 months do not make appropriate vergence movements in response to prisms placed in front of one eye. Whether this is because infants attend to the prism rather than to the distant target in such a circumstance has not been ruled out.

In our own studies of vergence (e.g., Hainline and Riddell, 1995), in which we measure both accommodation and vergence, the findings are more in line with the earlier laboratory results, with certain extensions in our understanding. Our measurements of infant angle λ (Riddell, Hainline, and Abramov, 1994), as well as average Hirschberg ratios (a ratio that conveys the degree of change in the location of a corneal reflection for a given real change in ocular alignment, dependent among other things on corneal curvature and axial length of the eyeball) for infants of different ages allows us to get a reasonable estimate of actual eye alignment, simultaneously with estimates of the plane of accommodation.

The first useful observation is that some of the flat accommodation behavior reported for the youngest infants is probably because they are simply not attending to the targets; a higher proportion of 1- and 2-month-old infants show neither vergence nor accommodation responses. As many of the same infants tested shortly before or after did vary behavior with target distance, the most likely explanation is that these infants are "asleep with their eyes open", a trance-like state familiar to anyone who works with young infants. Even after excluding the data from such infants, however, we still see infants who vary vergence but fail to change accommodation. However, as Fig. 3 illustrates, there are no obvious age differences in the slopes of the average vergence functions during infancy or adulthood, after the effect of the larger adult interpupillary distance which reduces vergence angle is taken into account. The systematic change in the intercept value is due to changes in angle λ with age, as discussed earlier. Unlike accommodation, the individual curves mirror well the group behavior; the data reported are based on data from a total of 297 infants 2 months and under. Individual infants' vergence response curves are almost always linear, and eye alignment at all distances is in almost all cases very appropriate (e.g., Fig. 4a).

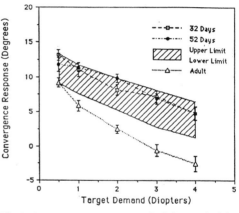

Fig. 3: Average vergence response for infants and adults. As in Fig. 1, hatched zone represents ages between 60 and 365 days. Convergence to nearer targets plots at bottom of scale, divergence to far targets plots up. Adult slope reflects larger interpupillary distance.

In contrast to problems reported in the clinical literature, we simply do not see large exotropias or other misalignments in normal infants under 2 months. However, we have observed that infants are very adept at correcting their ocular position to adjust for the fact that they are not being held directly on the target axis. If a subject is attending to a central target and the head is positioned off axis, one eye will have to turn less than the other to maintain single vision. If one fails to note the off-axis placement of the head, this difference in vergence angle between the two eyes could be confused with a tropia (see Fig. 4b). This behavior is what we think that Archer and colleagues may have, in fact, been observing in their informal observations of infant eye alignment. In our own work in which we give videofeedback to the person holding the infant about

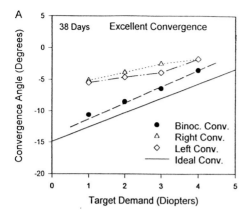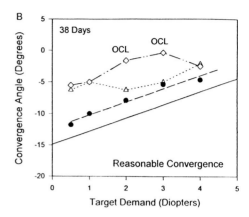

Fig. 4: A. Excellent binocular (closed symbols) and monocular (open symbols) vergence responses in a 38-day-old infant. As expected, the monocular slopes are half that for the binocular response function. The ideal vergence line reflects responses expected for the infant's specific interpupillary separation. B. Similar responses from another 38-day-old. The monocular response represents ocular adjustment for the fact that when the 2 and 3D photos were taken, the infant's head was off center to the left of the target ("OCL"). The excellent binocular vergence response demonstrates that the infant was able to make a visually-guided oculomotor correction for the position of the head with respect to the target.

whether the infant's head is supposed to be, we still find evidence in the pictures we take that infants are not always correctly positioned, usually because of problems holding a wiggling or floppy infant or inattention on the part of the holder. But in many of those cases (see Fig. 4b), infants adjust their eye position precisely, maintaining the correct binocular vergence angle. Besides verifying that this must mean that infants are actually attending the targets on those trials, it also implies that vergence behavior at this age is visually-driven, and that infants have a motive to control or at least contain diplopia.

The Development of Spatial Vision in Early Infancy

The critical question is whether we can find any plausible explanation for these differential rates of development for accommodation and vergence. I suggest that the answer may lie in the development of the visual system's response to different spatial frequencies. It is well established that compared with adults, infants have poor spatial vision (Banks and Crowell, 1993), although for the tasks that infants are normally required to do, it may be well not to emphasize these deficits too heavily (Hainline and Abramov, 1992). It is likely that the marked improvements in spatial vision and acuity in the first year of life are due to changes in factors such as outer segment length and a decrease in foveal cone spacing, although there is some disagreement about how to apportion the effects of these factors to explain improvements in spatial vision with age (see e.g., Wilson, 1993 vs. Banks and Bennett, 1988). The general developmental sequence seems to be a shift of a consistently shaped spatial contrast sensitivity function (CSF) upwards and to higher spatial frequencies with age, resulting in a shift in the peak of the CSF at middle spatial frequencies to more mature values by 4-5 months, and a continued improvement in acuity (the intercept of the CSF at higher spatial frequencies) for a longer period throughout the first year (e.g., Norcia, Tyler and Hamer, 1990). Yet, even with poor spatial vision, many details about the world remain accessible to infants, at least as long as those objects are reasonably close to the infant.

Fig. 5 attempts to show what we estimate a face may look like to an 2-month old infant, based on estimates of spatial contrast sensitivity at this age. The panel at the lower left shows a rendition of the face viewed at 30 cm; it is still noticeably a face. Note, however, that when the same face is moved further away, to 150 cm in the panel at the bottom right, it increases the high spatial frequencies that a young infant is insensitive to, causing the object to lose much of its information value as it becomes an ensemble of meaningless (to us at least) blobs. The point is that taking the development of spatial vision into account, viewing distance will probably influence what infants are able to respond to. We can see this in terms of gross attention; many researchers have commented on the fact that young infants in particular have a kind of attentional "zone" at close distances around them, rapidly losing interest as objects move out of this region. The suggestion here is that the same factors may influence more molecular

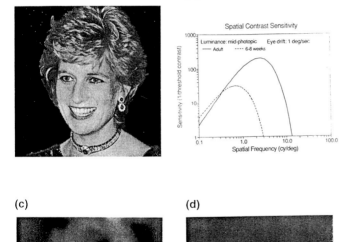

Fig. 5: Simulations of the appearance of a face (panel a) at 30 cm (panel c) and 150 cm (panel d) when viewed by a visual system with a 2-month-old's contrast sensitivity function, graphed in panel b and compared with an average adult function. (image processing assistance by Schuai Chen.)

behaviors such as accommodation and vergence.

The fact that vergence appears to be more robust to such a distance effect than accommodation may imply that vergence at this age depends more on low and intermediate spatial frequencies than higher spatial frequencies. The fact that vergence behavior seems relatively similar even in the youngest infants, and reaches "mature" infant levels (i.e., like that of older infants) relatively early suggests that the response function has a threshold, so that to have adequate vergence, the system needs to be just "good enough". That accommodation is slower to develop and more variable across infants implies that it may depend more heavily on the time course of a particular infant's CSF development. Unfortunately, while we have simultaneous data on infant vergence and accommodation, we do not have CSF or acuity values from these same infants, which would be necessary to elucidate these developmental relationships more precisely. Certainly, the data do allow the observation that linkages between vergence and accommodation seem much looser, if they exist at all in the early months of life (see Aslin and Jackson, 1979 for research on this topic with slightly older infants). Unfortunately, the current data do not allow us to determine if the data imply the non-existence of accommodation-vergence linkages, or simply a very large "dead zone" for accommodation until spatial vision reaches some criterion level. Further studies designed to address this particular issue will be needed to look specifically at the development of accommodation/vergence linkages.

It is clear that the accommodative response is less accurate at this age than will be the case later, but this may not matter for the level of vision that young infants possess. No one has as yet measured Panum's fusional area in infants, but we surmise from their lower contrast sensitivity that it is probably wider in infants than in adults (see Schor, Wood, and Ogawa, 1984). Tyler and colleagues (Liu, Tyler, Schor and Lunn, 1990) have shown that even in adults, low contrast dichoptic stimuli are able to be fused without significant binocular rivalry, so that the lowered contrast sensitivity may actually be an advantage for infants as the preconditions for mature binocular vision are being established. With significantly more sensitive spatial vision, infants might experience disorienting diplopia which could interfere with their ability to process visual information in an adaptive way. As it is, the vergence responses that we have measured are probably good enough for the infant to maintain sensory fusion from the conjoined motor responses of the two eyes.

The fact that accommodative error increases with viewing distance suggests that infants are not getting a sufficient error signal to control accommodation as targets move away. The intercept of their accommodation function at 1-2 months is about 3.5D, or an accommodative plane of about 28-30 cm. We have recently been collecting some pilot eye movement data on infants' vergence while looking at targets that move back and forth in depth in front of them (Hainline, Raptis, Raptis, and Riddell, 1995). There were some problems with the infrared illumination used in our recording that led to quite noisy data as we tracked the location of the two pupil centers over time, and we ran only a few infants around 1-2 months old, but several of the subjects showed evidence of appropriate vergence to a slowly moving target only while the target was at a distance nearer than about 30 cm; as the target receded further than this, tracking disappeared or changed to versional eye movements. As the target approached, one infant tracked it with vergences but only after it was closer than 30 cm; the other ignored the target when it approached, but verged appropriately out to about 30 cm when it was receding. While very preliminary, these data may mean that there are no serious differences in vergence to static versus dynamic targets, once attention differences due to target visibility and reaction time are taken into consideration. We plan to extend this line of work with a more precise eye tracker soon. Aslin's work on prism response also needs further replication, ideally in a situation in which the addition of prisms can be done without attracting the infant's attention.

The data reviewed here suggest that accommodation and vergence are part of a complex system of sensory/motor coordination even early in infancy. Motor responses such as vergence and accommodation cannot be interpreted apart from consideration of developments in sensory factors such as spatial contrast sensitivity. Attentional factors also play an important role, but are not independent; all other things being equal, infants stop paying attention as objects become less visible, so that provided the infant is not in an overall groggy state, near objects will be more visible and therefore more attended than distant objects because of their salience. While these facts create some complexity for the student of infant visual behavior, they also reveal an interesting and important system in which to study the establishment of sensory and motor linkages with relevance for explaining the problems of accommodation and vergence described in more mature individuals.

Acknowledgments:
This work was supported by NSF grant IBN-9319683 and PSC-CUNY grants 6664235, 6665462, and 666240. I thank my long-time collaborator Israel Abramov for the benefit of his knowledge of vision and visual development over many years, and Patricia Riddell and Jill Grose-Fifer for their insights and help specifically in the work on infant accommodation and vergence. I also thank the members of the Infant Study Center Staff who provided critical assistance in this work, and the parents of the infants whose data are reported here for their cooperation in allowing their children to take part in these studies.

References

Abramov, I., Hainline, L. and Duckman, R. (1990) Screening infant vision with paraxial photorefraction. *Optometry and Visual Science* 67: 538-545.

Archer, S. M. (1993) Detection and treatment of congenital esotropia. In K. Simons (ed.) *Early Visual Development, Normal and Abnormal* New York: Oxford University Press.

Aslin, R.N. (1977) Development of binocular fixation in human infants. *Journal of Experimental Child Psychology* 23:133-150.

Aslin, R.N. (1988) Normative oculomotor development in human infants. in G. Lennerstrand, G.K. von Noorden, and E.C. Campos (eds.) *Stabismus and Amblyopia*. London: Macmillan Press.

Aslin, R.M. and Jackson, R.W. (1979) Accommodative-convergence in young infants: Development of a synergistic sensory-motor system. *Canadian Journal of Psychology* 33:222-231.

Banks, M.S. (1980) The development of visual accommodation in early infancy. *Child Development* 51:646-666.

Banks, M.S. and Bennett, P.J. (1988) Optical and photoreceptor immaturities limit the spatial and chromatic vision of human infants. *Journal of the Optical Society of America, A* 5: 2059-2079.

Banks, M.S. and Crowell, J.A. (1993) Front-end limitations to infant spatial vision: Examination of two analyses. In K. Simons (ed.) *Early Visual Development, Normal and Abnormal* New York: Oxford University Press.

Daw, N. (1995) *Visual Development*. New York: Plenum Press.

Hainline, L. and Abramov, I. (1992) Assessing visual development: Is infant vision good enough? In C. Rovee-Collier and L.P. Lipsitt (ed.) *Advances in Infancy Research, Vol 7*. Norwood, NJ: Ablex. pp. 39-102.

Hainline, L., Raptis, E., Raptis, R. and Riddell, P.M. (1995) A comparison of static and dynamic vergence eye movements in human infants. *Investigative Ophthalmology and Visual Science* 36: S457.

Hainline, L., Riddell, P.M., Grose-Fifer, J., and Abramov, I. (1992) Development of accommodation and convergence in infancy. *Behavioural Brain Research* 49:33-50.

Hainline, L. and Riddell, P.M. (1995) Binocular alignment and vergence in early infancy. *Vision Research* 35: 3229-3236.

Haynes, H., White, P.L. and Held, R. (1965) Visual accommodation in human infants. *Science* 141:528-530.

Liu, L., Tyler, C.W., Schor, C.M. and Lunn, R. (1990) Dichoptic plaids: No rivalry for lower contrast orthogonal gratings. *Investigative Ophthalmology and Visual Science* 31: 526.

Norcia, A.M., Tyler, C.W. and Hamer, R.D. (1990) Development of contrast sensitivity in the human infant. *Vision Research* 30: 1475-1486.

Riddell, P.M., Hainline, L., and Abramov, I. (1994) Calibration of the Hirschberg test in human infants. *Investigative Ophthalmology and Visual Science* 35: 538-543.

Schor, C.M., Wood, I. and Ogawa, J. (1984) Binocular sensory fusion is limited by spatial resolution. *Vision Research* 24: 661-665.

Simons, K. (ed.) *Early Visual Development, Normal and Abnormal*. New York: Oxford University Press.

Slater, A.M. and Findlay, J.M. (1975) Binocular fixation in the newborn baby. *Journal of Experimental Child Psychology.* 20: 248-273.

Sondhi, N., Archer, S.M. and Helveston, E.M. (1988) Development of normal ocular alignment. *Journal of Pediatric Ophthalmology and Strabismus* 25: 210-211.

Thorn, F., Gwiazda, J., Cruz, A.V., Bauer, J.A. and Held, R. (1994) The development of eye alignment, vergence, and sensory binocularity in young infants. *Investigative Ophthalmology and Visual Science* 35: 544-553.

Vital-Durand, F., Atkinson, J. and Braddick, O. (eds.) (1996) *Infant Vision*. Oxford: Oxford University Press.

Wilson, H.R. (1993) Theories of infant visual development. In K. Simons (ed.) *Early Visual Development, Normal and Abnormal*. New York: Oxford University Press.

Accommodation and Vergence Mechanisms
in the Visual System
ed. by O. Franzén, H. Richter and L. Stark
© 2000 Birkhäuser Verlag Basel/Switzerland

The accommodative response to blur in myopic children

R. Held and J. Gwiazda

New England College of Optometry, Myopia Research Center, 424 Beacon Street, Boston, Massachusetts 02115, United States

Summary: Accommodation has been implicated in one way or another in the genesis of myopia. Two experimental tests of these claims are reported. In the first, the accommodative response function of myopic children was measured and compared to that of a comparable group of emmetropes. The response of the myopes to lens-induced demand was markedly less at near than that of emmetropes; that to real targets was slightly but significantly less. In the second test, change in the response function was measured over an interval of six to twelve months and correlated with accompanying variation in the refractive error. Myopes showed a significant correlation indicating that accommodative response was reduced during increasing myopia but returned to normal during stabilization of refractive error. Emmetropes showed no correlation between these measures. In conclusion: we see no evidence that excessive accommodation is a factor in the progression of myopia, but dioptric blur from reduced accommodation may be. Reduced accommodation in early myopes appears to result from reduced effectiveness of dioptric blur and subsequent retinoneural signals in activating the ciliary muscle.

Introduction

The increasing prevalence of myopia among urban youth in technologically advanced countries has brought this form of pathology to epidemic levels. Cultural activities thought to be responsible are the amounts of time devoted to reading and viewing television and computer screens (so-called nearwork). Schooling and gainful occupations increasingly demand reading, school work, and attention to fine visual detail. Each of these has been shown to be correlated with the increasing prevalence of myopia (Parssinen and Lyyra 1993, Zylbermann, et al. 1993).

As is well known, the performance of near work implies accommodation of the lens of the eye in order to achieve and maintain focus of the image on the retina in the normal (emmetropic) eye. Consequently, the process of accommodation has been implicated in the onset and development of school-age myopia, although its mode of action has remained obscure. One prevalent view has maintained that intensive near work with its strong demand on accommodation has some kind of mechanical or hydraulic effect, perhaps by increasing intraocular pressure, on growth of the eyeball so as to increase axial length (Oakley and Young 1975). By now this view has been largely discredited by animal experiments in which myopia is produced by form deprivation of eyes in which the accommodative system is essentially eliminated as a functioning entity (Schaeffel, et al. 1990) and by new evidence that myopes may in fact accommodate less effectively than emmetropes (Gwiazda, et al. 1993b, Gwiazda, et al. 1995). This report summarizes current findings from our research concerning the relation between accommodation and myopization.

In studies of a variety of animal species reared early in life with form deprivation, produced either by lid suture or translucent occluders, an elongated eyeball is produced with attendant myopia. This result has suggested, and appears to be partially confirmed by Robb (1977) in cases of early hemangioma and by Rabin et al (1981), that such a condition might also induce myopia in human observers. In principal, habitual blur from defocus could be produced in human observers by habitual misaccommodation. In fact, there have been suggestions in the literature that low accommodative gain with consequent underaccommodation at near is a condition observed in myopes (Goss 1991). However, proof that dioptric blur resulting from accommodative defocus is involved in the progression of myopia must come from a longitudinal study of the development of human refraction and the accompanying status of accommodation. A beginning in that direction was made in our laboratory as an accompaniment of psycho-physical and behavioral studies of the development of aspects of the vision of human infants made over the course of the last twenty years (Gwiazda, et al. 1993a).

Recognizing the need for an assessment of refraction before coming to conclusions about the visual capacities of infants, we normally had such measurements taken in conjunction with the measures of vision. Any significant deviation from emmetropia could then be taken into account and not confounded with purely visual function. As these subjects grew out of infancy, we had them return periodically, although with decreasing frequency, in order to follow their post-infantile development. The measurements taken continued to include refractions such that by the 1990s we had subjects whose refractions had been prospectively measured over periods ranging up to twenty years. Within the last few years we have summarized and published these data showing the progression of emmetropization during the first five or six years followed by an increasing incidence of myopia in our sample of the population (Gwiazda, et al. 1993a).

Measurements of refraction were taken by near-retinoscopy (Mohindra 1977) during the first three years of life, after which non-cycloplegic distance retinoscopy was used. The essential findings are as follows: We see a high incidence of astigmatism ranging up to 60% in infants but declining over the first few years to the population mean for adults between 5 and 10% (Gwiazda, et al. 1985). Mean refraction in spherical equivalents rose from slightly negative {probably resulting from residual accommodation inherent in the near refraction procedure in these very young infants (Thorn, et al. 1996)} to plus 0.5 D within the first year and remaining stable up to approximately 7 years, after which there ensues a steady drop resulting from an increasing tendency to myopia. As had been found by previous investigators, the scatter of refractive errors is greatest in the early months of life but decreases steadily until about six years of age. This reduction of error is attributed to an active process of emmetropization tending to correct for refractive errors irrespective of their direction. After the sixth year the scatter increases largely as a result of the increasing tendency to myopia.

These developmental data reveal that the refractions of individuals repeated over the years tend to maintain themselves in similar positions within their annual distributions. As a consequence, Pearson correlations between refractions measured at one year of age and those measured at six years and later range between 0.65 and 0.75. Thus, knowledge of early refractive error and its distribution in the population yields reasonably good predictions of later error during the school years and of the potential for developing myopia. In addition, the data show that the likelihood of myopia developing in children is related to its incidence in the parents of these children. In our sample, with two myopic parents 42 % of the children became myopic in later childhood, with one myopic parent 22.5% became myopic, and with no myopic parent only 8% became myopic (Gwiazda, et al. 1993a). With both knowledge of early refraction and of parents' refraction, predictions can be made of the population of children at-risk for development of myopia. Such knowledge will be invaluable when treatment becomes established. In addition to its practical significance, the implication of parental refraction is consistent with a long-suspected genetic factor influencing susceptibility to myopia (Grice, et al. 1996). Recent data on a correlation between axial lengths of the eyeballs of parents and children (Zadnik, et al. 1994), constitute further confirmation. This evidence in no way reduces the signficance of the experiential factor of exposure to blur which modulates emmetropization and its deviations as revealed most strongly by the animal studies of induced myopia (Schaeffel, et al. 1988, Norton and Siegwart 1995). In a number of species, including monkeys, lenses of various powers and signs have been fixed over the eyes of young animals. In response, the growth (axial length) of these eyes has been shown to be modulated so as to compensate for the induced defocus (Hung, et al. 1995). Epidemiologic evidence that increased nearwork is correlated with the prevalence of myopia appears to confirm the relevance of visual experience in producing human myopia (Zylbermann, et al. 1993).

Knowing the refractive histories of individuals allows us, at least in principle, to relate the status of their accommodative function to refraction and its changes. To date, two major studies along these lines have been performed. In the first of these a group of our longitudinally-followed children was tested for their accommodative response functions and their data categorized into emmetropes and myopes for analysis of results (Gwiazda, et al. 1993b). In the second study, the changes occurring between the results of two successive tests of accommodative function taken within six months to a year were compared to co-occuring refractive changes (Gwiazda, et al. 1995).

The Accommodative Response in Young Myopes and Emmetropes

The accommodative response function relates dioptric power of the eye and its changes resulting from accommodation to demand for focusing. Demand for focus and its change can be produced by several means. The following tests utilized real targets at varying distances (Fig 1B) and a target at a fixed distance with lenses interposed between it and the eye (Fig 1A). The accommodative response was measured using a Canon R-1 Autorefractor (Canon Europa N.V., Amstelveen, The Netherlands), an optometer that allows targets to be viewed at any distance through an infrared reflecting mirror. Testing was done monocularly on the right eye while the left eye was covered with an occluder. Children wore their best subjective refraction for 4.0m and were seated in front of the device with head held by means of chin and head rests so that the right eye viewed the display. At each level of target demand at least three successive measurements were taken at intervals of 5 to 10 seconds. Target presentations at successive levels of demand were separated by 5 to 30 seconds. The child was instructed to keep each target as clear as possible and to report any persistent blurring. As shown in Figure 1B, the real targets were 3 x 3 arrays of 20/100 (Snellen) letters. Target distance was varied from 4.0 to 0.25 m in seven steps. Size of the target letters was varied so as to keep the visual angle constant. In the second procedure, shown in Figure 1A, negative lenses ranging from plano to -4.0 D in 0.5 D steps were placed in front of the eye as the child fixated the target at 4.0 m.

In the first of two experiments, 64 members of our longitudinal group were tested. At the time of testing 16 had become myopic by 0.50 D or more while 48 remained emmetropic (between -0.25 and +0.50 D). Ideally the response curve should have a slope of 1.0 when focusing power exactly matches demand. As shown in Figure 2A, the mean slopes (emmetropes: 0.78; myopes: 0.88) using real targets show the increasing lag of accommodation for increasingly near targets exhibited by typical observers. The myopes show a small but significantly increased lag compared to the emmetropes on the two nearest targets. In addition, the slope of the function for myopes is slightly but significantly less than that for the emmetropes. Nevertheless, the overall performance of the myopes on this test is barely distinguishable from that of emmetropes.

176

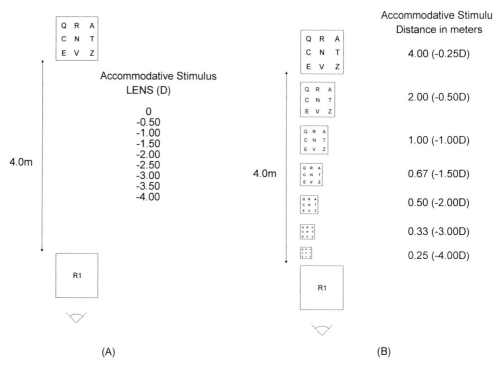

(A)   (B)

Figure 1 Optically corrected monocular observer views letter array target through the R1 which measures the accommodative response of the eye under two conditions: (A) Remote target viewed through a sequence of lenses beginning with the least power. (B) Targets set at a sequence of distances beginning with the most remote. Visual angle subtended by target remains constant at all distances.

The accommodative response functions of Fig 2B, taken with the addition of negative lenses, show increased lag of accommodation, compared with those of Fig 2A. Differences between emmetropes and myopes are seen for both procedures although the lens addition yields much larger differences between the magnitudes and slopes of the functions. Both methods of testing reveal the relative weakness of the accommodative response in myopes.

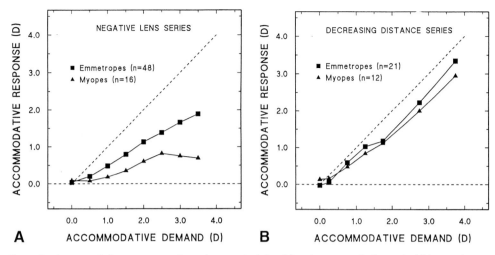

Figure 2 Accommodative responses whown by myopic (triangle) and emmetropic (square) children under two conditions: (A) As in Figure 1A (B) As in Figure 1B.

Since the accommodative response functions of myopes shown in Figure 2A appear close to normal, the response of the ciliary musculature must be near normal contrary to arguments of Avetisov (1979). Consequently, comparison of the two methods reveals that the principal weakness stems from inadequate response to dioptric blur of the system intervening between retina and ciliary innervation. The proximal cues entailed in testing with real targets apparently mask the weakness of response to blur as such. Apart from that weakness of the myopes, a number of other factors may account for the high degree of accommodative lag in testing with the lenses. These include the fact that lens-induced blur is in general a less adequate stimulus for accommodation than real targets and emmetropic observers show great variability in their responses to accommodative demand. In addition, a number of subjects, defined as emmetropes, may well have been incipient myopes who already show reduced accommodative responses.

The results of this study lead us to conclude that in early myopes, the accommodative response to dioptric blur signifying near objects is reduced compared to that of emmetropes of comparable age. Moreover, the reduction apparently results from a failure to couple the neuronal response to a dioptrically blurred retinal image to the available focusing response of the ciliary muscle in order to bring the image into focus.

Concomitant Changes in Refractive Error and Blur-driven accommodation

The results reported above show reduced accommodative response in early myopes. We should also like to know when in the course of development of myopia does the normal accommodative response show this decline: either before or during ? Our own data (Gwiazda, et al. 1995) show normal accommodation in children when their degree of myopia stabilizes. Pilot data on chronic myopics also show normal accommodation. Consequently, we believe that the reduced accommodative responses during the development of myopia must be followed by improvement after stabilization of myopia. As a first step in attempting to relate these concomitant changes, the accommodative response functions were measured twice in each of 63 subjects (40 emmetropes, 23 myopes) separated in time by six months to one year. The measuring procedure was that described above in the case of the lens-induced dioptric demand but the range of demand was increased to a maximum of 10D. The measure of accommodative efficiency was taken as the area under the accommodative response function, which is to a first approximation, proportional to its slope and power at the cutoff (the maximum accommodative response to demand after which accommodation drops to its resting state).

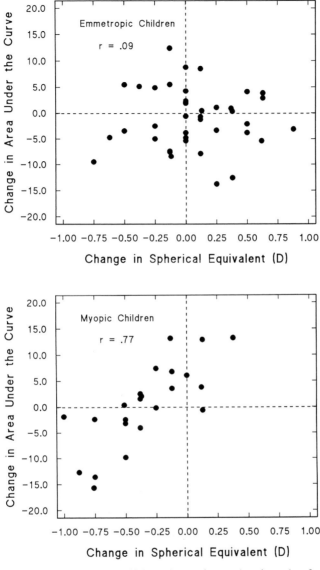

Figure 3 Scatter plots relating accommodative efficiency (area under curve) to change in refractive error (spherical equivalent: equal to spherical error plus one-half of cylindrical error). Upper plot: emmetropes, Lower plot: myopes.

The results are shown in Figs 3A and 3B. The changes in refractive error ranged between -0.75D and +0.90D among the emmetropes and between -1.00D and +0.40D among the myopes. Among the emmetropes no particular relation was seen between concomitant accommodative and refractive changes (Pearson r=0.09). However, the data for myopes show a correlation of 0.77 between these values with the regression line crossing the zero axis at approximatly -0.35 D.

The results imply a relation between increasing myopia and reduced accommodative response.

Conclusions

The tendency to show **reduced** accommodation during the development of **increasing** myopia adds to the evidence against accommodative strain as a causal factor in the genesis of myopia. Inadequate accommodative response to demand implies a blurred image produced by defocus. Since we have no evidence that such defocus occurs before the onset of myopia, but does occur during increasing myopia, we tentatively conclude that such blurring is doubtful as a cause of its onset, but may contribute to its progress. We must also consider the possibility that some as yet unknown factor is common to both increase of myopia and reduction of accommodation. This reduction appears to result from reduced effectiveness of dioptric blur and subsequent retinoneural response in activating the ciliary muscle in accommodation.

Acknowledgements
Support for all phases of this research was provided by NIH grants EY01191 and EY02621.
Joseph Bauer, Kenneth Grice, and Frank Thorn have participated in the performance of this research and/or the preparation of the manuscript.

References

Avetisov, E. S. (1979) Unterlagen zur entstehungstheorie der myopie 1. Mitteilung. Die rolle der akkommodation in der entstehung der myopie. *Klin. Mbl. Augenheilk.* 175:735-740.
Goss, D. (1991) Clinical accommodation and heterophoria findings preceding juvenile onset of myopia. *Optometry and Vision Science* 68:110-116.

Grice, K., DelBono, E., Haines, J., Wiggs, J. and Gwiazda, J. (1996) Genetic analysis of pedigrees affected by juvenile onset myopia. *Investigative Ophthalmology and Visual Science (Supplement)* 37:S1002.

Gwiazda, J., Bauer, J., Thorn, F. and Held, R. (1995) A dynamic relationship between myopia and blur-driven accommodation in school-aged children. *Vision Research* 35:1299-1304.

Gwiazda, J., Mohindra, I., Brill, S. and Held, R. (1985) Infant astigmatism and meridional amblyopia. *Vision Research* 25:1269-1276.

Gwiazda, J., Thorn, F., Bauer, J. and Held, R. (1993a) Emmetropization and the progression of manifest refraction in children followed from infancy to puberty. *Clinical Vision Sciences* 8:337-344.

Gwiazda, J., Thorn, F., Bauer, J. and Held, R. (1993b) Myopic children show insufficient accommodative response to blur. *Investigative Ophthalomology and Visual Science* 34:690-694.

Hung, L., Crawford, M. and Smith, E. (1995) Spectacle lenses alter eye growth and the refractive status of young monkeys. *Nature Medicine* 1:761-765.

Mohindra, I. (1977) A non-cycloplegic refraction technique for infants and young children. *Journal of the Americam Optometric Association* 48:518-523.

Norton, T. and Siegwart, J. (1995) Animal models of emmetropization: matching axial lengths to the focal plane. *Journal of the American Optometric Association* 66:405-414.

Oakley, K. and Young, F. (1975) Bifocal control of myopia. *American Journal of Optometry and Physiological Optics* 52:758-764.

Parssinen, O. and Lyyra, A. (1993) Myopia and Myopic Progression Among Schoolchildren: A Three-Year Follow-Up Study. *Investigative Ophthalmology and Visual Science* 34:2794-2802.

Rabin, J., van Sluyters, R. C. and Malach, R. (1981) Emmetropization: a vision-dependent phenomenon. *Investigative Ophthalmolgy and Visual Science* 20:561-564.

Robb, R. (1977) Refractive errors associated with hemangiomas of the eyelids and orbit in infancy. *American Journal of Ophthalmology* 83:52-8.

Schaeffel, F., Glasser, A. and Howland, H. (1988) Accommodation, refractive error and eye growth in chickens. *Vision Research* 28:639-657.

Schaeffel, F., Troilo, D., Wallman, J. and Howland, H. (1990) Developing eyes that lack accommodation grow to compensate for imposed defocus. *Visual Neuroscience* 4:177-183.

Thorn, F., Gwiazda, J. and Held, R. 1996. Using near retinoscopy to refract infants. In *Infant Vision,* edited by F. Vital-Durand, J. Atkinson and O. Braddick:Oxford University Press. 113-124.

Zadnik, K., Satariano, W., Mutti, D., Sholtz, R. and Adams, A. (1994) The Effect of Parental History of Myopia on Children's Eye Size. *JAMA* 271:1323-1327.

Zylbermann, R., Landau, D. and Berson, D. (1993) The Influence of Study Habits on Myopia in Jewish Teenagers. *Journal of Pediatric Ophthalmology and Strabismus* 30:319-322.

Accommodation and Vergence Mechanisms
in the Visual System
ed. by O. Franzén, H. Richter and L. Stark
© 2000 Birkhäuser Verlag Basel/Switzerland

The role of muscarinic antagonists in the control of eye growth and myopia

Neville A. McBrien

Cardiff University of Wales, P.O. Box 905, Cardiff, CF1 3XF, Wales, United Kingdom

Summary. The success of cycloplegic agents such as atropine in stopping the progression of juvenile myopia in adolescent children has given strong support for a major role for accommodation in the control of ocular growth and myopia. This paper presents data from two recent studies which demonstrate that atropine is effective in preventing axial myopia development via a nonaccommodative mechanism and that the effect is mediated through an M_1 muscarinic receptor process.

Introduction

A role for accommodation as a causative mechanism in the development of myopia has long been suggested. The increased prevalence of myopia in college-based samples and occupational groups with a large near-work component is well documented (e.g. Adams & McBrien, 1992). Although there is substantial epidemiologic evidence that some aspect of the near response in humans can lead to myopia this does not prove a cause and effect relationship, it simply indicates an association between near-work and myopia in humans. However, findings from animal studies of refractive development provide direct evidence that restricting vision to close viewing results in the development of myopia (e.g Young, 1963). Studies have attempted to implicate accommodation specifically, as opposed to just general near-work, by pharmacologically blocking the accommodative apparatus with cycloplegic agents such as atropine. In adolescent humans, daily administration of 1% atropine was found to prevent the progression of juvenile myopia in the treated eye (Bedrossian, 1979; Gimbel, 1973). In mammalian animal models of refractive development, atropine administration has also been found to prevent the development of experimentally induced myopia in certain species of monkey (e.g. Young, 1965; Raviola & Wiesel, 1985) and tree shrew (McKanna & Casagrande, 1981). These studies all concluded that atropine effectively prevented or retarded the progression of myopia due to its cycloplegic effect on smooth ciliary muscle, which blocked the accommodative function of the eye. The above findings have given strong support to a role for accommodation in the development of myopia.

However, recent studies of refractive development in animal models have provided evidence that questions the role of accommodation as a major causative mechanism in the development of myopia. Findings suggest that the visual signals controlling eye growth and refractive development may proceed directly from the retina to the choroid and/or sclera and not through higher central visual processing centres. Studies have demonstrated that blocking communication between the eye and higher centres, either by optic nerve section in chicks (e.g.Troilo et al, 1997) or monkeys (Wiesel & Raviola, 1977), or by blockade of retinal ganglion cell action potentials in tree shrews (Norton, Essinger & McBrien, 1994) or chicks (McBrien et al, 1995; Wildsoet & Wallman, 1995) does not prevent the development of, or recovery from, experimentally induced myopia. Further evidence against a major role for accommodation in experimental myopia is given by studies which demonstrate that myopia can be induced in a species that does not possess a functional accommodation system (McBrien et al, 1993) or where accommodation has been blocked by bilateral lesioning of the Edinger-Westphal nucleus (Schaeffel et al, 1990).

The apparent conflict between the evidence implicating local ocular control of postnatal eye growth and the findings demonstrating the success of atropine in preventing the progression of axial myopia, questions the proposed accommodative mode of action by which atropine is suggested to prevent myopia. This paper discusses the results of two recent studies which address the questions of whether atropine is working via a nonaccommodative mechanism (McBrien, Moghaddam and Reeder, 1993) and secondly through which muscarinic receptor subtype it produces its inhibitory effect on myopia development (Cottriall & McBrien, 1996).

In primate and mammalian models of myopia, atropine invariably blocks the accommodative function of the eye via muscarinic receptors in the ciliary muscle. As a result, in these species it has been difficult to differentiate accommodative and nonaccommodative effects of atropine on myopia development. However, in chick the ciliary muscle is striated and contains predominantly nicotinic receptors (Pilar et al, 1987), thus atropine (a muscarinic receptor antagonist) should not produce cycloplegia or pupil dilation. Therefore, we investigated whether atropine was effective in preventing experimentally induced myopia via a nonaccommodative mechanism in chick. We then went on to test in a mammalian model whether an M_1 selective muscarinic antagonist could inhibit myopia without reducing accommodative amplitude.

Materials and Method

Experiment 1

A detailed description of the methodology is given in the original publication (McBrien et al, 1993) and only a brief description will be given here. Chicks were allocated to one of six experimental groups on the basis of whether they were monocularly deprived (MD) and whether they received intravitreal injections of either atropine or saline. Each of the six groups contained eight chicks. On day 6 after hatching, chicks were given an intravitreal injection of 7μl of phosphate buffered atropine (calculated concentration at retina, 250-300μmol/l, equivalent to 0.01%) or phosphate buffered saline in one eye under halothane (2-3.5%) anaesthesia. Chicks in the MD groups then had a translucent occluder placed over the injected eye. This initial injection procedure was repeated every 48 hours over an 8 day period. The occluders were replaced at each injection. As a control for the intravitreal injection one group of chicks (sham-injected, MD) underwent MD and anaesthesia on the same basis as injected MD chicks but without having a needle inserted in the vitreous. Another group of chicks underwent neither injections or occlusion but were housed in identical conditions (normal-open). Thus the six groups were sham-injected MD, saline-injected MD, atropine-injected MD, atropine-open, saline-open and normal-open.

Eight days after the first injection chicks were anaesthetized (Ketamine/xyalzine) and refractive and biometric measures were taken. All refractive and structural measures were taken after cycloplegia using 5 drops of vecuronium bromide (2mg/ml) spaced at 1 minute intervals. Refraction was measured by streak retinoscopy, corneal curvature by a modified one-position keratometer and axial ocular dimensions by A-scan ultrasound.

To be certain that atropine did not reduce the accommodative amplitude in chick a control experiment was conducted. A separate group of chicks were given intravitreal injections of either atropine (n=5) or saline (n=5). One hour after injection chicks were anaesthetized and baseline measures of refraction and ocular biometry were taken for injected and contralateral control eyes. To pharmacologically induce accommodation carbachol 10% (nicotinic and muscarinic agonist) incorporated in a 2.5% agar button was topically applied to the injected eye via corneal iontophoresis. Because of the rapid and relatively short-lived response to pharmacological stimulation of chick striated ciliary muscle, all measures of refractive and axial dimensions were measured within 10 minutes after topical application of carbachol.

Experiment 2

To address the possibility that muscarinic antagonists are effective in preventing axial elongation and myopia via a retinal mechanism we investigated whether a selective M_1 muscarinic receptor antagonist was effective in preventing myopia in a mammalian model of refractive development without causing cycloplegia (Cottriall & McBrien, 1996). Atropine is a broad-band muscarinic antagonist that binds to all five identified muscarinic receptors. More selective muscarinic antagonists that have affinity for specific muscarinic receptors (Goyal, 1989) make it possible to target a particular class of receptors. It is known that M_1 receptors are usually (although not exclusively) found in neural tissue (e.g. retina), whereas M_3 receptors are found in smooth muscle (e.g. mammalian ciliary muscle).

A similar design to the atropine study in chick was adopted with six groups of tree shrews (n=6 each group) with one eye receiving either the M_1 selective antagonist pirenzepine or the saline vehicle and either monocularly deprived of pattern vision with a translucent occluder (MD) over the treated eye or left binocularly open. Pirenzepine was administered on a daily basis due to its elimination half life in ocular tissues of 4-5 hours (unpublished data). The drug or vehicle was given on an alternate subconjunctival/topical (to reduce insult to conjunctiva) regime for a total of 12 days. A 10% pirenzepine dose was chosen based on preliminary results and was injected in a total volume of 75µl giving a daily dose of 17.7µmol. Injections were given under Halothane (2-3.5%) anaesthesia. On the first day of treatment axial dimensions were also recorded using A-scan ultrasound and, if the animal was in an MD group, they had a head mounted spectacle frame fitted in which the translucent occluder was placed. The goggle could be removed each day to facilitate injections and be reattached immediately afterwards. After 12 days of treatment a full set of refractive and biometric measures were recorded on both eyes of each animal. Measurements of corneal radius, ocular refraction and axial ocular dimensions were taken as previously described (McBrien et al 1993).

To determine whether pirenzepine, at the dose used, had any effect on ocular accommodation a control study was performed. Tree shrews were given a subconjunctival injection of either pirenzepine (n=3) or saline vehicle (n=3) under halothane anesthesia. Ninety minutes after injection baseline measures of ocular refraction and axial dimensions were taken on both eyes of animals. To maintain a sufficient pupil diameter for ultrasound measures after treatment with carbachol, animals were pre-treated with 5% phenylephrine and 2% ephedrine combined in a 2.5% agar gel button and given iontophoretically. Refractive and ocular measures were repeated after full mydriasis to control for any drug-induced changes. To stimulate accommodation, carbachol 20% was administered to the treated eye iontophoretically, again incorporated in a 2.5% agar gel button. Previous studies have shown that this is a supramaximal dose to induce accommodation. Pupil diameter was recorded for the first 10 minutes after carbachol administration before refractive and ocular dimension measures were again taken. Measurements were repeated at 30 and 60 minutes after carbachol administration.

Results

Experiment 1

Figure 1A shows the refractive differences between treated and control eyes for all six groups of chicks. While in sham-injected MD and saline-injected MD chicks the deprived eye developed a relative myopia compared to its contralateral eye of -18.5 ± 2.9 D and -20.9 ± 1.2D respectively atropine-injected MD chicks developed a relative myopia of only -2.8 ± 1.5 D. The difference between the atropine treated MD chicks and control MD groups was found to be highly significant ($p<0.001$). Thus, the muscarinic antagonist atropine, administered as described, almost completely blocks experimentally induced myopia in chicks. Figure 1B shows that atropine blocks experimentally induced myopia by blocking the axial elongation associated with myopia. While the deprived eyes of sham-injected and saline-injected MD chicks underwent an average axial elongation of 1.00 ± 0.14 mm and 1.04 ± 0.06 mm respectively when compared to the contralateral control eye, the deprived eyes of atropine-injected MD chicks underwent an axial elongation of 0.21 ± 0.05 mm.

Intravitreally injected atropine was found not to cause a reduction in the direct pupil response to light (Figure 2A). Corneal iontophoresis of carbachol 10% induced the same degree of accommodation (9.9 ± 1.6 D versus 9.1 ± 1.4 D; $p=0.7$) and increase in lens thickness ($p=0.77$) in atropine-injected and saline-injected chicks (Figure 2B). Thus atropine does not reduce accommodative amplitude in chick.

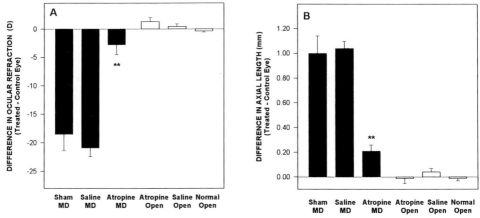

Figure 1. **A**. Differences in ocular refraction between deprived and open control eyes in monocularly deprived (MD) chicks and between injected and contralateral control eyes or left and right eyes of binocularly open chicks. Cycloplegic refraction as measured by streak retinoscopy. **B**. Differences in axial length between deprived and contralateral control eyes in monocularly deprived (MD) chicks and between injected and contralateral control eyes or left and right eyes of binocularly open chicks. Error bars = 1SEM. (Adapted from McBrien et al,1993).

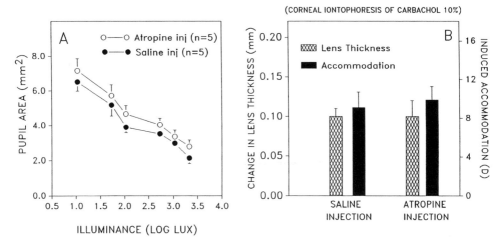

Figure 2. **A**. Constriction of the pupil area in response to increasing light levels on chick eyes after intravitreal injections of atropine or saline. There was no significant difference in pupil area at any light level between atropine- and saline-injected chick eyes (P>0.17). **B**. Change in lens thickness and amount of carbachol-induced accommodation in chick eyes after intravitreal injection of atropine or saline (pre-post injection). No significant difference in lens thickness changes (p=0.77) or induced accommodation (p=0.7) were observed between atropine- and saline-injected chick eyes. N=5 in each group. Error bars=1 SEM. (Adapted from McBrien et al, 1993).

Experiment 2

Refractive differences for the six experimental groups of tree shrews are shown in Figure 3A.
While sham-injected (-13.2D) and saline-injected (-14.1D) MD tree shrews developed similar high
degrees of myopia in the deprived eye relative to the contralateral control eye after 12 days,
pirenzepine-injected MD tree shrews developed only -2.1D of myopia. The observed differences
in induced myopia between MD groups was highly significant (p<0.001). The major structural
correlate to the induced myopia in sham-injected and saline-injected MD animals was vitreous
chamber elongation. Elongation of the vitreous chamber in the deprived eye, relative to the
contralateral open eye, of the pirenzepine-treated MD animals (0.05 ± 0.04mm) was significantly
less than either the sham-injected (0.24 ± 0.02mm) or saline-injected (0.29 ± 0.01mm) MD
animals (p<0.001, Fig 3B).

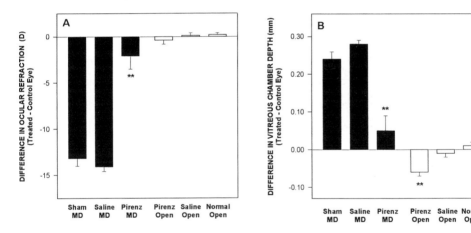

Figure 3. **A**. Differences in cycloplegic ocular refraction between deprived eyes and open control eyes in monocularly
deprived (MD) tree shrews and treated and control eyes or right and left eyes of non-deprived animals. **B**. Differences
in vitreous chamber depth between deprived and open control eyes in monocularly deprived (MD) tree shrews and
between treated and control eyes or right and left eyes of binocularly open animals. Vitreous chamber depth measured
with A-scan ultrasonography. Error bars = 1 SEM. (Adapted from Cottriall and McBrien, 1996).

The level of carbachol-induced accommodation in pirenzepine-treated (7.2 ± 0.3D) and saline-
treated (7.8 ± 0.2D) animals was not significantly different (p=0.13). Histologic evaluation of the
retina and sclera at the light microscopic level (X1000) revealed no apparent toxic effect on either

neural retina or sclera of tree shrew eyes that received 12 days of pirenzepine treatment compared to saline-treated eyes (see Cottriall & McBrien, 1996).

Discussion

As atropine was shown not reduce the accommodative amplitude of the striated ciliary muscle of chick and yet was found to almost completely prevent the development of experimentally induced axial myopia the findings demonstrate that atropine is working via a nonaccommodative mechanism in inhibiting myopia. It does this by preventing the axial elongation of the vitreous chamber associated with pattern deprivation myopia, although there is also some reduction of anterior segment depth. It has also been found that atropine reduces the increased axial elongation and myopia associated with defocusing negative lenses (Wildsoet, McBrien & Clarke, 1994).

The results also demonstrate that the M_1-selective muscarinic antagonist, pirenzepine, was effective in markedly reducing induced myopia in a mammalian model, namely tree shrew, and it did so through a nonaccommodative mechanism. These findings demonstrating the success of pirenzepine in inhibiting axial myopia without causing cycloplegia in a mammalian model specifically implicate an M_1 receptor mediated mechanism, which has also been suggested by earlier studies of the effects of pirenzepine and other selective muscarinic receptor antagonists in chick myopia (e.g. Stone, Lin and Laties, 1991; Leech, Cottriall and McBrien, 1995; Rickers and Schaeffel, 1995).

Atropine has been found to effectively reduce or prevent axial myopia in other animal models of myopia (monkey, tree shrew) and has been shown to prevent the progression of human juvenile myopia (e.g. Bedrossian, 1979). It seems reasonable to assume that the mode of action by which atropine prevents the excessive elongation of the eye and myopia is similar in the various species investigated. The above findings clearly demonstrate that, contrary to previous hypothesis (e.g. McKanna & Casagrande, 1981; Young, 1981), the mechanism by which atropine reduces or prevents myopia is nonaccommodative. If muscarinic antagonists (both broad-band and M_1-selective) are not working via an accommodative mechanism then the question arises as to how they are producing their inhibitory effect on myopia development. It is known that M_1 receptors are predominantly, although not exclusively, found in neural tissue which in the eye would implicate the retina. Acetylcholine is a retinal neurotransmitter (e.g. Masland, 1980; Hutchins,

1987) and the retina contains a cholinergic amacrine cell population (Masland et al, 1987; Vaney, 1984) that includes muscarinic acetylcholine receptors (e.g. Vanderheyden et al, 1988). It is known that other retinal neuromodulators, such as dopamine, are reduced in axial myopia and that dopamine agonists, such as apomorphine, can also inhibit the development of experimentally induced myopia (Iuvone et al, 1991). It therefore seems a good possibility that atropine affects ocular growth by altering retinal neurotransmission or even by direct action on scleral cells as previously suggested (Marzani et al, 1994).

Summary

The widely held clinical belief that accommodation has a major causative role in the development of myopia is to a considerable extent based on work which showed that cycloplegic agents such as atropine were effective in preventing the progression of juvenile myopia in adolescent children. The data presented in brief above clearly demonstrates that muscarinic antagonists are effective in inhibiting axial myopia progression via a nonaccommodative mechanism. It is therefore not surprising that attempts to reduce accommodation using non-pharmacological approaches, such as bifocals, have been unsuccessful in reducing the progression of myopia (e.g. Parssinen et al, 1989). Although cholinergic antagonists do not produce their inhibitory effect on myopia development via an accommodative route, this does not mean that accommodation has no role in refractive development. Studies have found that when communication between the eye and brain are stopped (e.g. by optic nerve section) then emmetropization is not achieved (Troilo et al, 1987), with the suggestion that brain mediated mechanisms, such as accommodation, may be important for normal refractive development.

Acknowledgements
The author thanks Dr Charles Cottriall for his assistance and critical comments on an earlier version of the manuscript. Supported by grants from the Wellcome Trust and the Biotechnology and Biological Sciences Research Council.

References

Adams, D.W and McBrien, N.A. (1992) Prevalence of myopia and myopic progression in a population of clinical microscopists. Optometry and Vision Sciences 69:467-473.
Bedrossian, R.H. (1979) The effect of atropine on myopia. American Journal of Ophthalmology 86:713-717.
Cottriall, C.L. and McBrien, N.A. (1996) The M_1 muscarinic antagonist pirenzepine reduces myopia and eye enlargement in the tree shrew. Investigative Ophthalmology & Visual Science 37:1368-1379.
Gimbel, H.V. (1973) The control of myopia with atropine. Canadian Journal of Ophthalmology 8:527-532.
Hutchins, J.B. (1987) Review:Acetylcholine as a neurotransmitter in the vertebrate retina. Experimental Eye Research

45:1-38.

Iuvone, P.M., Tigges, M., Stone, R.A. Lambert, S. and Laties, A.M. (1991) Effects of apomorhine, a dopamine receptor agonist, on ocular refraction and axial elongation in a priamte model of myopia. Investigative Ophthalmology & Visual Science 32: 1674-1677.

Leech, E.M., Cottriall, C.L. and McBrien, N.A. (1995) Pirenzepine prevents form deprivation myopia in a dose dependent manner. Ophthalmic and Physiological Optics 15:351-356.

Marzani, D., Lind,G. J., Chew, S.J. and Wallman, J. (1994) The reduction of myopia by muscarinic antagonists may involve a direct effect on scleral cells. Investigative Ophthalmology & Visual Science 35:(Suppl) 2545.

Masland, R.H. (1980) Acetylcholine in the retina. Journal of Neurochemistry 1:501-518.

Masland, R.H., Mills, J.W. and Hayden, S.A. (1987) Acetylcholine-synthesizing amacrine cells:Identification and selective staining by using radioautography and fluorescent markers. Proceedings of Royal Society (Biol) 223:79-100.

McKanna, J.A. and Casagrande, V.A. (1981) Atropine affects lid-suture myopia development: Experimental studies of chronic atropization in tree shrews. Documenta Ophthalmologica 28:187-192.

McBrien, N.A., Moghaddam, H.O. and Reeder, A.P. (1993) Atropine reduces experimental myopia and eye enlargement via a nonaccommodative mechanism. Investigative Ophthalmology & Visual Science 34:205-215.

McBrien, N.A., Moghaddam, H.O., New, R. and Williams, L.R. (1993) Experimental myopia in a diurnal mammal (*Sciurus carolinensis*) with no accommodative ability. Journal of Physiology 469:427-441.

McBrien, N.A., Moghaddam, H.O. Cottriall, C.L. Leech, E.M. and Cornell, L.M. (1995) The effects of blockade of retinal cell action potentials on ocular growth, emmetropization and form deprivation myopia in young chicks. Vision Research 35:1141-1152.

Norton, T.T., Essinger, J.A. and McBrien, N.A. (1994) Lid-suture myopia in tree shrews with retinal ganglion cell blockade. Visual Neuroscience 11:143-153.

Parssinen, O., Heminki, E. and Klemetti, A. (1989) Effect of spectacle use and accommodation on myopic progression:final results of a three year randomised clinical trial among schoolchildren. British Journal of Ophthalmology 73:547-551.

Pilar, G., Nunez, R., McLennan, I.S. and Meriney, S.D. (1987) Muscarinic and nicotinic synaptic activation of the developing chicken iris. Journal of Neuroscience 7:3813-3826.

Raviola, E. and Wiesel, T.N. (1985) An animal model of myopia. New England Journal of Medicine 312:1609-1615.

Rickers, M and Schaeffel, F. (1995) Dose-dependent effects of intravitreal pirenzepine on deprivation myopia and lens-induced refractive errors in chickens. Experimental Eye Research 61:509-516.

Schaeffel, F., Troilo, D., Wallman, J. and Howland, H.C. (1990) Development eyes that lack accommodation grow to compensate for imposed defocus. Visual Neuroscience 4:177-183.

Stone, R.A., Lin, T. and Laties, A.M. (1991) Muscarinic antagonist effects on experimental chick myopia. Experimental Eye Research 52:755-758.

Troilo, D., Gottlieb, M.D. and Wallman, J. (1987) Visual deprivation causes myopia in chicks with optic nerve section. Current Eye Research 6:993-999.

Vanderheyden, P., Ebinger, G. and Vauquelin, G. (1988) Characterization of M1-and M2-muscarinic receptors in calf retina membranes. Vision Research 28:247-250.

Vaney, D.I. (1984) "Coronate" amacrine cells in rabbit retina have a "starburst" dendritic morphology. Proceedings of Royal Society (Biol) 220:501-508.

Wiesel, T.N. and Raviola, E. (1977) Myopia and eye enlargement after neonatal lid fusion in monkeys. Nature 266:66-68.

Wildsoet, C.F., McBrien, N.A. and Clarke, I.Q. (1994) Atropine inhibition of lens-induced effects in chick:evidence for similar mechainsm underying form deptivation and lens-induced myopia: Investigative Ophthalmology & Visual Science 35:(Suppl) 3774.

Wildsoet, C.F. and Wallman, J. (1995) Choroidal and scleral mechanisms of compensation for spectacle lenses in chicks. Vision Research 35:1175-1194.

Young, F.A. (1963) The effect of restricted visual space on the refractive error of the young monkey eye. Investigative Ophthalmology 2:571-577.

Young, F.A. (1965) The effect of atropine on the development of myopia in monkeys. American Journal of Optometry & Archives of American Academy of Optometry 59:828-841.

Young, F.A. (1981) Primate myopia. American Journal of Optometry & Physiological Optics 58:560-566.

Accommodation and Vergence Mechanisms
in the Visual System
ed. by O. Franzén, H. Richter and L. Stark

Accommodation, age and presbyopia

Kenneth J. Ciuffreda, Mark Rosenfield, John Mordi and Hai-Wen Chen.

Department of Vision Sciences, State University of New York, State College of Optometry, 100 East 24th Street, New York, New York 10010. U.S.A.

Summary: Static and dynamic aspects of accommodation were investigated with regard to age and the development of presbyopia in two relatively large populations. The majority of accommodative parameters were invariant with age, which in most cases suggested normalcy of neuromotor control. For those parameters which decreased, this decline was attributed to age-related crystalline lens saturation effects. These results are discussed in relation to current theories of presbyopia.

Introduction

One of the inevitable consequences of living is aging. Physiological systems typically demonstrate a linear decline in output with increasing age. This mean output value approaches zero under optimal conditions at approximately 150 years of age (Ordy, 1975) as illustrated in Figure 1.

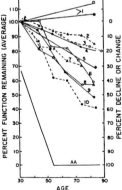

Figure 1. Selected physiological functions and their change with age. Key: 1 = blood pressure, 2 = conduction velocity, 3 = basal metabolic rate, 4 = standard cell water, 5 = hand grip strength, 6 = glomerular filtration rate, 7 = cardiac index, 8 = vital capacity, 9 = renal plasma flow, 10 = maximal breathing rate, and AA = amplitude of accommodation. Adapted from Ordy (1975).

The accommodative system also exhibits age–related changes. For example, the amplitude of accommodation, i.e., the maximum accommodative output or response, has been the parameter which has received most attention. However, its decline is much more precipitious (Figure 1). The amplitude of accommodation approaches zero at approximately 52 years of age (see Ciuffreda-in press, for a detailed review). The uniqueness in decline of this important basic and clinical accommodative parameter has been attributed to multiple factors, in addition to the basic physiological aging process (Pierscionek and Weale, 1995). Its value is highly correlated with the normal middle–aged signs and symptoms of clinical presbyopia (Ciuffreda-in press).

Although the amplitude of accommodation has been the primary focus of research attention over the past century, only relatively recently have some of the other monocular and binocular static and dynamic components of accommodation been studied with respect to their age–related changes (Ciuffreda-in press). This has followed the introduction of a number of bioengineering models describing both the accommodative and vergence systems and their interaction. However, these have not been comprehensive investigations involving all of the accommodative and related oculomotor components. Furthermore, some have either used small sample sizes or have had methodological problems.

Thus, the purpose of our investigation was to conduct a comprehensive study in this area using a relatively large population and a model of the oculomotor system as a guide. This chapter summarizes our results over the past several years (Ciuffreda et al., 1993; Ciuffreda, in press; Ciuffreda and Tannen, 1995; Hokoda et al., 1991; Mordi, 1991; Mordi and Ciuffreda, in press, in preparation; Rosenfield and Ciuffreda, 1990; Rosenfield et al., 1995, in preparation).

Materials and Methods

Subjects

Two subject populations were tested which only differed in their age range. One (n = 30) was 21–50 years of age, whereas the other (n = 42) was 21–70 years of age. The smaller population was divided into six age subgroups: 21–25, 26–30, 31–35, 36–40, 41–45, and 46–50 years of age with 5 subjects in each subgroup, whereas the larger population also had a group of subjects (n=12) between 51 and 70 years of age. Subjects were derived from the faculty, staff, and student body of the optometry college. Each was prescreened and found to be free from any systemic, neurological, or ocular disease; furthermore, none was not taking any drugs or medications that could compromise either accommodative or vergence function. Each subject had a corrected visual acuity of 20/25 or better and normal binocular vision.

Apparatus and Procedures

Several devices were used to obtain the measurements. Details of the equipment and experimental protocols have been provided elsewhere (Ciuffreda et al., 1993; Ciuffreda, in press; Ciuffreda and Tannen, 1995; Hokoda et al., 1991; Mordi, 1991; Mordi and Ciuffreda, in press, in preparation; Rosenfield and Ciuffreda, 1990; Rosenfield et al., 1995, in preparation).

1. Haploscope–optometer (Ciuffreda and Kenyon, 1983; Mordi and Ciuffreda, in press): Several subjective measurements were obtained with this device. These included the stimulus accommodative convergence to accommodation (AC/A) ratio, the response AC/A ratio, the stimulus convergent accommodation to convergence (CA/C) ratio, tonic vergence, vergence adaptation, proximally–induced vergence, and proximally–induced accommodation.

2. Dynamic infrared optometer (Mordi, 1991; Mordi and Ciuffreda, in press): Several dynamic measures were obtained with this device. These included latency, time constant, peak velocity/amplitude relation ("main sequence"), microfluctuations, accommodative controller gain (open–loop gain), slope of the accommodative stimulus/response function (closed–loop gain), objective depth–of–focus, and tonic accommodation.

3. Static infrared optometer (Canon Autoref R–1) (Matsumura et al. 1983; Berman et al., 1984; McBrien and Millodot, 1985): Two objective measures were obtained with this device, namely tonic accommodation and accommodative adaptation.

4. Badal system (Mordi, 1991): A Badal–based optical system using a split–field target arrangement was used to assess the subjective depth–of–focus.

In addition, the accommodative amplitude was assessed using the standard clinical push–up technique (Ciuffreda, in press).

Model

A recently developed, comprehensive static model of the interactive accommodative and vergence systems was used as a guide for our static assessment (Hung et al, 1996) (Figure 2). It included all four accommodative components (Ciuffreda, in press), i.e., blur, vergence, proximal and tonic accommodation.

However, no current dynamic model of accommodation exists which includes all of the parameters of interest. Nevertheless, the following dynamic accommodative parameters were studied: latency, time constant, peak velocity/amplitude relation ("main sequence"), and microfluctuations of accommodation.

Results

The overall results are qualitatively summarized in Table 1. Of some surprise was the finding that just over one–half (10 out of 18) of the static and dynamic parameters, which were all within population norms, remained invariant with age. Of the remaining 8 that changed with age, 6 decreased and 2 increased.

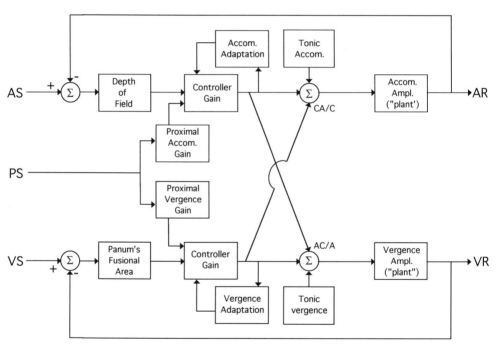

Figure 2. Hung-Ciuffreda-Rosenfield comprehensive static model of accommodation and vergence.
From Hung et al. (1996).

For those 8 parameters that changed with age, their annual rates of change are presented in Table 2. The largest annual dioptric change (0.34D) was found for the accommodative amplitude. This confirmed the numerous earlier investigations of this important clinical parameter (Ciuffreda-in press).

Table 1 Oculomotor parameters and age.

Increase	Decrease	Constant
Subjective depth-of-focus	Accommodative amplitude	Objective depth-of-focus
Accommodative latency	Tonic accommodation	Open-loop gain
	Accommodative microfluctuations	Closed-loop gain
	Accommodative adaptation	Accommodative time constant
	CA/C ratio	Stimulus AC/A ratio
	Proximally-induced accommodation	Response AC/A ratio
		Peak velocity / amplitude relation
		Proximally-induced vergence
		Tonic vergence
		Vergence adaptation

Table 2. Rate of parameter change per year with increasing age.

Parameter	Annual rate of change
Subjective depth-of-focus	0.027 D
Accommodative latency	2.5 msec
Accommodative amplitude	0.34 D
Tonic accommodation	0.04 D
Accommodative microfluctuations	–
Accommodative adaptation	0.034 D
CA/C ratio	0.006 D/?
Proximally-induced accommodation	0.008 D

Discussion

The results of our comprehensive investigation have clearly demonstrated that not all accommodative and related oculomotor parameters changed with age. In fact, the majority did not. However, the basic and clinical implications related to those parameters that did change, as well as those that did not, are of equal importance.

Parameters That Increased With Age

Two parameters increased with age (see Tables 1 and 2). The increase in accommodative latency was attributed to the normal age–related, neurosensory slowing of information processing (Mordi, 1991; Mordi and Ciuffreda, in press). All forms of reaction time increase by a few milliseconds per year throughout one's lifetime (Ordy, 1975). In contrast, the increase in subjective depth–of–focus was attributed to a perceptual adaptation phenomenon, which can be conceptualized as a "tolerance for error" to defocus blur (Mordi, 1991; Mordi and Ciuffreda, in press). As one approaches the pre–presbyopic years (i.e., around 35 years of age), the frequency with which the full range of near objects becomes slightly more defocused and may be perceived as being slightly blurred gradually increases due to the progressive and concurrent decline in accommodative amplitude. To minimize this occurrence, one might envision a gradual elevation in the subjective (perceptual) blur threshold with age.

Parameters That Decreased With Age

Six parameters decreased with age (see Tables 1 and 2). However, in all cases, it is believed that the neural controller signals remained *normal and constant with age*. Rather, the gradual and progressive loss in motor responsivity was attributed to an age–related saturation phenomenon of the crystalline lens and related peripheral structures (i.e., the "plant") that participate in the accommodative process (Ciuffreda et al., 1993; Ciuffreda, in press; Ciuffreda and Tannen, 1995; Hokoda et al., 1991; Mordi, 1991; Mordi and Ciuffreda, in press, in preparation; Rosenfield and Ciuffreda, 1990; Rosenfield et al., 1995, in preparation). Thus, the various systems' initial neural signals, as well as the final inputs to the accommodative apparatus, did not change with age, but rather the overall crystalline lens–related responsivity declined with age. It is important to note that all six parameters had a maximum response value (for the age range tested) around 20 years of age, and declined progressively with age to reach a zero value at approximately 50 years of age. This time course is remarkably similar to the widely-demonstrated reduction in accommodative amplitude, which strongly suggests that this parameter is the common underlying factor for the decline in the various accommodative components.

Parameters That Remained Constant With Age

Each of the 10 parameters that remained invariant with age had an underlying neurological/neuromotor aspect (see Table 1) (Ciuffreda et al., 1993; Ciuffreda, in press; Ciuffreda and Tannen, 1995; Hokoda et al., 1991; Mordi, 1991; Mordi and Ciuffreda, in press, in preparation; Rosenfield and Ciuffreda, 1990; Rosenfield et al., 1995, in preparation). For example, the lack of change in the peak velocity/amplitude relation suggested that the midbrain neural step controller signal remained constant and appropriate for its correlated response amplitude (Mordi and Ciuffreda, in press). It should be noted that for this as well as the other parameters, it is assumed that one is operating over the linear accommodative response region (Ciuffreda, 1991; Ciuffreda-in press). Similarly, constancy of the response AC/A ratio suggested that the neural crosslink gain did not change, whereas constancy of the stimulus AC/A ratio implied that the innervation to accommodation and convergence were also invariant with age (Ciuffreda et al., in press). Furthermore, constancy of the objective depth–of–focus suggested that threshold blur detection and the related neurologically–based reflex–driven accommodation (Fincham, 1951) did not exhibit any age–related change (Mordi and Ciuffreda, in press). With respect to vergence, the results demonstrated that this system remains relatively robust with age (Morgan, 1960; Ciuffreda et al., 1993; Rosenfield et al, in preparation). This is consistent with earlier findings (Ciuffreda et al., 1993; Ciuffreda, in press).

Theories of Presbyopia

The finding that the response AC/A ratio does not exhibit any significant age-related change has important implications with regard to current theories of presbyopia (Ciuffreda, in press). For example, the *Helmholtz-Hess-Gullstrand Theory* attributes all of the loss in accommodation to biomechanical changes in the lens and lens capsule. It states that the amount of ciliary muscle contraction required to produce a unit change in accommodation remains *constant* with age. Thus, with increasing age, the progressive decline in the amplitude of accommodation would be entirely due to lenticular factors. Additionally, as a corollary, the response AC/A ratio would not change with age. In contrast, the *Donders-Duane-Fincham Theory* attributes all of the age–related loss of accommodation to the ciliary muscle. It states that the amount of ciliary muscle contraction required to produce a unit change in accommodation progressively *increases* with age. Thus, with increasing age, the reduced accommodative amplitude results from progressive weakening of the ciliary muscle. Additionally, as a corollary, the response AC/A ratio would progressively increase with age. The present results clearly demonstrated constancy of the response AC/A ratio with age and therefore support the *Helmholtz-Hess-Gullstrand Theory* of presbyopia (Ciuffreda et al., in press; Rosenfield et al, in preparation).

Acknowledgements
Supported by NIH EY08817, Schnurmacher Institute for Vision Research, and Essilor International.

References

Berman, M., Nelson, P. and Cade, B. (1984) Objective refraction: Comparison of retinoscopy and automated techniques. *American Journal of Optometry & Physiological Optics* Opt. 61; 203-209.

Ciuffreda, K.J. (1991) Accommodation and its anomalies. *In*: W.N.Charman (ed.): *Vision and Visual Dysfunction:Visual Optics and Instrumentation*, Vol. 1. MacMillan, London, pp. 231-279.

Ciuffreda, K.J. (in press) Accommodation, pupil, and presbyopia. *In*: W.J. Benjamin and I.M.Borish (eds.): *Clinical Refraction: Principles and Practice*, Saunders, Philadelphia.

Ciuffreda, K.J. and Kenyon, R.V. (1983) Accommodative vergence and accommodation in normals, amblyopes, and strabismics. *In*: C.M. Schor and K.J.Ciuffreda (eds): *Vergence Eye Movements: Basic and Clinical Aspects* Butterworths, Boston, pp. 101–173.

Ciuffreda, K.J., Ong, E. and Rosenfield, M. (1993) Tonic vergence, age and presbyopia. *Ophthalmic and Physiological Optics* 13; 151–154.

Ciuffreda, K.J., Rosenfield, M. and Chen, H.W. (in press). The AC/A ratio, age and presbyopia. *Ophthalmic and Physiological Optics*

Ciuffreda, K.J., Tannen, B. (1995). *Eye Movement Basics for the Clinician*. Mosby Yearbook, St. Louis.

Fincham, E.F. (1951). The accommodation reflex and its stimulus. *British. Journal of Ophthalmology* 35; 381–393.

Hokoda, S.C., Rosenfield, M. and Ciuffreda, K.J. (1991) Proximal vergence and age. *Optometry and Vision Science* 68; 168–172.

Hung G.K., Ciuffreda, K.J. and Rosenfield, M. (1996) Proximal contribution to a linear static model of accommodation and vergence. *Ophthalmic and Physiological Optics* 16; 31-41.

Matsumura, I., Maruyama, S., Ishikawa, Y., et al. (1983) The design of an open view autorefractor. *In*: G.M. Breinin and I.M. Siegel (eds), *Advances in Diagnostic Visual Optics*, Springer-Verlag, Berlin, pp. 36-42.

McBrien, N.A. and Millodot, M. (1985) Clinical evaluation of the Canon Autoref R-1. *American Journal of Ophthalmic and Physiological Optics* 62; 786-792.

Mordi, J.A. (1991) Accommodation, age and presbyopia, Ph.D. dissertation, State University of New York, State College of Optometry, New York, New York, U.S.A.

Mordi, J.A. and Ciuffreda, K.J. (in press) Static aspects of accommodation in presbyopia, *Vision Research*

Mordi, J.A. and Ciuffreda, K.J. (in prep.) Dynamic aspects of accommodation in presbyopia.

Morgan, M.W. (1960) Anomalies of the visual neuromuscular system of the aging patient and their correction. *In:* M.J.Hirsch and R.E.Wick (eds) *Vision of the Aging Patient*. Chilton, Philadelphia, pp. 113-45.

Ordy, J.M. (1975) Principles of mammalian aging. *In:* J.M. Ordy and K.R. Brizzee (eds) *Neurobiology of Aging*. Plenum Press, New York, pp. 1–22.

Pierscionek, B.K. and Weale, R.A. (1995) Presbyopia – a maverick of human aging. *Archives of Gerontology and Geriatrics* 20; 229–240.

Rosenfield, M. and Ciuffreda, K.J. (1990) Accommodative convergence and age. *Ophthalmic & Physiological Optics* 10; 403–404.

Rosenfield, M., Ciuffreda, K.J. and Chen, H.W. (1995) Effect of age on the interaction between the AC/A and CA/C ratios. *Ophthalmic & Physiological Optics* 15; 451–455.

Rosenfield, M., Ciuffreda, K.J. and Chen, H.W. (in prep). Oculomotor adaptation and age.

Chromatic stimulus for accommodation to stationary and moving targets

P.B. Kruger, S. Mathews*, K.R. Aggarwala, L. Stark, S. Bean, J. Lee and S. Cohen.

Schnurmacher Institute for Vision Research, State University of New York, 100 E 24th Street, New York NY 10010 USA.
*Texas Tech. U.H.S.C. Dept. of Ophthalmology, 3601 4th Street, Lubbock, Texas 79430 USA.

Summary. Experiments using moving targets (sinusoidal and sum-of-sine motion) suggest that the longitudinal chromatic aberration (LCA) of the eye provides directional information at edges that specifies ocular focus and drives reflex accommodation (Kruger, Mathews, Aggarwala, Sanchez, 1993). Recently two studies used stationary targets to stimulate accommodation and both concluded that chromatic aberration has no effect on accommodation when the target is stationary (Bobier, Campbell and Hinch, 1992; Kotulak, Morse and Billock, 1995). The issue was addressed in three experiments: First, subjects (8) viewed a stationary target (3.5 cycles per degree square-wave grating) through a 3 mm artificial pupil at three dioptric stimulus levels (0 D, 2.5 D, 5 D) in a Badal stimulus system. There were 3 experimental conditions including Normal LCA, Reversed LCA and Monochromatic (550 nm; 10 nm bandwidth) and the resting position of accommodation was monitored while subjects viewed the grating target through a pinhole pupil. The target remained stationary during randomized 40-second trials, and accommodation was monitored continuously. Most subjects accommodated accurately in the normal condition at all three stimulus distances, a few subjects (3 of 8) had difficulty maintaining focus in monochromatic light at the near (5 D) or far (0) stimulus distances, and most subjects (7 of 8) could not maintain focus at both the near and far distances with LCA reversed. When the target is stationary it must be considerably closer or further away than the (tonic) resting level to demonstrate the effect of LCA.

In the second experiment the same subjects (8) viewed the grating target as it moved sinusoidally toward and away from the eye (1.5 to 2.5 D at 0.2 Hz) under the same three experimental conditions. As in previous dynamic experiments, accommodative gain was reduced by approximately 50% in monochromatic light, and accommodative tracking (gain) was poor with LCA reversed.

In the third experiment subjects (12) viewed a stationary sine-wave grating (3.88 c/deg) through a pinhole pupil (0.75 mm). The sine-wave grating was projected at high luminance by a video-projector, and imaged in the Badal stimulus system. The grating target simulated the effects of static defocus and LCA on the contrast of the "red", "green" and "blue" components of the "white" grating for a 3 mm diameter pupil. There were three simulated defocus conditions: myopic focus (1.0 D); hyperopic focus (1.0 D); and a control condition that simulated the effect of defocus without LCA (the three chromatic components of the grating had the same contrasts). The subject viewed a fixation target for 10.24 seconds followed by 10.24 seconds during which one of the defocused sine-wave simulations appeared. Ten of 12 subjects accommodated in the appropriate direction to randomized open-loop simulations of under- and over-accommodation.

The results support the hypothesis that the relative contrast of long- middle- and short-wavelength components of the retinal image with the same spatial frequency and spatial phase reliably indicates ocular focus and drives reflex accommodation for both stationary and moving targets.

Introduction

The standard model of accommodative control is that accommodation is a contrast-maximizing closed-loop negative feedback system. In this view the eye accommodates to maximize or optimize luminance contrast of the retinal image. Blur from innacurate focus reduces contrast of the retinal image, and the effect of defocus on contrast depends on the spatial frequency content of the target. The effect of defocus blur is best illustrated by the modulation transfer function (MTF) of the eye. The functions in Figure 1 were calculated using the method of Smith (1982) for a 3 mm pupil in monochromatic light and various amounts of defocus-blur. For an in-focus retinal image spatial contrast is transferred well at low and intermediate spatial frequencies, and the reduction in contrast at high spatial frequencies is due mostly to diffraction. Blur of the retinal image also results from scattered light, but we concentrated on the effects of innacurate focus and chromatic aberration. Defocus has no effect on contrast at low spatial frequencies (<1 c/d), and a small amount of defocus has a large effect on contrast at high spatial frequencies (10 c/d). Chromatic aberration of the eye complicates the issue. Dispersion by the ocular media produces a large amount of longitudinal or axial chromatic aberration which measures more than 2 diopters across the visible spectrum. In addition to the effects of defocus-blur on the modulation transfer function of the eye, chromatic difference of focus alters MTF as a function of wavelength in focus.

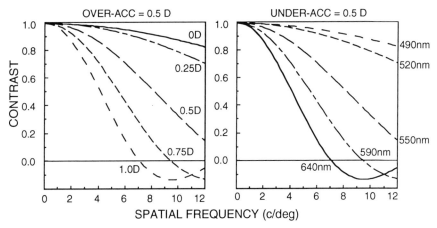

Figure 1. Modulation transfer functions for a 3 mm pupil and defocus from zero to 1 diopter in 0.25 D steps. Focus is referenced to 550 nm light with over-accommodation (0.5 D) on the left and under-accommodation (0.5 D) on the right.

Figure 1 illustrates the effects of both defocus and chromatic difference of focus (LCA) on MTF of the eye for 5 conditions of focus (0, 0.25 D, 0.5 D, 0.75 D and 1.0 D). In addition, the figures represent under- and over-accommodation in the amount of a half diopter with focus referenced to 550 nm light. In the left side (over-accommodation) 550 nm light is focussed 0.5 D in front of the retina and 640 nm light is in focus, and on the right side (under-accommodation) 550 nm light is focussed 0.5 D behind the retina and 490 nm light is in focus. At low spatial frequencies, contrast of the retinal image is the same for all spectral components of the image, and defocus-blur has no effect. However at moderate and high spatial frequencies, contrast is different for each spectral component of the image. At intermediate spatial frequencies (3 to 5 c/d) the difference in contrast is approximately 20 percent between spectral components that focus a half diopter apart. Under-accommodation is specified by higher contrast for the short-wavelength (blue) component of the image than for the long-wavelength (red) component, and over-accommodation by higher contrast for the long-wavelength component than for the short-wavelength component. A comparison of the contrast of spectral components of the retinal image specifies focus behind or in front of the retina. At high spatial frequencies small amounts of defocus have a large effect on contrast, and the effects of chromatic aberration are particularly pronounced. The effects of defocus and LCA can be *simulated* by superimposing red, green and blue sine-wave gratings of the same spatial frequency and the same spatial phase, but having different contrasts. Dynamic simulations of the effects of defocus and LCA, in which the contrasts of the red, green and blue grating components change independently to simulate sinusoidal changes in focus, drive accommodation as expected (Kruger et al., 1995). Static simulations of dioptric blur are used here to drive accommodation (experiment 3).

Methods

In the first two experiments subjects viewed a real square-wave grating target in a Badal stimulus system. The target remained stationary in the first experiment, and it oscillated sinusoidally toward and away from the eye in the second experiment. The target was a 3.5 c/d square-wave grating (Ronchi ruling) in a 6 degree field, viewed through a 3 mm artificial pupil. The square-wave grating was illuminated from behind by broadband "white" light to give a mean target luminance of 100 cd/m2. An interference filter could provide narrowband monochromatic light at 550 nm at the same luminance as the white target, and a specially designed lens doublet could be imaged in the subject's pupil to reverse the normal longitudinal chromatic aberration of the eye. The reversing lens has no effect on contrast sensitivity, depth-of-focus, or lateral chromatic aberration measured through the stimulus system. Accommodation was monitored continuously by an infrared recording optometer, and stimulus presentation and data collection were under computer control.

There were 3 chromatic conditions, including chromatic aberration *normal* and *reversed*, and a *monochromatic* condition. In the first experiment the target distance was fixed at 0, 2.5 or 5 diopters of stimulus to accommodation during trials that lasted 40.96 seconds. The 3 chromatic conditions and 3 stimulus distances were presented in random order to 8 subjects. In the second experiment the target oscillated sinusoidally toward and away from the eye at 0.2 Hz between 1.5 and 2.5 diopters of stimulus. Three trials were run for each condition on the same 8 subjects. In addition, the resting position of accommodation was monitored for 40 seconds with a pinhole imaged in the subject's pupil, while subject viewed the square-wave grating target. The subject's left eye was aligned with the apparatus by viewing an infrared image of the subject's pupil (and Purkinje Image I) on a video-display. Final adjustment of the lateral position of the subject's eye in the apparatus was made by the subject, while viewing the square-wave grating target with red (632 nm dominant wavelength) and blue (486 nm dominant wavelength) wratten filters superimposed over the top and bottom halves of the grating field. The subject aligned the red and blue gratings by adjusting the horizontal position of the bite-plate assembly. This vernier alignment method positioned the visual achromatic axis of the eye close to the optic axis of the Badal stimulus system throughout each trial.

In the third experiment the target *simulated* the effects of defocus and LCA on the contrast (modulation = L_{max}-L_{min}/L_{max}+L_{min}) of red, green and blue components of a static sine-wave grating. The grating simulated the effects of static defocus and LCA on the contrast of the "red", "green" and "blue" components of the "white" grating for a 3 mm diameter pupil, using the method of Smith (1982). A similar method has been described previously for simulations of sinusoidally moving gratings (Kruger et al., 1995). In the present experiment twelve subjects viewed a stationary sine-wave grating (3.88 c/deg) with a pinhole (0.75 mm in diameter) imaged in the plane of the subject's natural pupil. The sine-wave grating was imaged on a diffuse screen at high luminance (max = 240 candelas per meter[2]) by a video-projector (Sharp XGE-800U), and the grating was imaged in the Badal stimulus system. At 3.88 c/d, contrast (modulation) of the simulated image is transferred through the pinhole to the retinal image without degradation by diffraction, defocus or aberrations. Retinal illumination was 50 trolands. There were three simulated defocus conditions: 1 D of myopic focus or over-accommodation; 1 D of hyperopic focus or under-accommodation; and a control condition that simulated the effect of defocus without chromatic aberration (the three chromatic components of the grating had the same contrasts in the control condition). The subjects aligned themselves on the visual achromatic axis of their eye before each trial while viewing a pair of vertical red-green vernier lines. The eye was maintained in position during each trial by the investigator, while viewing the eye and Purkinje Image I on an infrared display. Trials lasted 20.48 seconds. The subject viewed a low contrast fixation target for the first 10.24 seconds of the trial, and the fixation target changed to one of 3 sine-wave grating simulations for the remaining 10.24 seconds of the trial. Six trials were run for each of the three conditions in counterbalanced order, and accommodation data from the six trials for each condition were averaged for each subject.

Results

Figure 2 (left) shows data from three subjects in the normal condition, for 40-second trials at 0, 2.5 and 5 diopters. Response traces are shown for more than one target distance for each subject, although recordings were made one trial at a time. The first subject accommodates relatively accurately at all 3 dioptric stimulus distances, and this subject performed this way under all 3 chromatic conditions. This was the only subject that could maintain focus for the full 40 seconds in both the monochromatic and reversed conditions. The other 2 subjects in Figure 2 also accommodated accurately in the normal condition at the 3 stimulus levels, but subject 3 shows large erratic oscillations of accommodation when the target is at the near (5 D) distance. Subjects 2 and 3 had much more difficulty maintaining focus in the monochromatic and reversed conditions (bottom and right of Fig. 2). In monochromatic light subject 2 over-accommodates at all distances, the data show large oscillations of accommodation with the target at the near distance (5 D), and the oscillations are absent at the intermediate and far distances. Subject 3 could not maintain focus at the near distance for more than a few seconds in monochromatic light. If a single measure of refractive state had been taken at the beginning of the trial one might conclude that the subject accommodates accurately in monochromatic light. When chromatic aberration was reversed, both subjects 2 and 3 had difficulty maintaining focus. The data show that subject 2 could maintain focus for 15 or 20 seconds, albeit with large unstable fluctuations in focus, and that subject 3 could not maintain focus at all with LCA reversed.

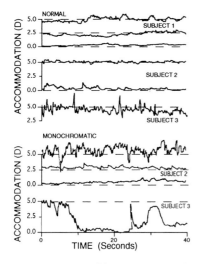

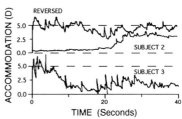

Figure 2. Data from three subjects for 40-second trials in which the target was stationary at 0, 2.5 and 5 D in white light (normal and reversed) and in monochromatic light.

In summary, most subjects accommodated accurately in the normal condition, some had difficulty in monochromatic light, and the majority were unable to accommodate effectively in the reversed condition. The data from each 40-second trial were sampled at 100 Hz and averaged, and in Figure 3 the mean accommodative responses from each subject (8) are presented as standard stimulus-response plots.

Accurate accommodation at all distances would give a straight line with a slope of 45 degrees. The first subject shows this type of pattern -- accommodation was relatively accurate at all distances for all 3 stimulus conditions. The intermediate resting position of accommodation is indicated by the horizontal line, which for this subject is close to the intermediate stimulus distance of 2.5 D. Subjects accommodated most accurately when the target was close to the resting position of accommodation. The first subject over-accommodated in the reversed condition (dotted line) when the target was closer or further away than the resting position of accommodation. Most subjects had more difficulty at the near and far stimulus distances. As an example, subject 3 over-accommodates in the normal condition at near and far. Accommodation is poor in monochromatic light (dashed line) at near, and in the reversed condition accommodation is poor at both near and far distances. Subject 4 shows similar difficulty in monochromatic light and with chromatic aberration reversed at all 3 stimulus distances. The stimulus-response functions for the remaining 4 subjects suggest relatively accurate accommodation in both the normal and monochromatic conditions. With chromatic aberration reversed, accommodation was less accurate at the near or far stimulus distances. To summarize, 7 of 8 subjects had difficulty maintaining focus in monochromatic light or when chromatic aberration was reversed, especially when the target was considerably closer or further away than the subject's tonic resting position.

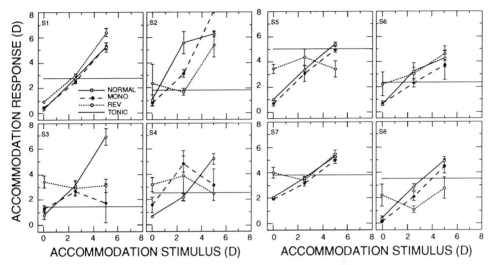

Figure 3. Stimulus-response plots for accommodation (8 subjects) under 3 conditions (normal, monochromatic and reversed).

The data in Figure 4 are from one subject for one sinusoidal trial of each chromatic condition. The amplitude of the sinusoidal response was reduced in monochromatic light compared to white light (normal); and when chromatic aberration was reversed the amplitude and form of the tracking response was poor. The data from each 40-second trial were Fourier analyzed (FFT) to estimate gain and phase-lag of the response, and the data from 3 trials for each subject were vector-averaged.

208

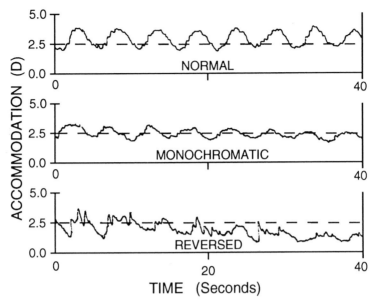

Figure 4. Data from one subject for three 40-second sinusoidal trials in each experimental condition (normal, monochromatic and reversed).

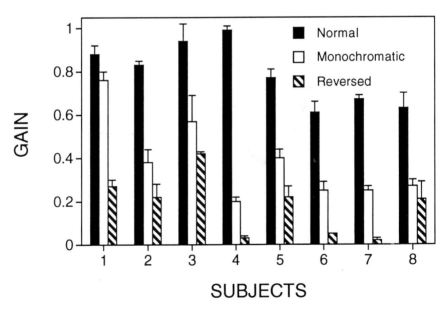

Figure 5. Dynamic gains for 8 subjects with longitudinal chromatic aberration normal, reversed and eliminated by illuminating the target with monochromatic light.

209

Figure 5 shows dynamic gains for 8 subjects under three conditions. For the group of subjects gain is significantly smaller for the monochromatic and reversed conditions than for the normal condition, and the monochromatic and reversed conditions also are significantly different from each other (paired t-tests: p≤0.001).

The data on the left side of Figure 6 are averaged data from a subject who responded well to the simulations of under- and over-accommodation. When the target simulated under-accommodation, there was a long latency (approximately twice as long as usual) before accommodation responded with a positive step change in accommodation, followed by a slow ramp change in accommodation at a rate of one tenth of a diopter per second. On the other hand, when the target simulated over-accommodation, focus moved to the far point of the eye (optical infinity) following a long latency, and accommodation remained at the far point for the remainder of the trial. In the control condition accommodation remained approximately the same throughout the trial. The simulation was viewed through a pinhole pupil (open-loop) so that the subjects' accommodation was close to the subject's resting position of accommodation (approximately 0.5 D) during the first half of the trial and in the control condition. Ten of 12 subjects accommodated in a similar manner to the open-loop simulations of under- and over-accommodation. The data on the right of Figure 6 are from a subject that responded poorly to the simulation. This subject has a higher resting level of accommodation (approximately 1.5 D) than the other subject. Two of 12 subjects responded in this manner.

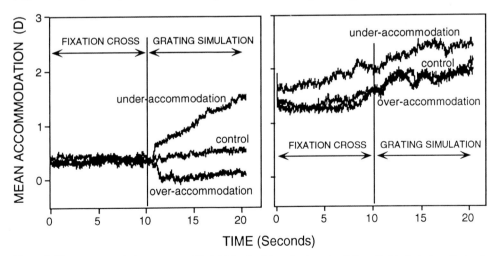

Figure 6. Mean accommodation responses (6 trials) for 2 subjects for two simulations that include the effects of defocus blur and LCA (1 D under-accommodation and 1 D over-accommodation), and a control simulation that includes the effects of defocus blur (1 D) without the effects of LCA. During the first 10.24 seconds the target was a fixation cross, and during the second 10.24 seconds the target was a grating simulation.

The simulation data were analyzed by calculating the mean accommodative response during the first 10.24 seconds of the trial, while the fixation target was present, and the mean response during the second 10.24 seconds, while the grating simulation was present. The change in mean response from the fixation target to the grating simulation was used as the main parameter in the subsequent analysis. Comparisons were made for each subject using t-tests, between the myopic and hyperopic simulations, and between the control condition and each of the two defocus simulations. Six subjects showed a significant difference between the myopic and control conditions; seven showed a significant difference between the hyperopic and control conditions; and 10 of 12 subjects showed a significant difference between the myopic and hyperopic simulations (significance levels ranged between $p < 0.05$ and $p < 0.01$).

Discussion

The method of oscillating targets along the z-axis of the eye provides a sensitive method for examining the sensory aspects of accommodation with minimal intrusion from voluntary accommodation. In the first experiment subject 1 probably used voluntary accommodation to maintain focus during the stationary trials. His voluntary accommodation was examined by asking him to view the grating through a pinhole pupil while focusing for near, far, and intermediate distances, and by accommodating rapidly from near to far and back every 2 seconds. The subject had excellent voluntary accommodation and no previous accommodation training. The dynamic stimulus clearly revealed this subject's accommodative sensitivity to chromatic aberration, while the static stimulus did not. Some subjects can use voluntary accommodation to maintain focus for short periods of time, and there are subjects with remarkably good voluntary control. In the experiment of Bobier, Campbell and Hinch (1992) stationary targets were used together with a subjective method (stigmatoscopy) to measure accommodation. The subjective method is not likely to detect the large oscillations of accommodation, sudden loss of focus, or intermittent and innacurate focus that is evident in continuous recordings of accommodation (Figure 1). Kotulak, Morse and Billock (1995) monitored accommodation continuously, but they used a stationary target positioned at an intermediate distance (1 D) to stimulate accommodation, and found no effect of eliminating LCA. The present results (Figure 3) show that the target should be considerably closer or further than the intermediate resting position of accommodation in order to demonstrate the influence of LCA.

We conclude that the visual-accommodative system monitors the contrast of spectral components of the retinal image, whether the target is stationary or moving. In these experiments we have concentrated on the chromatic stimulus for accommodation from LCA in broadband white light. But it is important to recognize that accommodation continues to operate to some extent in narrowband monochromatic light. Moreover, the response continues even in the absence of blur feedback and this raises the possibility that monochromatic aberrations, the angle of incidence of light at the retina, and perhaps a decentered Stiles-Crawford function, play a role in accommodative control (Kruger, Mathews, Katz, Aggarwala and Nowbotsing; 1997). The long latency and slow time course of the response to the present simulations of dioptric blur are in line with data from previous simulations of polychromatic blur that show larger temporal phase-lags to sinusoidal simulations than to real sinusoidally moving targets (Kruger et al., 1995). This supports the notion that the **contrast** of spectral components of the retinal image may not be the only aspect of dioptric blur that indicates focus and drives accommodation (Kruger et al., 1997).

Besides reflex accommodation, the present results may have implications for the process of emmetropization. There is consensus that an in-focus retinal image is an important part of the process (Rabin, Van Sluyters and Malach; 1981) but the optical stimuli that control emmetropization remain obscure (Wallman, 1995). Recently Bartmann and Schaeffel (1992) suggested that luminance contrast could provide the stimulus for emmetropization without the need for information from chromatic aberration, in line with the standard view that luminance contrast is the stimulus for accommodative control. Considering the present results, it would be a mistake to discount the effects of ocular aberrations as stimuli for emmetropization.

Finally, the effects of chromatic aberration have not been incorporated into theories of early visual processing and edge detection. As an example, simulated defocus blur at textured occluding edges provides directional information that contributes significantly to the perception of depth, and such perceptual sensitivity varies widely among subjects (Marshall, Burbeck, Ariely, Rolland and Martin; 1996). It is possible that dioptric information helps distinguish between occluding edges, edges from shadows, and edges that are due to changes in the material (color) of surfaces, and that dioptric information supplements the pictorial information that specifies the layout and substance of the environment.

212

Acknowledgements
The research was supported by the National Eye Institute of NIH (RO1 EYO5901; F32 EYO6403; T35 EYO7079), the Schnurmacher Institute for Vision Research (95-96-069; 96-97-076) and the Research Foundation of the State University of New York.

References

Bobier W.R., Campbell M.C.W. and Hinch M. (1992) The influence of chromatic aberration on the static accommodative response. *Vision Research* **32**, 823–832.

Kotulak J.C., Morse S.E. and Billock V.A. (1995) Red-green opponent channel mediation of control of human ocular accommodation. *Journal of Physiology (London)* **482**, 697-703.

Kruger P.B., Mathews S.M., Aggarwala K.R. and Sanchez N. (1993) Chromatic aberration and ocular focus: Fincham revisited. *Vision.Research* **33**, 1397–1411.

Kruger P.B., Mathews S. M., Aggarwala K.R., Yager D. and Kruger E.S. (1995) Accommodation responds to changing contrast of long, middle and short spectral-waveband components of the retinal image. *Vision Research* **35**, 2415-2429.

Kruger P.B., Mathews S. M., Katz M., Aggarwala K.R. and Nowbotsing S. (1997) Accommodation without feedback suggests directional signals specify ocular focus. *Vision Research, accepted.*

Marshall J.A., Burbeck C.A., Ariely D., Rolland J.P. and Martin K.E. (1996) Occlusion edge blur: a cue to relative visual depth *Journal of the Optical Society of America A* **13**, 681-8.

Rabin J., Van Sluyters R.C. and Malach R. (1981) Emmetropization: a vision-dependent phenomenon. *Investigative Ophthalmology and Visual Science* **20**, 561-564.

Smith G. (1982) Ocular focus, spurious resolution and contrast reversal. *Ophthalmic and Physiological Optics*, **2**, 5-23.Wallman, J. (1993) Retinal control of eye growth and refraction. *Progress in Retinal Research*, **12**, 133-53.

Error signal detection and dynamics of accommodation

T. Yamada, K. Ukai and S. Ishikawa

Department of Ophthalmology, School of Medicine, Kitasato University
1-15-1 Kitasato, Sagamihara, as the error signal in the negative feedback system of accommodation control.
However, it is difficult to estimate the amount of defocus from the contrast reduction of a retinal image, because
both *Kanagawa 228, Japan*

Summary: Usually, defocus is considered defocus and the spatial structure of the target determine the contrast. To investigate whether the accommodation control system can perceive the amount of error signal (defocus) from the contrast reduction of a retinal image, step responses for various amplitudes of stimuli were measured. The data obtained were analyzed using an exponential curve fitting method in which not only the time constant, but also the final value of the curve, could be estimated from a portion of the exponential curve. The near-to-far (or intermediate) step responses had a common path and showed that the curve initially went towards a certain position around the dark focus of the subject and was suddenly clipped around the correct focus position. The time constant and the destination of the curve were independent from the stimulus step amplitude. These results diverged from the step responses of fusional vergence. The destination of the exponential curves in vergence coincided with the target position for any combination of stimulus. This fact suggests that the accommodative control system does not refer to the amount of error signal when error is large and that the fusional vergence control system does refer to the amount of error signal.

Introduction

The dynamics of human ocular accommodation, such as latency time, reaction time, maximum velocity and transfer function have been reported (Tucker and Charman, 1979; Wildt *et al.*, 1974). Models of the control system, including closed-loop feedback control, have also been reported (Stark *et al.*, 1965; Krishnan and Stark, 1975). The characteristic of the accomodative control system that most sets it apart from other motor control systems of the human body is the lack of an odd error signal (Stark and Takahashi, 1965). This means that the system cannot detect whether the accommodative state is myopic or hyperopic to the stimulus based on the retinal image alone, and that the defocused retinal image does not contain the information for the direction of defocus. When the defocus is detected, the initial direction of the accommodative response should be given by cues other than the blur (e.g., Kotulak and Schor, 1986). Following the determination of the initial direction of correcting defocus, almost all authors (Stark *et al.*, 1965; Krishnan and Stark, 1975; Kotulak and Schor, 1986; Sun and Stark, 1990; Hung *et al.*, 1986; Hung and Ciuffreda, 1988; Schor *et al.*, 1992) have described models in which the human ocular accommodation system uses the amount of defocus as the error signal and operates to minimize this error.

Retinal image is affected not only by defocus, but also by the spatial structure of the target pattern. In other words, the amount of defocus cannot be determined by the retinal image alone. These calculations were carried out by Charman and Tucker (1977, 1978). In order to estimate the amount of defocus from the degraded retinal image, the inverse calculation, which is very complex, should be performed. The question arises as to whether or not the accommodative control system can perceive the amount of error signal claimed by nearly all the previously mentioned authors. The present work is an attempt answer this question by measuring dynamic responses for various step stimulus amplitudes and to compare the accommodation step responses with fusional vergence step responses.

Method

Accommodation

In order to present the target and to measure accommodation, a modified optometer was used. A personal computer was used for controlling target motion and processing the data obtained. The target has an asterisk-like shape and was viewed through a Badal optical system. Details of the apparatus and target condition have been reported elsewhere (Ukai *et al.*, 1983, 1986; Ukai and Ishikawa, 1989; Tsuchiya *et al.*, 1989; Woung *et al.*, 1993). The measurements were taken by shifting the target position in a stepwise manner between any two points including 0, 2, 4, 6 and 8 diopters above the accommodative far point, of each subject. Ten different stimuli of various amplitudes were performed. At least 10 step responses for each stimulus were measured using a repetitive stimulus between far and near positions. The target remained for at least 5 s at each position. The output of the optometer was sent to the personal computer through an A/D converter. The traces were analyzed using an exponential curve fitting method in which not only the time constant, but also the initially determined final value of the curve could be estimated from a part (about the first half) of the response. The values of the three coefficients were estimated using the data obtained from measurements.

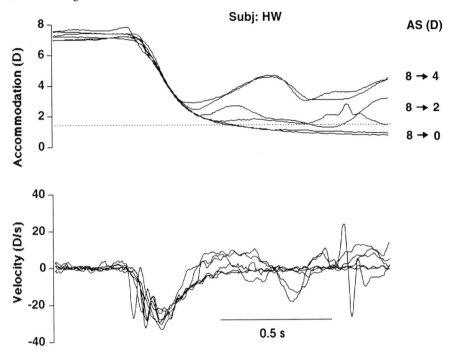

Figure 1 The upper figure shows an example of the accommodation step responses in subject HW, when the target moved from 8 D to 4, 2, or 0 D. Two traces for each stimulus were selected and superimposed. Dashed line indicates dark focus level of the subject. The lower figure shows velocity curves.(from Yamada and Ukai,1996)

Three healthy males were used as the subjects. Two out of three subjects were myopes. They wore no corrective lenses during experiments but the stimulus positions were adjusted to compensate for their refractive errors. They were instructed to see the target clearly. Prior to the examination, the dark focus of accommodation was measured using the same optometer. Their age, refraction and dark focus are listed in Table 1, together with the exact stimulus condition.

Vergence

In order to compare accommodative step response with vergence step response, a similar analysis of fusional vergence was performed. Step responses of fusional convergence and divergence were evoked by a computer-controlled phase-different haploscopic device. Stimulation was performed using a Liquid Crystal shutter and a personal computer. Two blurred bars were displayed alternatively on the TV monitor. When one bar was present, the other bar was turned off. Each bar was turned on and off every 16.7 ms. The opening of two Liquid Crystal shutters was synchronized with the appearance of the two bars. One eye could see only one bar. The subject was forced to fuse these two images of bars. Eye movements of convergence and divergence were recorded by a limbus tracking method. Disparities were selected from far, intermediate, and near conditions and stimulus was changed in a stepwise manner between any two points including 0, 2, 4, and 6 meter angles. Six different stimuli of various amplitudes were performed. The output of the limbus tracker was sent to the personal computer through an A/D converter, and the vergence component was obtained by subtracting the right-eye and left-eye position signals. Evoked vergence responses were analyzed using the same method as for accommodation.

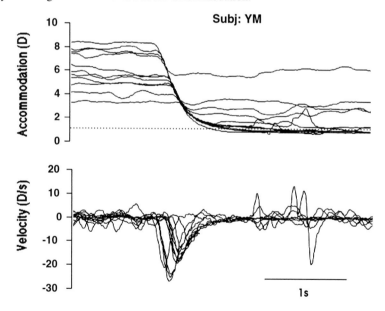

Figure 2 The upper figure shows another example of the accommodation step responses in subject YM, when the target moved from near to far (including intermediate). Near and far positions of the target are selected adequately. Horizontal positions of the traces are shifted so as to have the maximal common path. The dashed line indicates dark focus level of the subject. The lower figure shows velocity curves.(from Yamada and Ukai,1996)

Two healthy males and one healthy female were used as subjects. Their ages and phoria angles in prism diopters are listed in Table 2, together with the exact stimulus condition.

Table 1: Subjects, stimulus conditions and results

Subject / Age	stimulus		Exponential fitting		TC (ms) in
Refraction / Dark Focus			Destination (D)	Time Const (ms)	Reverse Movement
	from 8.00	to 4.00	1.65 ± 0.73	129 ± 19	368 ± 145
HW		2.00	0.87 ± 0.62	141 ± 34	532 ± 124
20 year-old		0.00	0.89 ± 0.46	170 ± 28	614 ± 202
Emmetropia	6.00	2.00	0.63 ± 0.38	133 ± 25	433 ± 129
DF: 1.44 D		0.00	0.57 ± 0.30	168 ± 45	702 ± 160
	4.00	0.00	*	211 ± 52	968 ± 602
	from 8.75	to 4.75	1.94 ± 0.17	79 ± 19	344 ± 99
TK		2.75	1.78 ± 0.16	121 ± 11	441 ± 129
19 year-old		0.75	1.53 ± 0.16	152 ± 8	1137 ± 533
Myopia (-3.25 D)	6.75	2.75	1.45 ± 0.09	119 ± 23	490 ± 99
DF: 1.11 D		0.75	1.25 ± 0.10	175 ± 45	822 ± 111
	4.75	0.75	1.17 ± 0.05	172 ± 12	489 ± 168
	from 8.50	to 4.50	1.19 ± 0.09	113 ± 4	475 ± 116
YM		2.50	1.33 ± 0.10	160 ± 17	730 ± 41
21 year-old		0.50	0.70 ± 0.13	156 ± 19	356 ± 54
Myopia (-1.50 D)	6.50	2.50	0.74 ± 0.09	146 ± 12	492 ± 148
DF: 1.20 D		0.50	0.99 ± 0.09	170 ± 13	498 ± 65
	4.50	0.50	0.98 ± 0.11	178 ± 23	308 ± 62

* could not be calculated due to irregular waveforms

Results

Accommodation

Figure 1 shows examples of the step responses in subject HW, when the target moved from 8 D to 4, 2, or 0 D. Two traces for each stimulus were selected and superimposed. Figure 2 shows examples of step responses in subject YM, when the target moved from near to far (including intermediate). Near and far positions of the target are selected adequately. Horizontal positions of the traces are shifted to have so as to have the maximal common path.

These two Figures show that the results of near-to-far (and intermediate) step responses are on the same path and independent from the initial and final target positions. The common path shows that the curve initially starts towards a certain position. Each curve was suddenly clipped around the correct focus position. Velocity curves are composed from 3 components especially in Figure 1, i.e., the increasing velocity phase (downward in Figures 1 and 2), then the gradually decreasing velocity phase, and the rapidly decreasing velocity phase. The first phase shows a limitation of muscle acceleration. The second phase shows the common path part. Curves in this phase are also exponential due to the characteristics of exponential curve. The third phase shows the clipping process.

When the target moved from far to near, the common path could not be found. The initial responses were often too small to reach stimulus position, sometimes too large, and rarely correct. When the initial response was too small, multiple-step waveforms were observed. When the initial response was too large, the response was clipped suddenly at approximatly the correct focus position. This sudden clipping is similar to the near-to-intermediate step response. Thus the amplitude in the initial fast component in accommodative step response seems to be randomly determined in far-to-near responses.

Table 1(from Yamada and Ukai,1996) shows the results calculated by the curve fitting method. Each value was calculated from 10 measurements. Four stimuli with an amplitude of 2 D were excluded from the analysis because an exponential curve.

If a double-step response was observed in the far-to-near response, the calculation portion was selected from the first step response. If a triple- or more multiple-step response was observed, the response was omitted from the calculation. The averages of the time constant for subjects HW, TK and YM are 159, 136 and 154 ms in the near-to-far accommodative response and 603, 621 and 477 ms in the far-to-near accommodative response, respectively. The ground averages of the destination of the curve in the near-to-far step responses are 0.92, 1.52 and 0.99 D in subjects HW, TK and YM, respectively. Destinations of the curve in the far-to-near accommodation were not calculated because they seemed to be determined randomly and had a large variation.

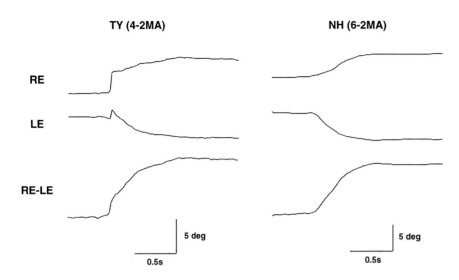

Figure 3 An example of the divergence step responses in two different subjects. *Left:* The response in subject TY when the target moved from 4 to 2 meter angles. *Right:* The response in subject NH when the target moved from 6 to 2 meter angles. RE: Right eye position. LE: Left eye position. RE-LE: Vergence component.

Table 2: Subjects, stimulus conditions and results

Subject age / phoria	stimulus		Divergence		Convergence	
			Destination (MA)	Time Const (ms)	Destination (MA)	Time Const (ms)
	from 6.00	to 4.00	3.23 ± 0.08	193 ± 36	6.60 ± 0.12	249 ± 92
TY		2.00	1.63 ± 0.21	188 ± 54	6.27 ± 0.80	219 ± 81
33 year-old		0.00	0.32 ± 0.41	302 ± 85	6.28 ± 0.84	261 ± 81
orthophoria	4.00	2.00	1.64 ± 0.48	149 ± 27	4.59 ± 0.35	230 ± 80
		0.00	0.31 ± 0.32	233 ± 68	4.49 ± 0.88	266 ± 116
	2.00	0.00	-0.13 ± 0.42	341 ± 101	2.53 ± 0.46	259 ± 80
	from 6.00	to 4.00	3.45 ± 0.17	275 ± 62	6.65 ± 0.89	253 ± 108
NH		2.00	1.68 ± 0.57	281 ± 109	6.55 ± 0.96	236 ± 128
26 year-old		0.00	0.21 ± 0.35	317 ± 63	6.51 ± 0.99	210 ± 64
-4 ² exophoria	4.00	2.00	1.62 ± 0.30	231 ± 28	4.89 ± 0.80	232 ± 107
		0.00	0.32 ± 0.68	300 ± 129	4.90 ± 0.83	297 ± 83
	2.00	0.00	-0.73 ± 0.28	199 ± 23	2.33 ± 0.88	248 ± 116
	from 6.00	to 4.00	3.05 ± 0.19	257 ± 14	6.25 ± 0.47	227 ± 113
JM		2.00	1.97 ± 0.28	283 ± 67	6.34 ± 0.94	259 ± 141
30 year-old		0.00	-0.22 ± 0.35	272 ± 79	6.20 ± 0.89	210 ± 56
-2 ² exophoria	4.00	2.00	1.10 ± 0.49	295 ± 62	4.58 ± 0.69	235 ± 128
		0.00	-0.13 ± 0.29	261 ± 43	4.65 ± 0.92	203 ± 65
	2.00	0.00	0.17 ± 0.33	276 ± 23	2.35 ± 0.67	251 ± 92

Vergence

Figure 3 shows examples of the divergence step responses in subjects TY and NH. The step response curve does not clip. The absence of clipping differs from accommodation step responses. Table 2 shows the results calculated by the curve fitting method. Each value was calculated from 10 measurements. If double-step response was observed in the far-to-near response, the calculation portion was selected from the first step response. The averages of the time constant for subjects TY, NH and JM are 234, 267 and 274 ms in divergence and 247, 246 and 231 ms in convergence, respectively. A good match to the exponential curve was obtained in the traces of vergence responses, and the destination of the exponential curves coincided with the target position for any combination of the stimulus with a stable value for the time constant.

Discussion

The time constant of the step response is a few times smaller in near-to-far responses than in far-to-near responses. The lens is believed to have a naturally accommodated shape. Conversely, the ciliary muscle is stretched in distant accommodation. The difference of the time constant between forcing accommodation and relaxing accommodation may reflect the elasticity of this organ. It may also reflect the difference in the control system of accommodation between forcing and relaxing. The destination of the common path does not seem to be a far point of accommodation, but a certain position around the dark focus of each subject. In the following part of this study we assumed that the destination is located at the dark focus although this hypothesis should be confirmed in further experiments involving many subjects. However, even if this assumption is not valid, the importance of the other findings is still maintained.

Step responses of accommodation have a common path and are independent from initial and final positions of the target during near-to-far responses until the curves are clipped at around the correct focus position. The time constant and destination of the curves are also independent from the initial and final position of the target. These results are schematically drawn in Figure 4. In the far-to-near stimulus, the situation is very similar, but the destinations are determined before the start of the response and seem to be random.

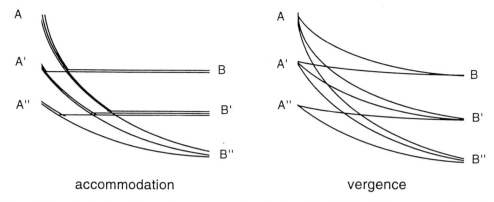

accommodation   vergence

Figure 4 Schematic drawing of traces of step responses from A, A' or A'' to B, B' or B'' in accommodation and vergence. Abscissa indicates time but the scales of both figure are not correct. Eight lines in each figure have the same time constant. If curves starting at A' and A'' in the left figure (accommodation) are shifted horizontally to the right, they have a common path with curves starting at A.(from Yamada and Ukai,1996)

This is in contrast to the case of the vergence control system. A good match to the exponential curve was obtained in the traces of vergence responses, and the destination of the exponential curves coincided with the target position for any combination of the stimulus with a stable value for the time constant. These results are also schematically drawn in Figure 4. This means that the amount of error signal was exactly detected in the fusional vergence control system.

In the present study, subjects could predict the movement of the target due to the repetitive stimuli. Even under such experimental conditions, subjects chose the way to accommodate toward a certain position which can be determined without any calculations rather that which corresponds to the target position. The amount of defocus was not used as the input of the accommodation control system. The discussion will be valid if the defocus is relatively large (here, the defocus is over 4 D).

In the latest model by Sun and Stark (1990), which has an open-loop component, the active part in tracking a target is provided by the closed-loop model. Using ramp stimuli, they reported that blur processing by the retina and cortex uses switching based threshold triggering. The ON state consists of a fairly high gain, closed-loop control system that minimizes error. The OFF state, which occurs when the error is small, provides for a quasi-open-loop state and a slow drift to a bias level, under the influence of a leaky integrator. The ON state of this model plays an active role in tracking a target.

Hung *et al.* (1986) and Hung and Ciuffreda (1988) recently developed a new dual mode model for vergence and for accommodation. They measured accommodative responses to ramp stimuli of various velocities (0.5-6.0 D/s). They found that the amplitude of the initial step movements increased with increasing ramp stimulus velocity and that the response approximately matched the instantaneous ramp stimulus amplitude in spite of the 360 ms delay in the feedback loop. It appears that the neural control process converted the error rate information into step amplitude information. This involves preprogramming of the necessary step control signals to generate an open-loop step movement. The dual-mode model incorporates an elaborate dual-mode controller determined by stimulus velocity: one mode generates smooth tracking of a slowly moving stimulus; the other, fast mode, produces a preprogrammed "catch-up step" in the ramp response, using both sampling and prediction mechanisms. In this model, the feedback loop is open during the fast component. The amplitude of the step movement is, however, preprogrammed before the start of the response and, for the accommodative model, the amount of defocus is used in preprogramming.

The present results suggest that the accommodation control system has two modes similar to the above models, but that the mechanism is different. Specifically, a fast movement for a catching-up of the target and a slow movement for maintaining the correct focus. The catching-up mechanism is neither an open-loop, nor a quantitative feedback system. The movement is determined before the response (similar to the model by Hung and Ciuffreda, 1988), but it is not based upon the amount of defocus. A certain position close to the far point or dark focus position, we assumed it to correspond to the dark focus position, is adopted as the initially determined destination in the near-to-far step responses and is randomly selected in the far-to-near-step responses. During the fast movement, the only thing that is monitored is whether or not the retinal image is blurred. Here the term blur is an ambiguous concept. Reduced contrast may correspond to it for the simplified target such as sinusoidal grating, and the broad gradient for the clear edge pattern. It is not clear wheter the detection of blur involves the higher activity of spatial vision or not. When the image is not considered to have potential blur, the fast movement is suddenly arrested, and the control is transferred to the slow, maintained holding mechanism. We can conclude that the control system of accommodation does not detect the amount of error signal when the error is large.

References

Charman, W. N. and Tucker, J. (1977). Dependence of accommodation response on the spatial frequency spectrum of the observed object. *Vision Research.* **17**, 129-139.

Charman, W. N. and Tucker, J. (1978). Accommodation as a function of object form. *American Journal of Optometry and Physiological Optics.* **55**, 84-92.

Hung, G. K., Semmlow, J. L. and Ciuffreda, K. J. (1986). A dual-mode dynamic model of the vergence eye movement system. *IEEE Transactions on Biomedical Engineering.* **33**, 1021-1028.

Hung, G. K. and Ciuffreda, K. J. (1988). Dual-mode behavior in the human accommodative system. *Ophthalmic and Physiological Optics.* **8**, 327-332.

Kotulak, J. C. and Schor, C. M. (1986). A computational model of the error detector of human visual accommodation. *Biological Cybernetics.* **54**, 189-194.

Krishnan, V. V. and Stark, L. (1975). Integral control in accommodation. *Computer Programs in Biomedicine.* **4**, 237-245.

Rashbass, C. and Westheimer, G. (1961). Disjunctive eye movements. *Journal of Physiology(London).* **159**, 339-360.

Schor, C. M., Alexander, J., Cormack, L. and Stevenson, S. (1992). A negative feedback control model of proximal convergence and accommodation. *Ophthalmic and Physiological Optics.* **12**, 307-318.

Stark, L. and Takahashi, Y. (1965). Absence of an odd-error signal mechanism in human accommodation. *IEEE Transactions on Biomedical Engineering.* **12**, 138-146.

Stark, L., Takahashi, Y. and Zames, G. (1965). Nonlinear servo analysis of human lens accommodation. *IEEE Transactions on Systems Science and Cybernetics.* **1**, 75-83.

Sun, F. and Stark, L. (1990). Switching control of accommodation: experimental and simulation responses to ramp inputs. *IEEE Transactions on Biomedical Engineering .* **37**, 73-79.

Tsuchiya, K., Ukai, K. and Ishikawa, S. (1989). A quasistatic study of pupil and accommodation aftereffect following near vision. *Ophthalmic and Physiological Optics.* **9**, 385-391.

Tucker, J. and Charman, W. N. (1979). Reaction and response times for accommodation. *American Journal of Optometry and Physiological Optics.* **56**, 490-503.

Ukai, K., Tanemoto, Y. and Ishikawa, S. (1983). Direct recording of accommodative response versus accommodative stimulus. In: *Advances in Diagnostic visual optics* (eds G. M. Breinin and I. M. Siegel), Springer-Verlag, Berlin, pp 61-68.

Ukai, K., Ishii, M. and Ishikawa, S. (1986). A quasistatic study of accommodation in amblyopia. *Ophthalmic and Physiological Optics.* **6**, 287-295.

Ukai, K. and Ishikawa, S. (1989). Accommodative fluctuations in Adie's syndrome. *Ophthalmic and Physiological Optics.* **9**, 76-78.

Wildt, G. J. v. d., Bouman, M. A. and Kraats, J. v. d. (1974). The effect of anticipation on the transfer function of the human lens system. *Optica Acta.* **21**, 843-860.

Woung, L.-C., Ukai, K., Tsuchiya, K. and Ishikawa, S. (1993). Accommodative adaptation and age of onset of myopia. *Ophthalmic and Physiological Optics.* **13**, 366-370.

Yamada, T. and Ukai, K.(1996). Amount of defocus is not used as an error signal in the control system of accommodation dynamics. *Ophthalmic and Physiological Optics.* **16**, in press.

Mode switching in control of accommodation

Lawrence W. Stark, Charles Neveu and V. V. Krishnan

University of California at Berkeley, School of Optometry 94720-2020

Summary

The block diagram describes the functioning of a four mode control system model for ACC that is switched among one or more of these operational modes. Blur controls certain switchings between modes dependent upon upper and lower thresholds for detection of blur. Blur is the "sufficient" neurological control signal; other clues to distance may supplement, but not supplant blur. Acceptance of a residual detectable blur error signal is characteristic of "hill climbing".

Control Mode I is the high gain closed loop operation of classical accommodation. Control Mode II is a slow drift. Three conditions must apply --- the forward loop must be opened, memory of the previous command signal must be forgotten, and there must be a preferred level of ACC, the "bias set level", toward which the system is pulled. The leaky integrator allows the system to forget the previous forward loop command. The bias set-level then determines the direction of drift; its slow velocity is set by the time constants of the leaky integrator.

Control Mode III, is termed "schematic control"; this includes such older terms as 'voluntary' ACC, instrumental myopia, and proximal ACC. An "hysteretic" effect, would always be seen, were it not for the action of schematic ACC. Control Mode IV, vergence-accommodation, depends upon a largely unmodifiable brain cross-link that plays an important role in normal vision.

The bioengineering control model may clarify aspects of the complex neurological ACC system; evidence from both experiments and simulations help to support the model.

Introduction

Early control studies of accommodation, ACC, were stimulated, and partly guided, by preceding studies of the pupil (Stark and Sherman, 1957) (Fig 1). Indeed, the late Dr. Fergus Campbell, to whose memory this paper is dedicated, and the senior author worked together for two summers at Yale University in the fifties, studying pupil noise (Stark, Campbell and Atwood, 1958).

Bioengineering and control theoretic approaches lead to certain points of view in studying a neurological control system. The concepts -- input and output functions, negative feedback, providing for error correction, the possibilities of oscillations with a feedback control system -- are all part of the thinking process of the experimenter. Arranging elements of a complex system into a severely constrained block diagram is another productive approach from control theory that has been applied to the ACC control system (Stark, Takahashi and Zames, 1965; Brodkey and Stark, 1967; Shirachi, Liu, Lee, Jang, Wong and Stark, 1978) (Fig 1).

Our aim is to present a control model that explains the manner in which the amount of blur, with its upper and lower thresholds, switch the control modes of ACC. Evidence from both experiments and simulations helps to support the model.

226

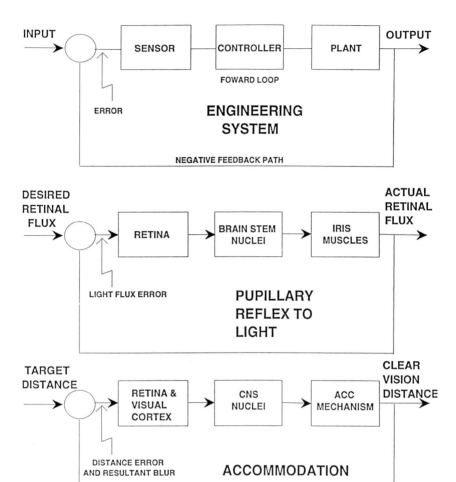

Fig1   Block Diagrams in Bioengineering

Classical control system, upper; early pupil control diagram, middle (Stark and Sherman, 1957); and early ACC control diagram, lower (Stark, Takahashi and Zames, 1965).

Blur Control of Switching Between Modes

Many phenomena can be captured in control models for accommodation (Stark, Takahashi and Zames, 1965; Brodkey and Stark, 1967; Shirachi, Liu, Lee, Jang, Wong and Stark, 1978); for any particular sets of behaviors it may be helpful to simplify the model as much as possible. A multiple mode ACC model has been developed over many years to capture the basic ideas of four-modes in the control system for ACC, as well as the nature of the blur and other functions that control the switchings between these modes (Krishnan and Stark, 1975; Sun and Stark, 1990; Neveu and Stark, 1995) (Fig 2).

Arguments against blur as the "sufficient" neurological control signal, from Fincham onward, have partially been put to rest by demonstration of (i) the even error nature of ACC responses in careful experiments and (ii) the fact that target blur and distance blur are equivalent, if control open loop, OL, and/or closed loop, CL, conditions are made equivalent (Phillips and Stark, 1977). Many believe that one or more of the multiple clues to target distance can substitute for blur, and it is likely that this can be accomplished via the schematic ACC pathway (Phillips, Shirachi and Stark, 1972) as, for example, when the ACC response predicts motion of the target. Optical aberrations with odd-error signal information can be documented in laboratory experiments; this may supplement but not supplant blur.

The even-error blur signal (Stark and Takahashi, 1965; Phillips and Stark, 1977) may be either perceived or not perceived by the sensory mechanisms (Fig 2). If the blur signal is too small, it falls below the lower threshold (Campbell, 1957; Kasai, Kondo, Sekiguchi and Fujii, 1971). "Hill climbing" is an important aspect of an even error signal in that the attempt is to minimize blur, not remove it completely; thus some minimum level of blur or error is always accepted. Again, if the blur signal is too great, it falls above the upper threshold. In both cases, with blur too small or too great, the system reverts to an "open loop" state. In the block diagram (Fig 2), this is represented by switch 1 (S1) being in the open state; note the heavy arrows leading from blur levels to the switches S1 and S2.

Control Mode I -- High Gain. In Mode I, blur error is passed through S1, in its closed state, to the high-gain controller, which then drives the ciliary muscle and other elements of the "plant" to produce CVD (Fig 2). The seven early responses (Fig 3) to step changes of distance show fast, quite adequate gain in the ACC response signal. This documents an interesting difference between the accommodative control system and the pupil control of retinal flux, in that the pupil responses are limited by the low gain of the pupil.

Control Mode II -- Drift. In Mode II, three conditions must apply --- the forward loop must be opened, memory of the previous command signal must be forgotten, and there must be a preferred level, the bias set level toward which the system may drift. The leaky integrator (second control block in Fig 2, before the summer) allows the system to forget the previous forward loop command. The bias set-level, here considered as an average value of noise at the final common path nucleus, then determines the direction of drift.

MODEL for SWITCHING CONTROL of ACCOMMODATION

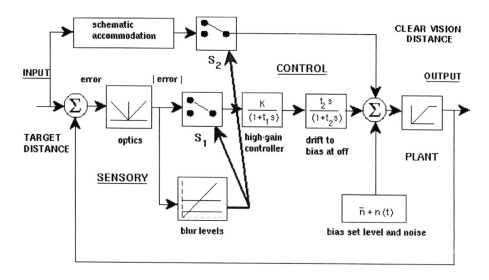

Fig 2   Multi-Mode Switching Control ACC Model
  Many aspects of ACC have been omitted so that the functions of the elements can be clearly identified (Modified from Krishnan and Stark, 1975; Sun and Stark, 1990; Neveu and Stark, 1995). Input is target distance (TD), and output is clear vision distance (CVD) measured in diopters, or inverse meters. Organization of the loop subtracts CVD from TD and yields distance error. Very importantly for ACC system, odd-error distance error is converted by optics of the eye into an even-error signal.

The speed at which the system drifts to the bias set-level is elegantly documented in the last response of Figure 3, following "light off", that serves to open the loop. Again, here the ACC system shows its difference from the pupil reflex to light, as well as its similarity to the vergence system, where drift to the phoria position is a well-known clinical and experimental phenomenon (Krishnan & Stark, 1977; Hung, Semmlow & Ciuffreda, 1986).

Three Ways to Open the Loop. Besides the maneuver of shutting off the light, also called night myopia (Figure 3), it is also possible to remove a signal completely under conditions of adequate illumination without any structured image information (fig 4A, 4B); this is called empty-field myopia (Whiteside, 1957). Also, when the target distance is clamped to the CVD by a carefully designed experimental apparatus (Phillips and Stark, 1977), allowing a perceived blur but setting the target to be slaved to the ACC output., the system will be in an open loop condition, and drift to the bias set-level (Figure 4A; 4B). Although the time constants for the drifts resulting from these three ways of opening the loop may vary for a single subject, and from subject to subject, they generally fall in the range of 5 to 15 seconds, and are very different from the fast, high gain dynamical responses seen in closed-loop ACC (Figs 3 and 4B).

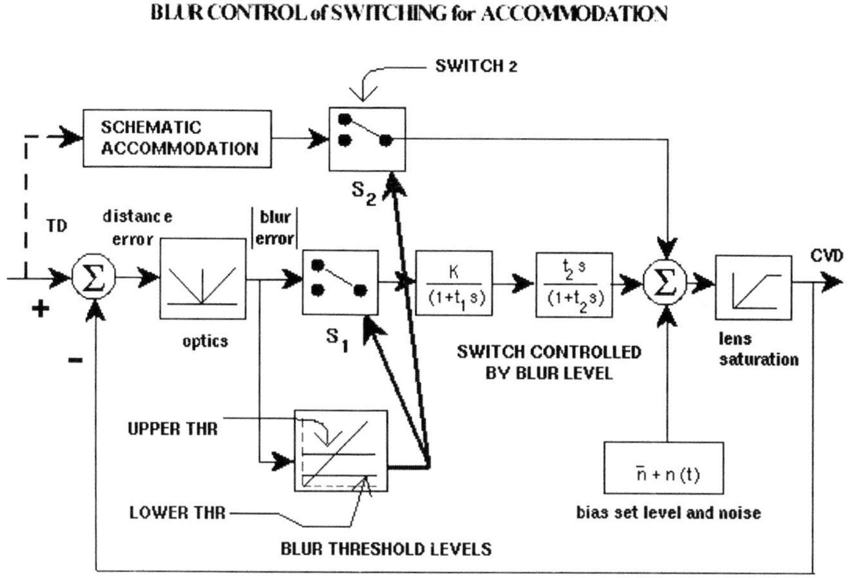

BLUR CONTROL of SWITCHING for ACCOMMODATION

Fig 3   Leaky Integrator in ACC
    Experimental ACC Mode I responses to step stimuli demonstrate rapid, high gain with excellent control properties. Next when "light off" removes target, the system goes into Mode II, with slow drift to bias set level (upper) (modified from Krishnan and Stark, 1975).
    Simulated ACC responses in both Mode I and Mode II operating conditions, lower (Neveu and Stark, 1995). Excellent fit of simulation to experimental data affords evidence in favor of ACC model in Fig 2; overshoots are a consequence of our oversimplified model.

Control Mode III -- Schematic Control. ACC, controlled by the brain, has access to many sorts of information and misinformation other than blur. Prediction of regularly spaced target motions leads to decreased ACC latency (Phillips, Shirachi and Stark, 1972), but as Mark Twain warned, "Never make predictions, especially about the future." The subject may know where the target is even in a dark room, if, for example, he holds it in his hand. He may also choose between two targets, both visible, that are at different distances, thus changing his ACC regard "at will." He may mistakenly feel that the target is physically located at the entry lens of an optical instrument, and thus produce a high degree of ACC for a target that is actually at optical infinity. In all of these cases, the subject is using a 'spatial schema' to represent the three-dimensional world, and to control his CVD within that schema. Since "free will" is a speculative philosophical notion, we have elected to use the term "schematic ACC" for older terms such as 'voluntary' ACC, instrumental myopia, and proximal ACC (Fig 2). These older forms of schematic accommodation have been long studied (e.g., Hung *et al.*, 1996) extra-ocular 'schematic'movements exist; an early model was applied by Young (1977) to smooth pursuit.

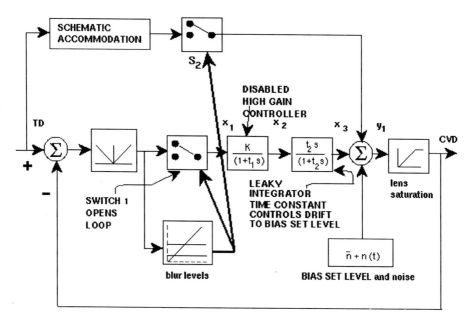

Fig 4   Mode II: Drifting to Bias Set Level

Experimental responses showing slow drifts in Mode II operation to bias set level for conditions:--- N, night or dark; EF, empty field; and OL, open loop (upper and lower). For comparison, see high gain fast, Mode I response, CL (lower) (modified from Phillips, S., Ph.D. Dissertation, Univ. of Calif. at Berkeley, 1974).

When target blur is greater than the upper threshold, schematic ACC may or may not come into play, depending on the subject's notion of target whereabouts (See control path in Fig 2, and its switch, S2, that enables it to control CVD). Experiments showing the action or inaction of schematic ACC can be performed using the push-down stimulus, rather that the common clinically-used push-up stimulus (Fig 5). In trial 1, the subject is unaware that the target is at 7 diopters, and remains at the bias set-level of approximately zero diopters, until the push-down stimulus has reached 3 diopters. At this point, the blur lies below the upper threshold, and control Mode I, high-gain ACC, can occur. On trial 2, the subject now predicts where the target is located, and schematic ACC is indeed used in this instance to bring the CVD as close as possible (see lens saturation block in Fig 2, representing the pre-presbyopic limit).

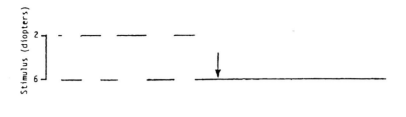

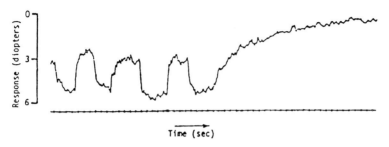

Time (sec)

Fig 5   Schematic ACC Experiments
   Two push-down (in diopters) trials follow one another and are displayed as time functions for both stimulus and response. In the first, the blur is too great and the ACC systems remains at the bias set level until the target approaches 3 diopters and the blur is less than the upper threshold. In the second, the subject now knows the target is likely to be very close, and uses schematic ACC to produce a maximum response (limited by his pre-presbyopic amplitude) (modified from Neveu and Stark, 1995) .

Hysteresis   These last experiments (Fig 5) can also be replotted as input-output relationships, with time as an implicit function only (Fig 6, see arrows). This shows an hysteretic-like effect in ACC, due to the upper blur threshold, analogous to magnetic hysteresis (Fig 7,left). An ACC hysteresis curve has been arranged similarly (Fig 7,right). The hysteretic effect in ACC would always be seen, were it not for the action of schematic ACC acting as in Trial 2. That is, the response to a 6D stimulus would depend upon the history, what path had been followed -- Path I (Figure 7,left) or the horizontal path along the abscissa (Trial 2) (Figure 7,left). It takes considerable experimental maneuvering to block schematic ACC, and yet document the hysteretic response of ACC (Neveu and Stark, 1995).

Control Mode IV: Vergence Accommodation(VG--ACC)   Demonstration of ACC-VG (Krishnan, Shirachi and Stark, 1977; Myers and Stark, 1990) is beyond the scope of the present paper; suffice it to say that this cross-link plays an important role in normal vision where in a target jump, the dispacity usually removes the target from the fovea and the triadic response depends upon blur-resistant disparity sensing mechanisms. The largely unmodifiable nature of the gain of the cross-link has been understood since the days of Donders and Hering.

232

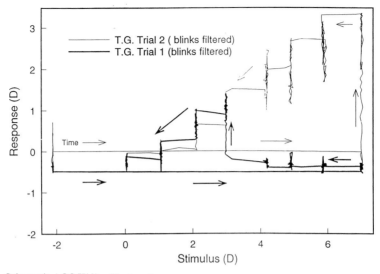

Fig 6 Schematic ACC Hiding Hysteresis
Same experiment as in Fig 5, but plotted as Response as a function of Stimulus. Note that same stimulus (7D) in trial 1 yields no response, while in trial 2, schematic ACC provides for a maximum response. Thus there are two different responses to the same stimulus depending upon the history of the system. (Modified from Neveu and Stark, 1995).

Discussion

Not Dark Focus Nor Tonic ACC. The bias set-level is referred to clinically as "tonic ACC." This refers to a myopia, such as night myopia or empty-field myopia, produced by the bias set-level, when the system's initial condition is at infinity or zero diopters. Thus, for a nominal bias set level of ID and an ACC ciliary muscle amplitude of 12D, 8% of the muscle force would be active.

The term 'tonic' expresses an implicit belief that the ACC mechanism, in particular, the ciliary muscle, has a tonic component. No such tonic muscular element has been found and distinguished from a phasic component, even though some muscles, as the extra-ocular muscles do indeed have such tonic and phasic components. What seems valid is that a neurological signal, the bias set level, is controlling ACC at this incorrectly named "tonic level". It would reduce some of the mystery and false attribution of aspects of ACC if the correct control (neurological) term were used.

The term "dark-focus" is doubly incorrect, since it does not depend uniquely on darkness, nor is it produced by an active focusing, but rather by an absence of focusing or of error control by the ACC system. A recent careful study, unfortunately using the term "dark focus," is by Jiang *et al.* (1996).

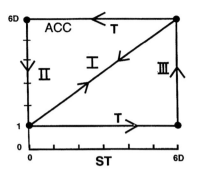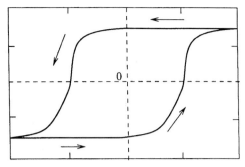

Fig 7   Classical and ACC Hysteresis

Magnetism with memory (*husteros*) of the history of the stimulus permits different responses at the same stimulus level; this is the basis of computer magnetic memories from tape to core to disc (right). ACC hysteresis plotted in similar fashion (left). Note that Mode II in return path demonstrates leaky memory and dynamic loss of hysteresis. Vertical Mode III path is a mechanism for hiding the upper threshold from clinical observers and enabling them to interpret Mode III, schematic ACC, incorrectly as CL Mode I ACC.

Bioengineering And Explicit Models   The different control modes of ACC have clarified some of the usual and unusual aspects of this complex neurological system. The terminology introduced and indeed the block diagram formulation enable one to define explicitly and talk clearly about these interesting aspects of ACC. The control approach also enables one to simulate the proposed model (Fig 3B) and show that these elements are sufficient to characterize and reproduce many behaviors seen under a variety of experimental conditions. Any system lying within the human brain is likely to have many complex phenomena not yet revealed or understood, and also may yield alternate explanations of behaviors, as those described above, that we think we now understand. In any case, the control approach clarifies the level of discourse, and may speed new, more expanded and less mistaken models in the future.

Acknowledgements

We thank Dr. Gerard Obrecht and Dr. Christian Miege, Scientific Monitors, Essilor Corporation, Paris and Professor Masaru Miyao, Adviser to the Tomey Corporation, Nagoya for support of our accommodation research. We also thank Professor John L. Semmlow for contentious but friendly discussion, and Professors Fuchuan Sun, Kenneth J. Ciuffreda and Derek Hendry, and others who have worked with Dr. Stark on accommodation over the years.

References

Brodkey, J. and Stark, L.W. (1967) Accommodative Convergence: An Adaptive Nonlinear Control System. *IEEE Transactions on Systems Science and Cybernetics* SSC-3: 121-133. {63}

Campbell, F.W. (1957) The Depth of Field of the Human Eye. *Optica Acta* 4: 157-164.

Hung, G.K., Semmlow, J.L. and Ciuffreda, K.J. (1986) A Dual-Mode Dynamic Model of the Vergence Eye-Movement System. *IEEE Transactions on Biomedical Engineering* BME-33.

Hung, G.K., Ciuffreda, K.J., and Rosenfield, M. (1996) Proximal contribution to a linear static model of accommodation and vergence. *Ophthalmic and Physiological Optics* 16(1):31-41.

Jiang, B.C. and Woessner, W.M. (1996) Dark focus and dark vergence: an experimental verification of the configuration of the dual-interactive feedback model.*Ophthalmic and Physiological Optics 16*(4):342-347.

Kasai, T., Kondo,K., Sekiguchi, M. and Fujii, K. (1971) Influence of the Depth of Focus on the Human Eye Accommodation. *Japanese Journal of Medical Electrical and Biological Engineering 9*(1): 28-36.

Krishnan, V.V., Shirachi, D. and Stark, L.W. (1977) Dynamic Measures of Vergence Accommodation. *American Journal of Optometry and Physiological Optics* 54: 470-473. {169}

Krishnan, V.V. and Stark, L.W. (1975) Integral Control in Accommodation. *Computer Programs in Biomedicine* 4: 237-245. {144}

Krishnan, V.V. and Stark, L.W. (1977) A Heuristic Model for the Human Vergence Eye Movement System. *IEEE Transactions in Biomedical Engineering* BME-24: 44-49. {161}

Myers, G.A. and Stark, L.W. (1990) Topology of the Near Response Triad. *Ophthalmic and Physiological Optics* 10: 175-181. {335}

Neveu, C. and Stark, L.W. (1995) Hysteresis in Accommodation. *Ophthalmic and Physiological Optics* 15:3: 207-216. {337}

Phillips, S. R. (1974) Ocular Neurological Control Systems: Accommodation and the Near Response Triad. Ph.D. Thesis, University of California,. U.S.A.

Phillips, S. R., Shirachi, D. and Stark, L.W. (1972) Analysis of Accommodative Response Times Using Histogram Information. *American Journal of Optometry* 49: 389-401. {119}

Phillips, S. R. and Stark, L.W. (1977) Blur: A Sufficient Accommodative Stimulus. *Documenta Opthalmologica* 43: 65-89. {166}

Shirachi, D., Liu, J., Lee, M., Jang, J., Wong, J. and Stark, L.W. (1978) Accommodation Dynamics: 1. Range Nonlinearity. *American Journal of Optometry and Physiological Optics* 55: 631-641. {182}

Stark, L.W. , Campbell, W. and Atwood, J. (1958) Pupil Unrest: An Example of Noise in a Biological Servomechanism. *Nature* 182: 857-858. {8}

Stark, L.W. and Sherman, P.M. (1957) A Servoanalytic Study of Consensual Pupil Reflex to Light. *Journal of Neurophysiology* 20: 17-26. {4}

Stark, L.W. and Takahashi, Y. (1965) Absence of an Odd-Error Signal Mechanism Human in Accommodation. *IEEE Transactions on Biomedical Engineering* BME-12: 138-146. {35}

Stark, L.W., Takahashi, Y. and Zames, G. (1965) Nonlinear Servoanalysis of Human Lens Accommodation. *IEEE Transactions on Systems Science and Cybernetics* SSC-1: 75-83. {44}

Sun, F. and Stark, L.W. (1990) Switching Control of Accommodation: Experimental and Simulation Responses to Ramp Inputs. *IEEE Transactions on Systems Science and Cybernetics* 37:1: 73-79. {334}

Troelstra, A., Zuber, B. and Miller, D. (1964) Accommodative Tracking: A-Trial-and-Error Function. *Vision Research* 4: 585-594, {33}

Whiteside, T.C.D. (1957) *The Problems of Vision in Flight at High Altitude*. Pergamon Press Ltd., London.

Young, L.R. (1977) Pursuit Eye Movement: What is Being Pursued? in Control of Gaze by Brain Stem Neurons. *In*: Baker and Berthoz (eds): *Developments in Neuroscience, Vol. I*. Elsevier-North Holland Press, Amsterdam, pp.29-36.

A Modified Control Model for Steady-State Accommodation

Bai-Chuan Jiang

University of Houston, College of Optometry, Houston, Texas 77204-6052, USA

Summary: Previous control models of steady-state accommodation have difficulty explaining the effect of the accommodative stimulus conditions on the accommodative response function. In this study, we suggest a model, in which an accommodative sensory gain (ASG) acting as a linear operator is added to simulate the sensory part of the accommodation system. Results derived from the model show that the sensory part, which may degrade the accommodative error (or blur) signal, not only affects the slope of the accommodative response function but also increases the system's effective threshold (ET) to the blur signal. An experiment was conducted to estimate the ET values for eight emmetropic subjects.

Introduction

Control system theory has been used to describe the steady-state accommodation for decades (Westheimer, 1963; Stark et al., 1965; Toates, 1972; Hung and Semmlow, 1980). Figure 1 is an example of the models (Hung and Semmlow, 1980). The input to the accommodative system, accommodative stimulus (AS), is the desired lens power in diopters, i.e. the value needed for optimal focus at the retina. The output, accommodative response (AR), is the actual value of accommodation of the lens, which may or may not equal the input. At the first summing point of the feed-forward loop, actual accommodation (i.e., AR) is subtracted from desired accommodation (i.e., AS), and this gives a dioptric, accommodative error (AE=AS-AR, called defocus or blur). The error signal has to exceed a threshold value in order to generate a signal from the central neural controller to drive the plant. This threshold is represented by a nonlinear operator in the model, called the dead zone (or deadspace, DSP). The effectiveness of the controller is represented by the accommodative controller's gain (ACG). In the model, there are also inputs from the vergence system (i.e., vergence accommodation, CA) and the tonic posture of accommodation (called accommodative bias, ABIAS). These sum with the signal from the controller at the second summing point to drive the plant and create an AR. Assuming the plant is a unity gain element with a saturation property, Hung and Semmlow (1980) attributed all of the signal degradation in the feed-forward loop to the ACG.

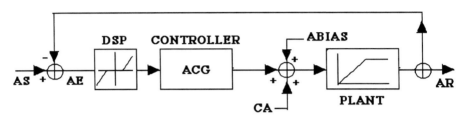

Figure 1. Block diagram of static model of the accommodative system. Adapted from Hung and Semmlow (1980).

This negative-feedback control model simulates the static accommodative stimulus-response relationship as described in following formulas:

$$AR = (AS - AR\ DSP) * ACG + ABIAS \qquad \text{for } |AS - AR| = |AE| \geq DSP \qquad (1.1)$$

$$AR = ABIAS \qquad \text{for } |AS - AR| = |AE| < DSP \qquad (1.2)$$

Here, the vergence loop is assumed to be open, i.e., CA=0. The DSP operator is an ideal mathematical function that approximates the threshold of the controller. If the AE is larger than DSP, minus (-) is used; if the AE is less than -DSP, plus (+) is used in equation (1.1). On the other hand, if the absolute value of the AE is less than the positive threshold value, then the output from the DSP operator is zero as represented in equation (1.2). In a diagram of the accommodative stimulus-response function, the point, at which the accommodative stimulus and response are equal, is located where the stimulus-response curve crosses over the unit ratio line (called the cross-over point). In previous studies (Ward and Charman, 1985; Rosenfield et al. 1993) researchers were not aware of the nonlinear property of the DSP and used equation (1.1) to determine the cross-over point. This has led a wrong prediction that the cross-over point is equal to ABIAS with a shift of DSP*ACG. Since the AE equals zero at the cross-over point, equation (1.2) has to be used, which predicts that the AR at the cross-over point is equal to ABIAS.

Experimental results reported by Ward and Charman (1985) supported this prediction. The results obtained from 5 subjects showed that the cross-over point values were very close to the ABIAS for each subject under the conditions with 5 different pupil diameters.

One of major parameters describing the relationship between the AR and the AS is the slope of the accommodative response function. In order to find how the slope relates to other parameters of the model, we rearrange equation (1.1) under the condition where:

$$AR = \frac{ACG}{1+ACG} * AS \pm \frac{ACG}{1+ACG} * DSP + \frac{1}{1+ACG} * ABIAS \qquad (2)$$

We let:

$$K = \frac{ACG}{1+ACG} \qquad (3)$$

here, K represents the slope of the accommodative response function.

Substituting K into (2), we have

$$AR = K * AS \ K * DSP + (1-K) * ABIAS \qquad (4)$$

Equation (3) indicates that any change of K has to be due to the ACG's change. But, this is not always true. For example, K becomes low when the pupil size becomes small (Ripps et al., 1962; Hennessy et al., 1976; Ward and Charman, 1985). Based on the current model, we have to assume that the pupil size influences the ACG. However, from geometrical optics we know that the pupil size changes the depth-of-focus of the eye, which many researchers equate with the DSP (Hung and Semmlow, 1980; Rosenfield et al., 1993). If this were the case, DSP's change due to pupil size change would not affect the K as anticipated by the current model (equation 3). But evidence shows that pupil size does cause a change of K.

The model also fails to account for other effects of the stimulus condition on the slope of the accommodative response function, such as the target's luminance (Johnson, 1976), blurring (Heath, 1956), spatial frequency and contrast (Owens, 1980; Ciuffreda and Hokoda, 1983; Ciuffreda and Rumpf, 1985). Based on the current model, we have to assume that these conditions affect the ACG but not the DSP. We are confronted with a similar problem discussed above, i.e., how we can relate the ACG to these stimulus conditions. The purpose of this study is to modify the static model of accommodation so it can account for the problems described above.

Modeling

As suggested in the previous literature, the effectiveness of a blur signal driving the controller of accommodation can be affected by pupil size, target luminance, blurring, spatial frequency and contrast. In contrast to the current model in which stimulus conditions are predicted to change the ACG, we suggest that effects of stimulus conditions take place before the AE signal reaches the DSP operator. As a first order approximation, we add an additional linear operator with gain (ASG, accommodative sensory gain) before the DSP in the feed-forward loop of the accommodation system to account for sensory responses induced by stimulus conditions. A diagram of the modified model is shown in figure 2. The ASG would decrease when the pupil becomes small or when luminance reduces. In general, we assume that the ASG represents the degradation of the blur (or error) signal caused by all sensory factors that occur before the signal reaches the DSP. Here, the DSP is assumed to represent the threshold of the motor controller for accommodation. Based on this modification, we present the following equations to describe the relationship between the steady-state accommodative response and the stimulus under vergence open-loop condition:

$$AR = [(AS - AR)*ASG \pm DSP] * ACG + ABIAS \quad for \left| AS - AR \right| = \left| AE \right| \geq DSP \tag{5}$$

or

$$AR = \frac{ASG * ACG}{1 + ACG * ACG} * AS \pm \frac{ASG * ACG}{1 + ASG * ACG} * \frac{DSP}{ASG} + \frac{1}{1 + ACG * ACG} * ABIAS \tag{6}$$

Equation (6) can be simplified if we let:

$$K' = \frac{ASG * ACG}{1 + ACG * ACG} \quad i.e., \quad ASG * ACG = \frac{K'}{1 - K'} \tag{7}$$

here, K' represents the slope of accommodative stimulus-response function in the modified model.

Substituting K' into (6), we have

$$AR = K' * AS \pm K' * \frac{DSP}{ASG} + (1 - K') * ABIAS \qquad (8)$$

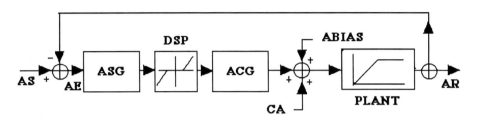

Figure 2. Block diagram of the modified control model of the accommodative system. A sensory operator with gain ASG is added in front of the DSP to simulate the function of the sensory part in the accommodative system.

Equation (8) is similar to equation (4). However, equation (8) indicates that the slope of the accommodative stimulus-response function (K') is affected by both the ASG and the ACG of the accommodative system. The ASG not only affects the slope, but also changes the threshold for the AE as shown in equation (5). As an operational definition, the term   is defined as the "effective threshold" (ET), which combines the sensory response of the system to varying stimulus conditions with the DSP (the motor controller threshold). Therefore, a lower ASG sets the ET high (with a constant DSP), and results in a larger AE needed to drive the motor controller. An alternative form of equation (8) is

$$\pm \frac{DSP}{ASG} = \pm ET = A - \frac{AR}{K'} + \frac{1-K}{K} * ABIAS \qquad (9)$$

This equation shows that the ET can be estimated from measurements of AS, AR, K', and ABIAS. The ET is an important parameter as it provides information on how the optical system of the eye and the sensory part of the accommodative system limits the error (or blur) signal which drives the motor controller. In order to verify our formula, we determined the ETs of eight emmetropic subjects from accommodative response function data and measures of dark-focus of each subject.

Methods

Eight volunteer emmetropes (average age, 25.0 years) participated. All subjects had at least 20/20 acuity, normal binocular vision and accommodative facility. Refractive errors of these subjects ranged from +0.50 to -0.50 D (average, -0.08 D). Subjects with an astigmatic refractive error greater than 0.75 D were excluded. After experimental procedures were described. each subject gave informed consent.

The accommodative stimulus/response function and the dark-focus (ABIAS) were objectively measured with a Canon R-1 infrared optometer. During each session, accommodative responses to 6 stimuli ranging from 0.25 to 6.00 D (presented in random order) were measured. For the accommodative function, the target was 4 white 20/100 E's with a luminance of 27 cd/m2 displayed on a dark computer screen in a dark room. The computer screen was physically moved to the positions representing the various dioptric demands and the physical size of the target was appropriately scaled. The subject's ABIAS was estimated as accommodative response after 10 minutes in the dark. At each demand, the accommodative response was measured at least 5 times and averaged.

Accommodative responses of the right eye were measured and all viewing was monocular with the right eye. Natural pupils were used throughout the study. Based on readings from the Canon optometer, accommodative responses were calculated as the sphere-equivalent power (i.e., sphere power + 1/2 cylinder power). Sessions were repeated 3 or 4 times for each subject. All data represent averages over the sessions.

Results

The average dark-focus (ABIAS) for the eight emmetropes was 0.71 ± 0.23 (S.E.) D. Average accommodative response function data are shown in figure 3. The slope of the accommodative response function is 0.96 ± 0.05 (S.E.) averaged across individual slopes. Each subject's slope of the accommodative response function was obtained by fitting a regression line to the response data measured at the four stimulus demands (1, 2, 4, and 6 D).

The effective threshold (ET) defined in equation (9) can be calculated from each individual's AS, AR, K', and ABIAS. Since we have four pairs of stimulus and response data corresponding to the 4 stimulus demands, we obtain four ETs named as ET(1), ET(2), ET(3), and ET(4), respectively, for each subject. The average ET(.)s across the subjects are shown in figure 4. ANOVA test results show that the difference between the four ET(.) averages is not significant (F(3, 28)=0.91, p=0.45). The average ET across each subject and each stimulus demand is 0.42 ± 0.04 (S.E.) D, which is close to the depth-of focus value for normal subjects.

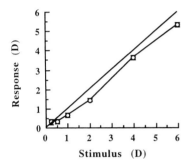

Figure 3. Averaged accommodative response function. The error bar represents ± 1 S.E..

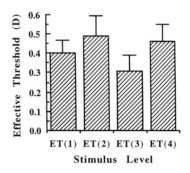

Figure 4. Averaged ET values obtained from 4 stimulus demands. The error bar represents +S.E..

Discussion

Current negative feedback control models of steady-state accommodation consist of a nonlinear element (DSP) and a linear element (ACG), both in the feedforward portion of the loop. This type of model is unable to successfully predict how pupil size and stimulus condition (target luminance, target spatial frequency and contrast) affect the accommodative response. To address these difficulties, we recommend adding an additional linear operator in front of the DSP to approximate the gain of the sensory part of the accommodative system. Here, the DSP is assumed to represent the threshold of the motor controller. It is uncertain where the exact location of the DSP is. It could be the cones and ganglion cells of the retina which are stimulated by the defocused image or the neural pathway from retina through the lateral geniculate body and cortical area 17 to the midbrain-oculomotor nucleus complex (Edinger-Westphal nucleus), where the blur-related signal is finally transformed into the motor command (Ciuffreda, 1991). The modeling results suggest that the DSP is affected by the ASG and becomes the ET, which represents at least how much AE is needed to drive the controller. In addition, the ASG contributes to the gain of the whole system, which can be considered the product of ASG and ACG. Therefore, the ASG affects the slope of the accommodative response function also.

In this study, we investigated only steady-state accommodation. From the equations derived, we are still unable to separate the ASG and the DSP. Most researchers believe the deadspace, DSP, is the subject's depth-of-focus, i.e., a perceptual threshold. Kotulak and Schor (1986) suggested that the motor threshold might be different from the perceptual threshold for blur. In this paper, we suggested an ET to emphasize that it is not perceptual, but sensory and motor related. Westheimer (1966) mentioned the difference between the DSP and the steady-state error of the accommodative response and pointed out that the DSP is a characteristic of the sensory system involved in the control of accommodation. Our results support his suggestion because the ETs calculated from different stimulus demands are the same, however, the errors of the accommodative response at different demands are very different.

Acknowledgments
I thank S. Morse, H. Bedell, and S. Patel for helpful comments on the manuscript. This work was supported by grant from the NEI (EY08862).

References

Ciuffreda KJ. (1991) Accommodation and its anomalies. in *Visual Optics and Instrumentation*, edited by Charman WN. CRC Press, Inc. pp.231-279.

Ciuffreda KJ, Hokoda SC. (1983) Spatial frequency dependence of accommodative responses in amblyopic eyes. *Vision Research*. 23:1585-1594.

Ciuffreda KJ, Rumpf D. (1985) Contrast and accommodation in amblyopia. *Vision Research*. 25:1445-1457.

Heath GG. (1956) The influence of visual acuity on accommodative responses of the eye. *American Journal of Optometry and Archives of American Academy of Optometry 33*:513-524.

Hennessy RT, Iida T, Shina K, Leibowitz HW. (1976) The effect of pupil size on accommodation. *Vision research* 16:587-589.

Hung GK, Semmlow JL. (1980) Static Behavior of accommodation and vergence: computer simulation of an interactive dual-feedback system. *IEEE Transactions on Biomedical Engineeering* (BME) 27:439-447.

Johnson CA. (1976) Effects of luminance and stimulus distance on accommodation and visual resolution. *Journal of the Optical Society of America* 66:138-142.

Kotulak JC, Schor CM. (1986) The accommodative response to subthreshold blur and to perceptual fading during the Troxler phenomenon. Perception 15:7-15.

Owens DA. (1980) A comparison of accommodative responsiveness and contrast sensitivity for sinusoidal gratings. *Vision Research* 20:159-167.

Rosenfield M, Ciuffreda KJ, Hung GK, Gilmartin B. (1993) Tonic accommodation: a review 1. Basic aspects. *Ophthalmic and Physiological Optics,* 13:266-284.

Ripps H, Chin NB, Siegel IM, Breinin GM. (1962) The effect of pupil size on accommodation, convergence, and the AC/A ratio. *Investigative Ophthalmology* 1:127-135.

Stark L, Takahashi Y, Zames G. (1965) Nonlinear servoanalysis of human lens accommodation. *IEEE Transactions on Systems Science and Cybernetics* (SSC) 1:75-83.

Toates FM. (1972) Accommodation function of the human eye. *Physiological Reviews* 1972; 52:828-863.

Ward PA, Charman WN. (1985) Effect of pupil size on steady-state accommodation. *Vision Research* 25:1317-1326.

Westheimer G. (1963) Amphetamine, barbiturates, and accommodative convergence. *Archives of Ophthalmology* 70:830-836.

Westheimer G. (1966) Focusing responses of the human eye. *American Journal of Optometry and Archives of American Academy of Optometry* 43:221-232.

Accommodation and Vergence Mechanisms
in the Visual System
ed. by O. Franzén, H. Richter and L. Stark
© 2000 Birkhäuser Verlag Basel/Switzerland

Binocular Accommodation

L. Marran and C.M. Schor

University of California at Berkeley, School of Optometry, Berkeley CA, 94720-2020 U.S.A.

Summary: Accommodation is normally described as a yoked consensual response but there are natural conditions that present stimuli for unequal accommodation of the two eyes such as uncorrected anisometropia, asymmetric convergence and unequal sclerosis of the ocular lenses. Prior studies have reported that the eyes can accommodate unequally (aniso-accommodation), in excess of 1D, to near targets viewed in asymmetric convergence (Rosenberg, Flax, Brodsky &Abelman, 1953). Yet other studies have reported only small aniso-accommodation responses (less than half a diopter) to anisometropic stimuli in primary gaze (Stoddard & Morgan, 1942; Ball, 1952) and in asymmetric convergence (Ogle, 1937; Spencer & Wilson, 1954). We have renewed this topic of investigation to determine whether aniso-accommodation does indeed exist and what stimulus conditions evoke it.

Aniso-accommodation was measured both subjectively with a haploscopic stigmascope and objectively with a binocular SRI Dual Purkinje eyetracker and optometer. Aniso-accommodation was measured in symmetrical convergence at 1 meter and 20 cm in response to lenses and in asymmetrical convergence at 16.7cm and in horizontal gaze eccentricities of 21 degrees, inducing a 0.75 D aniso-accommodative stimulus. Significant differences in the accommodative response of the two eyes during binocular viewing (aniso-accommodation) were possible when unequal accommodative stimuli were presented to the two eyes either with a lens before one eye or as a result of the spatial geometry of eccentric viewing.

The existence of aniso-accommodation suggests that the accommodative control mechanism has some independence for the two eyes. This may be mediated by the separate pre-ganglionic pools of the paired Edinger-Westphal nuclei which innervate the ciliary body of each eye. Aniso-accommodation may respond to the typically anisometropic refractive error of the newborn and may play a role in obtaining and maintaining isometropia and emmetropia throughout life.

Introduction

Binocular vision, in which there is predominant overlap of the two visual fields and foveal specialization of the retina, provides the benefits of stereoscopic depth perception and enhanced visual resolution. These sensory functions require that both retinal images be clearly focused and aligned on corresponding retinal regions of the two eyes. Oculomotor functions which support this form of binocular vision are simplified by yoked and coordinated voluntary movements of the two eyes. These yoked movements are described by Hering's law in which activity of one set of extraocular muscles is automatically associated with actions of another set of muscles in the contralateral eye. This reflex serves to maintain image alignment on corresponding retinal regions as gaze is altered from one direction to another.

This yoking behavior is also evident in the consensual accommodative response of the two eyes (Campbell & Westheimer, 1960), producing clear retinal images of bifoveally viewed objects. If accommodation were uncoupled in the two eyes, it would mainly benefit conditions in which the two eyes viewed independent objects. This situation mainly occurs in animals that lack foveas and have predominantly non-overlapping visual fields, or those with multiple foveas that lie in monocular crescents. For example, the chicken has only a 20 degree binocular overlap of the visual field and is capable of independent control of accommodation in the two eyes (Schaeffel, Howland & Farkas, 1986) and many rafters have independent accommodation (Glasser, Pardue, Andison & Sivak, 1996). However, even animals with single, well developed foveas and a larger extent of the binocular visual field encounter natural conditions that result in unequal retinal image blur of the two eyes. The visual system has two potential strategies to respond to these conditions. The first is a suppression of the blurred component of the binocular image (Fahle, 1982; Schor, Landsman & Erickson 1987). The second strategy is an unequal accommodative response of the two eyes. The first mechanism is well documented in human and is exploited in certain contact lens corrections of presbyopes where one eye is fitted with a far lens and the other with a near lens (Schor et al 1987). In this circumstance, the near eye is suppressed when viewing far targets and the far eye is suppressed while viewing near targets. Stereopsis is maintained at both viewing distances even though its threshold is slightly elevated (Schor & Heckmann, 1989). An alternative strategy which does not sacrifice stereo sensitivity is to accommodate unequally in the two eyes in response to anisometropic stimuli. An important question is whether there is any degree of uncoupling of binocular accommodation in the predominantly consensual accommodative response in humans.

Asymmetric Convergence

Anisometropic stimuli to accommodation occur naturally in asymmetric convergence. Combined version and vergence movements produce asymmetric convergence upon targets that are physically closer to the abducted eye than to the adducted eye. The unequal viewing distance produces an aniso-accommodative stimulus which increases with both the proximity of the target and with horizontal eccentricity. Does the visual system automatically correct these unequal stimuli to accommodation in asymmetric convergence?

Prior studies of aniso-accommodation to asymmetric vergence stimuli have produced unequivocal interpretations. Rosenberg et al. (1953) found large interocular differences of accommodation to a large aniso-accommodation stimulus (1.3 D) produced by a near target (12 cm viewing distance) at a horizontal eccentricity of 20 degrees. They observed aniso-accommodative responses of up to 2.91 D. In contrast, Ogle (1937) reported only small differences in accommodative responses of the two eyes to asymmetric vergence stimuli. Similarly, Spencer and Wilson (1954) found small aniso-accommodative responses, supporting Ogle's findings. However, analysis of the spatial geometry of the visual stimulus conditions in these latter experiments leads to different interpretation of their results. The viewing distances and gaze eccentricities used by Ogle (1937), and by Spencer and Wilson (1954), provided rather small differences in the accommodative stimuli to the two eyes (0.14 and 0.33D, respectively). Interestingly, the aniso-accommodative responses reported in these studies closely matched their stimuli (0.25 and 0.34D, respectively). These analyses suggest that aniso-accommodation responded appropriately to asymmetric convergence stimuli in all three studies.

These results are not conclusive because they lack the necessary controls to insure an aniso-accommodative response. It is possible that subjects eliminated aniso-blur, not by aniso-accommodation, but by suppressing the most blurred of the two retinal images. This form of binocular blur suppression has been demonstrated to occur regionally between corresponding retinal areas (Schor et al .,1987) and to allow unmatched dichoptic components of the two retinal images to remain visible to the two eyes. Accommodation was measured subjectively in prior studies of asymmetric convergence with a haploscopic stigmascope. This instrument presents a small spot or stigma to each eye, superimposed with beam splitters on a binocularly viewed target. Thus, suppression could eliminate blurred components of the binocular target while the separate stigma seen by each eye would be unsuppressed. This presents the subject the opportunity to accommodate consensually to the target viewed by the eye whose stigma is adjusted while suppressing blur of the non-measured eye. These studies only presented the stigma to one eye at a time such that the resulting measures of accommodative response could represent the monocular accommodative response of the eye viewing the stigma. When the stigma was presented to the contralateral eye, a different consensual response could could have occurred. Although Ogle used dichoptic cues that provided potential suppression cues, these were presented 4 degrees from the foveal fixation target, a location where visual acuity and blur detection would be slightly compromised, and as stated earlier, dichoptic blur cues do not prevent the possibility of regional blur suppression between corresponding retinal areas.

Anisometropia

Another stimulus condition for aniso-accommodation is naturally occurring anisometropia. Several studies have investigated the binocular accommodative response to lens-induced anisometropic stimuli. Stoddard and Morgan (1942) and Ball (1952) reported small aniso-accommodative responses to lens induced anisometropic stimuli presented in symmetrical convergence with average responses of 0.22D and 0.12D respectively, for aniso-stimuli ranging from 0.25D to 1.00D. Aniso-accommodation responses were as large as 0.50D in individual cases. As with the asymmetric convergence studies described above, the results were not conclusive because of the potential for monocular blur suppression and alternating consensual accommodation following the stigma target used to measure accommodation of one eye at a time. While Ball did use a dichoptic target, it consisted of a potentially rivalrous orthogonal line pattern which would encourage binocular suppression.

Unequal Binocular Progression of Presbyopia

A third natural stimulus condition for aniso-accommodation is the potential for unequal development of presbyopia in the two eyes. If the amplitude of accommodation does not regress at the same rate in the two eyes, or if there was unequal lens sclerosis, an emmetropic individual with isometropia at a far viewing distance would be anisometropic at near distances due unequal effective accommodation. The degree of anisometropia would increase with stimulus proximity. Consequently, as with asymmetric convergence, there would be reason for aniso-accommodation to be more pronounced at near viewing distances. To accomplish this, the aniso-accommodative response could be coupled to either the convergence or overall level of accommodation of the two eyes or to target proximity. Furthermore, one might expect that in asymmetric convergence, the aniso-accommodation response would be greater to aniso-lens stimuli that were consistent with spatial geometry (i.e., reduction of accommodation stimuli before the adducted compared to the abducted eye). The following experiments demonstrate that aniso-accommodation does indeed exist, that it can be trained to respond 0.75D, and that in many cases it responds more in near target conditions. However, it is not facilitated by aniso-stimuli that are consistent with the geometry of asymmetric convergence.

Materials and Methods

Binocular accommodation was measured in response to aniso-accommodative stimuli presented with lenses before one eye in symmetrical and asymmetrical convergence. Seven young adult subjects (ages 16-22 years) with normal binocular vision and stereo acuities of at least 20 arc sec participated in the study.

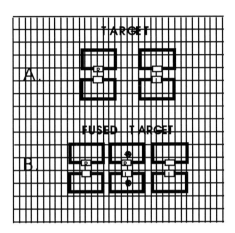

Figure 1. Fusion Target (A) andAppearance   (B)

Figure 1 illustrates the free-fusion target used in conjunction with monocular lenses to stimulate aniso-accommodation. Monocular dichoptic blur cues were presented in the binocular stimulus to accommodation by the dichoptically viewed letters (R and L) which provided subjects feedback for the accuracy of their accommodative response, while also serving as a binocular suppression check. The target was either crossed- or uncrossed fused in a manner similar to the free-fusion technique of an autostereogram. The grid background and the rectangular outline surrounding the letters served as a binocular stimulus for sensory and motor fusion. The target was viewed binocularly at a distance of either 20 cm or 1 meter . Target size and letter separation were adjusted to present the same convergence demand of 30 D for both near and far conditions and to maintain a constant visual angular subtense across both conditions. Target distance, and hence the accommodative stimulus around which the aniso-accommodative stimulus was manipulated, differed between the near and far target distances (5.0 D Vs 1.0 D, respectively). Plus lenses at the near distance and minus lenses at the far distance ranging from 0.5 to 3.0 D were placed in the spectacle plane before one eye to stimulate aniso-accommodation. Subjects were given 3 minutes to respond to the aniso-accommodative stimulus. Measurements began when both dichoptic letters appeared clear. If after 3 minutes they could not be cleared, then measures of accommodative response were made with the blurred images.

Conclusive demonstrations of aniso-accommodation require that the accommodative response of the two eyes be monitored simultaneously. This was accomplished subjectively with a haploscopic stigmascope apparatus as well as objectively with a binocular SRI Optometer mounted in a generation V dual Purkinje eye tracker apparatus.

Subjective Measurements of Aniso-Accommodation by Stigmatoscopy

Cross-hairs illuminated by two small light sources (stigma) were viewed haploscopically in a 10D Nagel Optometer. Subjects turned a knob to adjust the distance of each stigma from the Optometer lens until both sets of cross-hairs appeared clear simultaneously, at which point they were conjugate to each eye's retina. Two beam splitters, anterior to the spectacle plane of each eye, allowed the stigma to be superimposed on the binocular fusion target close enough to the dichoptically viewed letters so that both dichoptic letters and stigma could be viewed simultaneously. The stigma were cross-paired with the dichoptic letters so that each eye's stigma was positioned near the dichoptic letter viewed by the other eye. This prevented inadvertent measurement of the sequential *monocular* accommodative responses. If subjects had rapidly changed their consensual accommodative response as they looked back and forth between the two dichoptic letters, this would be revealed as an inverted pattern of results compared to the aniso-accommodative stimulus.

Objective Measures of Aniso-Accommodation with the SRI Optometer

Subjective measures of aniso-accommodation were verified with objective measures of lens-induced aniso-accommodation stimulated in symmetrical convergence at a 20 cm viewing distance with a binocular SRI infrared optometer and generation V binocular SRI dual Purkinje eye tracker system (Crane & Steele, 1978). Subjects' pupils were dilated with 2.5% phenylephrine hydrochloride. Aniso-accommodation was measured in response to steady-state and step changes in aniso-accommodative stimuli. Eye movements were monitored and trials were rejected in which artifacts could have occurred. Eye movements artifacts were minimized by instructions to maintain fixation at a point between two Nonius lines during the recording periods. (The dichoptic letters in the target were replaced by Nonius lines to minimize vergence eye movements.)

Results

Lens-Induced Aniso-Accommodation (Symmetrical Convergence)

A typical aniso-accommodative response function for the near (20cm) and far (1m) viewing distances in symmetrical convergence is shown for one subject in Figure 2. The horizontal axis represents the aniso-stimulus (0.5 to 3.0 D) and the vertical axis, the aniso-accommodative response. Across subjects the mean *slopes* of the aniso-accommodative response function for the near target ranged from 0.12 to 0.38 D (overall mean = 0.27 ($p < 0.05$). Mean slopes of the aniso-accommodative response function for the far target ranged from 0.03 to 0.27 (overall mean = 0.14). Thus, the average mean *response* that occurred with our largest stimulus (3.0 D) was 0.81D for the near target condition and 0.42D for the far target condition. The gain of the aniso-accommodative response for 4 of the 7 subjects was significantly reduced in the far compared to the near condition ($p < 0.05$).

Lens-Induced Aniso-Accommodation (Asymmetric Convergence)

Aniso-accommodation was tested in asymmetric convergence at 21^0 right or left horizontal eccentricity at the near viewing distance of 16.7 cm, measured from the midpoint between the two eyes.. The spatial geometry of this target presented a 0.75D aniso stimulus to accommodation. The direction and amount of this aniso-accommodative stimulus was manipulated by additional lenses, so that in each horizontal gaze position, there were five stimulus conditions: 0, +/-0.75, and +/-1.50. The zero stimulus value served as a control. The positive values and negative values represent aniso-accommodative stimuli with relatively more accommodative stimulus for the right and left eye, respectively. In right asymmetrical gaze, when the values were positive, the stimulus was consistent with spatial geometry, and one might expect facilitation of the response. A reduced response might be expected when the values were negative and the aniso-stimulus was greater for the left eye such that the stimulus was contrary to spatial geometry. The opposite sign convention would apply for left asymmetrical gaze. Only 3 of the 6 subjects showed significant aniso-accommodative responses in asymmetric convergence ($p < 0.05$). Of these 3 subjects, there was no consistent pattern of either facilitation or degradation. Thus the accommodative system does not appear to have an innate ability to make greater aniso responses to stimuli that are consistent with rather than opposite to the sign of naturally occurring spatial geometry.

Objective Measures of Aniso-Accommodation: Steady State Response

The average aniso-accommodative responses measured objectively with the SRI binocular optometers for the two steady state aniso-accommodation conditions (+1.00 over the right eye and +1.00 over the left eye) were -0.60 and 0.90 D. These all were significantly different than the iso-accommodative (plano) condition in which the response was 0.29 (paired t-test, p<0.001).

Objective Measures of Aniso-Accommodation: Step Response

A fifty second recording of the binocular accommodative response to a +1.50D step aniso-accommodative stimulus is shown in Figure 3. This is a representative trace of three 50s trials. A significant difference (p <0.01) in means of the aniso-accommodation (Right - Left eye) before the step response and after the completion of the step response was -0.65 D. The mean time for the aniso-accommodative response to begin was 11 seconds after the step stimulus was introduced. It took between 3.5 to 5 seconds for the response to be completed.

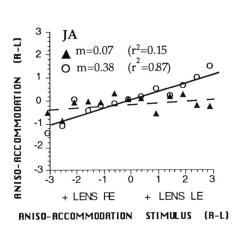

Figure 2. Aniso-Accommodation Response
to far (Δ) and near (o) targets

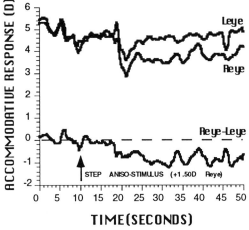

Figure 3. Aniso-Accommodation Step Response

Discussion

Target Distance and Aniso-Accommodation

These experiments demonstrate that: (1) aniso-accommodation of greater than 0.5 D is possible when a stimulus is presented with lenses placed before one eye and (2) in some subjects, the mechanism is more effective at near than at far viewing distances. We found that the aniso-accommodative response was significantly reduced for more than half of the subjects in the 1m viewing condition compared to the 20cm viewing condition (p<0.01). This result could indicate a coupling of the aniso-accommodative response to the overall level of accommodation, since convergence was held constant at the two stimulus distances. However, accommodative level does not explain why target distance had no effect on some subjects' aniso-accommodative response. Other possible factors include (1) mismatches between the vergence and accommodative stimuli in the far test condition and (2) proximal cues. If accommodative vergence interactions accounted for the difference, we would expect those subjects who were influenced by viewing distance to have different accommodative convergence/accommodation (AC/A) ratios than subjects who had equal aniso-accommodation ability at both viewing distances, but this was not the case. Proximity has been shown to affect both accommodation and its tonic aftereffects (Rosenfield & Gilmartin, 1991; Rosenfield & Ciuffreda, 1991). These studies point to a strong high level interaction between proximity and the accommodative response. Proximity may also contribute to the higher magnitude of the aniso-accommodative response that we found for near targets. Aniso-accommodation would be more important for near viewing tasks where detailed hand-eye coordination or fine stereopsis tasks usually occur. Target proximity may have a greater influence in triggering the response for some subjects than for others, just as individual differences in the proximal accommodative response have been demonstrated (Rosenfield & Gilmartin, 1991).

Reaction Time and Total Response Time

Although our objective data is limited, the contrast in reaction time and total response time of the aniso-accommodative response to that of consensual accommodation is remarkable. Reaction time for a 2.0 D *iso-accommodative* step stimulus ranges from 225-440 ms (Campbell & Westheimer, 1960). Reaction time to a 1.50 D *aniso-accommodative* step stimulus, as measured objectively with the SRI optometer, was about 11 seconds. Total response time (start to completion) for consensual accommodation ranges from 560 ms to 2 s (Tucker & Charman,

1979; Heron & Winn, 1989). In contrast, our objective measures of the total response time for aniso-accommodation is about 15 seconds. Subjective reports of target clarity after introduction of an aniso-accommodative stimulus during haploscopic measurement approximately confirm this response time. This long response time may explain the absence of an aniso-accommodative response to dynamically presented stimuli (Flitcroft, Judge & Morley, 1992).

Site and Mechanism of Neural Control of Aniso-Accommodation

In order for aniso-accommodation to occur, there must be separate pre-ganglionic pools in the Edinger-Westphal (E-W) nucleus that innervate the ciliary body in a given eye as has been demonstrated with retrograde HRP studies (Akert, Glicksman, Grob & Huber, 1980). Aniso-accommodation could be controlled prior to or at the site of the E-W nucleus in at least three different ways: (1) there could be independent gain adjustment of accommodation for each eye at the EW nucleus (possibly by excitatory or inhibitory interactions); (2) there could be two independent monocular pathways in parallel with a common binocular consensual pathway; (3) some combination of these two possibilities. Although these schemes could be implemented without monocular blur feedback with eye of origin information (e.g. by binocular contrast averaging and trial-and-error search for direction), it is more likely that a monocular error signal is used to direct the aniso-accommodative response.

Utility of Aniso-Accommodation

While some capability for differential accommodation exists, the human binocular accommodative response is predominantly consensual. Comparative studies suggest that the degree of consensuality of accommodation in other vertebrates corresponds to the degree of similarity between the two monocular images, which in turn is determined by the degree of binocular overlap of the two visual fields (Schaeffel et al, 1986) and the utilization of stereo-depth cues (Schaeffel & Wagner, 1991). Although binocular overlap does not exclusively produce identical retinal focus errors (e.g. asymmetric convergence and anisometropia refractive errors), there is sufficient similarity of the two ocular images to allow binocular blur suppression mechanisms to eliminate the differential blur of the two ocular images (Schor et al., 1987). In contrast to the high degree of binocular correlation in the two retinal images of frontal-eyed visual systems, lateral eyed animals rarely view the same object with both eyes so that the most common situation is unequal blur of the two retinal images. Binocular blur suppression is of no use under these circumstances and clarity of the two eyes' images would require independent control of the accommodation response of each eye. Why then, with our high degree of binocular overlap, do we have any capacity for aniso-accommodation? Is it simply the vestige of some obsolete mechanism or does it provide some useful function?

Aniso-accommodation could eliminate small amounts of unequal contrast in the two retinal images, caused by uncorrected refractive error or asymmetric convergence, and enhance fine stereo acuity. Stereo acuity is reduced by unequal image contrast of the two ocular images (Schor & Heckmann 1989) and is reduced two- and four-fold by 1D and 2D anisometropic blur respectively (Levy & Glick, 1974; Westheimer & McKee, 1980). Coupling the gain of aniso-accommodation with target distance is an optimal use of this function since the disparity associated with a given depth interval decreases with the square of viewing distance so that stereo is predominantly a near depth cue (Schor & Flom 1968). Its augmentation by aniso-accommodation is mainly needed at near viewing distances.

Aniso-accommodation could also provide the necessary independent accommodative feedback to guide the isometropization process of the developing eye. The typical neonatal eye is hyperopic and becomes emmetropic (zero refractive error) over the first six years of life, in a process called emmetropization. The prevalence of 1D or more of anisometropia declines during development. The prevalence is 17.3% in newborns (Zonis & Miller, 1974), 6.5%, in 1 year olds (Ingram, 1979) and 3.4% in 5-12 year olds (Flom & Bedell, 1985). Isometropization to optically-induced anisometropia has been demonstrated in monkeys (Hung, Crawford & Smith, 1995). The isometropization process has been suggested to depend on some form of accommodative response in kittens, where isometropization of surgically induced refractive errors only occurred if accommodation had been left intact (Hendrickson & Rosenblum, 1985). A similar isometropization process could exist in humans which would account for the reduction of the incidence of anisometropia during development.

Acknowledgments
This project was supported by a grant from the National Eye Institute of the National Institutes of Health (EY0-3532)

References

Akert K, Glicksman M, Grob P & Huber A (1980) The Edinger-Westphal nucleus in the monkey: a retrograde tracer study. *Brain Research,* 184; 491-498

Ball, E.(1952). A study in consensual accommodation. *American Journal of Optometry and Archives of American Academy of Optometry,* 561-574.

Campbell, F. W., & Westheimer, G. (1960). Dynamics of accommodation responses of the human eye. *Journal of Physiology,* 151, 285-295.

Crane, H.& Steele, C.(1978). Accurate three dimensional Eyetracker. *Applied Optics* ,17, 691-705.

Fahle, M.(1982). Binocular rivalry: suppression depends on orientation and spatial frequency. *Vision Research* ,22, 787-800.

Flitcroft, D., Judge, S.& Morley, J.(1992). Binocular interactions in accommodation control: effects of anisometropic stimuli. *Journal of Neuroscience* ,12, 188-203.

Flom, M., & Bedell, H. (1985). Identifying amblyopia using associated conditions, acuity and non-acuity features. *American Journal of Optometry & Physiological Optics,* 62, 153-160.

Glasser,A. Pardue, M.T., Andison, M.A. & Sivak, J.G (1996) Accommodation in Raptors. *Journal of Comperative Physiology A.* (submitted)

Hendrickson, P., & Rosenblum, W. (1985). Accommodation demand and deprivation in kitten ocular development. *Investigative Ophthalmology & Vision Science,* 26, 343-349.

Heron G, & Winn B. (1989) Binocular accommodation reaction and response times for normal observers. *Ophthalmic and Physiological Optics,* 9:176-83.

Hung, L., Crawford, M., & Smith, E. (1995). Spectacle lenses alter eye growth and the refractive status of young monkeys. *Nature Medicine,* 1(8), 761-765.

Ingram, R. (1979). Refraction of 1-year-old children after atropine cycloplegia. *British Journal Ophthalmology,* 63, 343-347.

Levy, N. S.& Glick, E. B.(1974). Stereoscopic perception and Snellen visual acuity. *American Journal of Ophthalmology* ,78, 722-724.

Ogle, K. N.(1937). Relative sizes of ocular images of the two eyes in asymmetric convergence. *Archives of Ophthalmology* , 1046-1061.

Rosenberg, R., Flax, N., Brodsky, B.& Abelman, L.(1953). Accommodative levels under conditions of asymmetric convergence. *American Journal of Optometry and Archives of American Academy of Optometry* , 244-254

Rosenfield, M., & Ciuffreda, K. (1991). Effect of surround propinquity on the open-loop accommodative response. *Invest Opthalmol Visual Science,* 32(1), 142-147.

Rosenfield, M., & Gilmartin, B. (1991). Effect of target proximity on the open-loop accommodative response. *Optometry & Vision Science* 67, 74-79.

Schaeffel F, Howland HC & Farkas L (1986) Natural accommodation in the growing chicken. *Vision Research* 26; 1977-1993

Schaeffel F & Wagner H (1991) Barn owls have symmetrical accommodation in both eyes, but independent pupilary responses to light. *Vision Research* 32; 1149-1155

Schor, C.M., & Flom, M.C. (1968) The relative value of stereopsis as a function of viewing distance, *American Journal of Optometry* 46(11):805-809.

Schor, C., Landsman, L. & Erickson, P. (1987). Ocular dominance and the interocular suppression of blur in monovision. *American Journal of Optometry and Physiological Optics,* 64, 723-730.

Schor, C., & Heckmann, T. (1989) Interocular differences in contrast and spatial frequency: Effects on stereopsis and fusion. *Vision Research,* 29, 837-847.

Spencer, R. W.& Wilson, W. K.(1954). Accommodative response in asymmetric convergence. *American Journal of Optometry and Archives of American Academy of Optometry* , 498-503.

Stoddard, K. B.& Morgan, M. W., Jr.(1942). Monocular accommodation. *American Journal of Optometry and Archives of American Academy of Optometry,* 460-465.

Tucker, J., & Charman, W. (1979). Reaction and response times for accommodation. *American Journal of Optometry & Physiological Optics,* 56(8), 490-503.

Westheimer, G.& McKee, S. P.(1980). Stereoscopic acuity with defocused and spatially filtered retinal images. *Journal of the Optical Society of America* ,70, 772-778.

Zonis, S., & Miller, B. (1974). Refractions in the Israeli newborn. *Journal of Pediatric Ophthalmology,* 11(2), 77-81.

Vertical vergence – normal function and plasticity

J. Ygge

Dept. of Ophthalmology, Karolinska Institute, Huddinge University Hospital, S-141 86 Huddinge, Sweden.

Summary: Eye movement recordings were made in three humans using the magnetic search coil system. The subjects were looking at targets that were mostly located in front of and close to one eye, thereby creating a disparity with a larger target in the close eye. We found that the eyes make rapid vertical eye alignment changes to the naturally occurring disparity. These alignment changes were very fast (up to a peak velocity of 30deg/sec) and 81% of the required change was completed within the saccadic pulse for downward saccades, whereas the corresponding figure for upward saccades was 47%. These alignment changes were independent of disparity cues since a corresponding disparity in the midline was introduced with the aid of a prism and no similar changes in vertical alignment were seen. The alignment change was also possible to alter by adapting the subjects by making them wear a prism in front of one eye for some time.

Introduction

It is a well known fact for most clinicians, that the vergence amplitudes for horizontal disparities are much larger than for vertical disparities (Houtman, Roze & Scheper, 1977; Kertesz, 1983). This gives rise to clinical problems in patients with suddenly appearing neuromuscular weaknesses in extraocular muscles that affect the vertical position of the eyes. This is mostly caused by a trochlear nerve palsy (Mottier & Metz, 1990) which gives rise to vertical fusional problems for the patients although the fusional amplitude necessary to overcome the problem is usually far less than what can be performed in the horizontal direction. The normal maximum vergence amplitude in the horizontal direction for an adult subject is about 14 prism diopters for convergence and about 6 prism diopters for divergence with a distance fixation target. However, the vertical amplitude is far less, only about 2.5 prism diopters, although a recent study (Sharma & Abdul-Rahim, 1992) has reported larger vertical amplitudes of about 4.6 prism diopters for normal subjects. Ellerbrock (Ellerbrock, 1949) could show even greater vertical vergence amplitudes by using larger targets and by separating the fusional targets more slowly. He reported a maximum of 16 prism diopters in normal subjects. Subjects with congenital fourth nerve palsy can often be diagnosed as having a congenital paresis instead of an acute palsy by measuring the vertical fusional amplitude. In congenital palsies it is often much increased compared to normal subjects and subjects with late onset fourth nerve palsies.

There are two instances where we use vertical vergence movements:
1. In the lens/prism induced disparity situation and
2. when the targets are located close to and in front of one eye.

When a subject is wearing spectacles with different lens powers, this gives rise to a disparity, aniseiconia. When wearing such glasses, the subject has to adapt to a situation where the images presented in front of the two eyes are of different sizes and thus, necessitates unequal eye movements when performing both saccades and smooth pursuit eye movements. The eye in which the spectacle has the larger power is presented with a larger image and therefore has to make larger eye movements than the eye with the less powerful lens. This change in image size between the two eyes requires some time for adaptation but has been shown to occur quite rapidly. This disconjugate or nonconjugate adaptation has been thoroughly investigated by several researchers (Erkelens, Collewijn & Steinman, 1989; Schor, Gleason & Horner, 1990; Oohira, Zee & Guyton, 1991; Lemij & Collewijn, 1991a-b, 1992; Schor, Gleason & Lunn, 1993).

The other instance when a subject has to perform vertical vergence movements is when the regarded object is placed close to and in front of one eye (Oohira, Zee & Guyton, 1991; Collewijn, 1994). When this occurs, the far eye views the target under a smaller angle and, subsequently, when performing vertical eye movements between targets located in front of and close to one eye, the eyes have to make different amplitude vertical movements, otherwise they end up at different targets and diplopia will probably be the result (Ygge & Zee, 1994, 1995).

The purpose of this study has therefore been to study the motor mechanisms that control the vertical eye alignment and furthermore, to investigate whether these mechanisms are able to undergo plastic changes.

Material and Methods

Three healthy emmetropic subjects were used in the present study. None of them had a history of diplopia or strabismus. The subjects had a visual acuity of at least 20/20 OU and no vertical misalignment. The horizontal phoria was less than 4 prism diopters for both near and distance viewing, and all had a stereoacuity of at least 60" as measured with the TNO test.

Eye movements were measured with binocular scleral annuli with the magnetic search coil technique. Horizontal as well as vertical signals from each eye was sampled at 500 Hz by a computer and the system noise limited the resolution to about 0.05 deg. The subjects head was stabilized with a bite bar during the experiments. The eye movements were calibrated monocularily in the same horizontal orbital position as when performing the actual experiments.

Two bars with three vertically-aligned LED's were used as stimuli. One of the bars had smaller LED's for near experiments (about 10 cm in front of the eyes) and the other had larger LED's for experiments at distance (115 cm from the eyes). The angle of horizontal vergence was about 3 degrees for the distance viewing and about 30 degrees for the near viewing. In some of the experiments the near LED bar was placed with the center LED aligned at the nasion and in some experiments in front of and close to one eye (fig. 1). The upper and lower LED's were then positioned at 10 degrees up and down, respectively, with regard to the close eye. When positioned in front of and close to one eye, the other eye had to converge about 30 degrees and thus, for geometrical reasons the experimental setup created a vertical disparity of about 1 degree between the two eyes. When the near eye thus had to perform a 10 degree saccade between two of the LED's, the other eye had to perform a saccade of about 9 degrees.

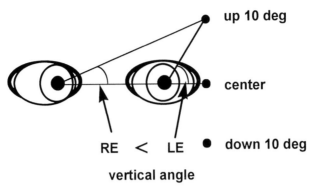

up 10 deg

center

RE < LE ● **down 10 deg**

vertical angle

Fig. 1. Schematic drawing of the experimental setup with the near LED bar placed close to and in front of the left eye. In this case the left eye has to perform a larger saccade when shifting the line of sight from the center LED to the up or down 10 degree LED, compared to the right eye, which thus has to perform an approximately 9 degree saccade to make the two eyes end up at the target.

To create an artificial vertical disparity a 2 D wedge-shaped prism was introduced and then taken away in front of one eye. This prism was positioned both base up and base down in different experiments.

The subjects performed vertical saccades between these LED's in blocks of four, starting from the center LED going up-down-down-up and back to the center. This cycle was repeated about 15 times in each experiment.

Results and Discussion

Vertical vergence movement to a prism-induced vertical disparity in the midline
When the subjects introduced the prism, they immediately experienced diplopia. In the case of near viewing, all three subjects were able to fuse the targets eventually, though when distance viewing was performed only one of the subjects was able to fuse the targets. The vertical vergence movements seen were very slow as can be seen in fig. 2.

The average mean velocities ranged from 0.16 to 0.35 degs/sec, and thus the vertical fusional process took several seconds for completion. The change in vertical alignment (Δ**VA**) in near viewing ranged between 59% ad 95% of that demanded by the prism (Table I).

Table I. Vertical motor fusional response to a prism-induced disparity. nr, no response, n, number of trials.

Subject	near			distance		
	ampl. (deg.)	vel. (deg/sec)	n	ampl. (deg.)	vel. (deg/sec)	n
1	80%	0.16	10	nr	-	-
2	76%	0.18	9	nr	-	-
3	59%	0.20	12	95%	0.26	6

There was no consistent difference between introducing the prism base-up or base-down. However, when the prism was removed the vertical vergence movements that restored the ocular alignment were generally faster (by about 44%) than those in response to the introduction of the prism, with a maximum value reaching 0.68degs/sec.

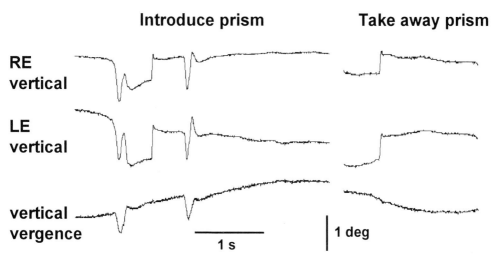

Fig. 2. Slow vertical vergence response to the introduction (left panel) and removal (right panel) of a 2 pD wedge prism oriented base-up in front of the left eye, during straight ahead fixation at distance. Note the difference in vertical vergence response, with a slower vergence movement when introducing the prism whereas a faster vergence response can be seen in association with a saccade when taking away the prism.

The general pattern for all subjects on removal of the prism was that both eyes made a conjugate saccade in the appropriate direction for the left eye (which had been having the prism in front) and then the right eye drifted back to the correct position. During the vertical vergence movements, small vertical as well as horizontal saccades were commonly seen and during these small saccades the vertical vergence movement often appeared to speed up (fig. 3).

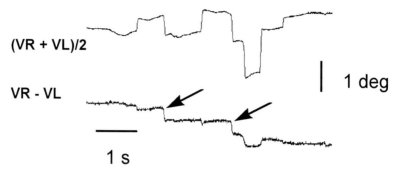

(VR + VL)/2

VR - VL

1 deg

1 s

Fig. 3. Vertical vergence response to a 2 pD wedge prism oriented base-down introduced in front of the left eye. Note the occurence of a increased vergence velocity in association with a vertical saccade (arrows). (VR+VL)/2 represents the vertical conjugate trace whereas the VR-VL represents the vertical vergence trace.

As has been shown in previous studies the human vertical fusional response to a prism induced disparity is very limited (Houtman, Roze & Scheper, 1981; Kertesz, 1983; Mottier & Metz, 1990; Sharma & Abdul-Rahim, 1992). In the present study, it could also be shown that the vertical vergence movements, as a response to a prism induced disparity were very slow and it seemed that the ΔVA when fusing targets at distance seems to be more ineffective since only one of the three subjects was able to perform it. The reason for why vertical vergence was more effective when performed in near viewing is not clear and we did not distinguish between other potential influences on vertical vergence such as accommodation, sense of nearness or influence of the background. Nevertheless, since naturally occurring vertical disparities are usually related to near viewing, it is not surprising that vertical vergence was better and more effective when targets were closer to the subject.

However, although the vertical vergence to a prism induced disparity was slow, we could sometimes note that the vertical vergence movements seemed to be facilitated by vertical as well as horizontal saccades, and sometimes even by blinks. Facilitation of horizontal vergence by vertical and horizontal saccades (Enright, 1984; Zee, Fitzgibbon & Optican, 1992) and by blinks (Peli, 1986) has recently been shown, and the findings in the present study seem to be analogous to these.

Conjugacy of normal vertical saccades

Vertical saccades made to targets in the median plane were found to be very conjugate. Only small intrasaccadic changes in vertical alignment (Δ**VA**; fig. 4) were found, although these changes were usually larger for far viewing than for near. As has been noted before (Collewijn, Erkelens & Steinman, 1988; Enright, 1989) (Zee, Fitzgibbon & Optican, 1992; Oohira, 1993), transient changes in horizontal alignment were also found during the vertical saccades. For upward saccades there was a consistent horizontal divergence (1.1-2.7 deg.) followed by a convergence during the pulse portion of the vertical saccade. On the other hand, during downward saccades a consistent convergence (0.8 -1.5 deg.) was found followed by a transient divergence during the vertical pulse of the saccade.

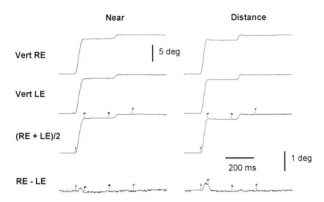

Fig. 4. Vertical vergence change (Δ**VA**) during vertical saccades in the midline at near and at distance from one of the subjects in the present study. "Vert RE" and "Vert LE", vertical position of the right and left eye, respectively. "(RE+LE)/2", conjugate vertical position of the eyes. "RE-LE", vergence vertical position (Δ**VA**) of the eyes. The 1 deg bar at the lower right corresponds to only the bottom trace of both panels, whereas the 5 deg bar in the top part of the figure corresponds to the upper three traces in both panels. In the lower two traces of both panels the "i" - "p" part corresponds to the saccadic pulse, the "p" - "s" and "s" - "f" part represent 160 msec periods each after the completion of the saccadic pulse. Note the transient Δ**VA** during saccades made on the midline and that the Δ**VA** is larger for distance targets.

Disconjugate vertical saccades with a natural lateral disparity

When performing this experiment, the targets were placed close to and in front of one eye, thereby introducing a vertical disparity of about 1 deg. There was a clear change in relative vertical alignment, Δ**VA,** during these experiments (fig. 5; Table II). Firstly, during the saccadic pulse there was a rapid change in Δ**VA** with a maximum velocity of about 20deg/sec for upward saccades and 30 deg/sec for downward saccades. This rapid change accounted for about 50% of the required Δ**VA** for upward and about 80% of the Δ**VA** for downward saccades (i-p in fig. 5

and in Table II). After the pulse portion of the vertical saccade was completed, there was still a ΔVA seen but this was much slower. During the first 160 msec following the vertical pulse (the s-p part in fig. 5 and Table II) the ΔVA had a mean velocity of about 3-7 deg/sec. This ΔVA was usually followed by a period (f-s in fig. 5 and Table II) with an even slower ΔVA with a mean velocity of about 0.1 - 0.0.6 deg/sec. This velocity is about the same as that of the the vertical vergence when introducing a vertical prism in front of one of the eyes without an accompanying saccade. The maximum ΔVA was reached after about 350 msec post pulse for downward saccades. For upward saccades this ΔVA continued for up to about 900 msec after the pulse was completed. The maximum ΔVA was always close to the demand calculated from monocular refixations. A small "sensory" component of vertical vergence could also be seen in some of the experiments but this component never exceeded 17% of the required ΔVA. The speed of vertical realignment was thus several orders of magnitude faster than that seen when vertical vergence is performed without an accompanying saccade.

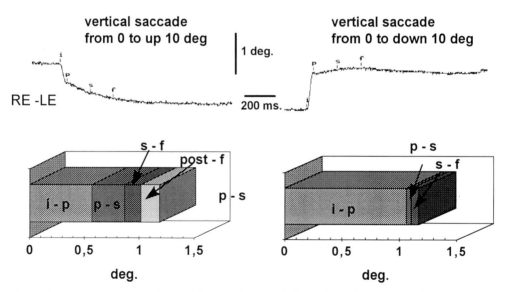

Fig. 5. Change in vertical vergence (RE-LE) during and after vertical saccades to the target bar in front of and close to the left eye in one of the subjects in the present study. In the two top panels the actual vertical vergence traces are shown for saccades from 0 to up 10 degrees (left panel) and from 0 to down 10 degrees (right panel). The horizontal histograms are drawn to represent the different parts of the vertical vergence change during different time intervals (for explanation of the different symbols, refer to figure legend 4). Note that more of the vertical vergence change is incorporated into the saccadic pulse for the 0 to down 10 degrees saccade than for the 0 to up 10 degrees saccade.

Thus the dynamics of the vertical vergence ΔVA during vertical saccades were quite different from the ΔVA to a prism induced response. Furthermore it could be concluded that much more, in fact almost double the ΔVA was included in the pulse portion of the saccade when the vertical movement was downward, where the ΔVA was 81% at mean, than upward where it was 47% at mean. This may reflect the fact that downward saccades are more frequently made when looking at a near target. Consequently, saccade performance is more likely to be optimized for the more commonly encountered visual circumstances.

Table II. Amplitude of the ΔVA in one representative subject (no. 1) in the present study.

Trial	demand	pulse	s-p	f-s	post-f	total ΔVA	sensory fusion
0 to up 10	-1.15	-0.46	-0.31	-0.13	-0.05	-0.95	-0.20
up 10 to 0	1.15	0.91	0.03	0.04	-0.02	0.96	0.19
0 to down 10	1.14	0.89	0.08	0.02	0.01	1.00	0.14
down 10 to 0	-1.14	-0.52	-0.29	-0.17	-0.07	-1.05	-0.09

Under different circumstances than in the present study the brain however seems to be able to adjust the vertical alignment of the eyes in the long-term, like the response to optics that require the changes in the vertical yoking of the two eyes (Zee & Levi, 1989; Schor, Gleason & Horner, 1990; Lemij & Collewijn, 1991a; Oohira, Zee & Guyton, 1991).

Disconjugate vertical saccades with a prism-induced disparity

The next question to be answered was if the brain was able to generate disconjugate vertical saccades to a prism-induced disparity with the targets on the midline, both near and at distance. The reason for this experiment was to know whether the rapid realignment of the eyes associated with making vertical saccades to targets near and close to one eye, was simply related to a linking of the vertical saccade with a vertical disparity, or if there was something unique about vertical saccades made to a naturally occurring disparity. The prism was placed in the upper visual field of one eye but sparing the line of sight when the eyes were looking straight ahead in the horizontal plane. This prism thus created a disparity for upward saccades which was similar to the one that was incurred naturally with the targets located near and close to one eye. The main finding was that no subject in the present study could use the prism induced disparity information to produce disconjugate vertical saccades. Neither the ΔVA during the saccade nor that during the following 160 msec after the saccade were affected by the prism induced disparity. However, two of the three subjects were able to produce a slow ΔVA after the saccade was completed and both those two subjects could eventually fuse the targets (fig. 6). The velocity of this Δ**VA** was in the range of the velocity seen in Δ**VA** when introducing a wedge prism in front of one eye as was described above. This ability was also most pronounced when he subjects were performing near viewing.

0 to up 10 deg.

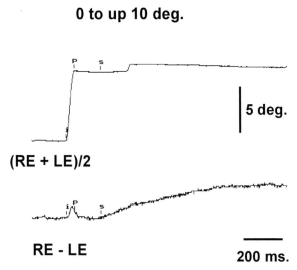

(RE + LE)/2

5 deg.

RE - LE

200 ms.

Fig. 6. Vertical saccades made to midline targets at near with a prism induced disparity (with a base up 2 pD wedge prism in front of the upper visual field of the left eye). For the upward saccade from 0 to up 10 degrees, note that there is first avery small transient change in vertical alignment of the eyes like that of naturally occurring saccades made on the midline without a prism. After the pulse and first part of the postsaccadic period is completed, there is a slow vertical vergence change continuing for several hundred msecs after the saccade.

From the up 10 degrees target where the eyes were already misaligned due to the prism, the eyes made disconjugate saccades back to the primary position. This change in vertical alignment during this return movement was very rapid, being mostly incorporated during the downward saccade and the immediate postpulse period (fig. 7). This response during the centripetal, resembled the change in vertical alignment that occurred with viewing the targets located near to the head and closer to one eye (cf. fig. 5) as was described above.

up 10 deg. to 0

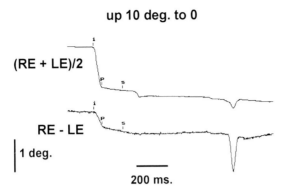

(RE + LE)/2

RE - LE

1 deg.

200 ms.

Fig. 7. Vertical saccades made to midline targets at near with a prism induced disparity (with a base up 2 pD wedge prism in front of the upper visual field of the left eye). For the downward saccade shown in this figure (compare with figure 6 for upward saccade) the vertical vergence change is much faster than for the upward saccade seen in fig. 6. In the present figure, the eyes are vertically misaligned due to the wedge prism when the vertical saccade is elicited from the up 10 degrees position.

Is there a need for disparity to make disconjugate vertical saccades to near targets?

Is the vertical disparity necessary or used to produce the disconjugate saccades that occurred with refixations between vertically displaced targets in front of and close to one eye? This question was addressed by placing a prism (oriented base down) in the upper visual field of the close eye (which naturally had to do a larger vertical eye movement) but sparing the line of sight in the horizontal meridian. This setup thus nullified the naturally occurring disparity induced when viewing a near target that is closer to one eye. The result from this experiment showed that the Δ**VA** both during the pulse portion of the saccade as well as during the first 160 msec postpulse of the saccade was the same as without the prism (fig. 8). But in this case the disconjugacies were inappropriate since it misaligned the eyes relative to demands of the prism. These unequal saccades were eventually followed by a slow vertical vergence movement that brought the eyes back to the correct alignment (fig. 8), given the demand of the prism. Centripetal saccades, however, back to the 0 deg horizontal position, where there was no prism, were associated with a considerable intrasaccadic change in vertical alignment, similar to that during natural viewing, as with and without the nullifying prism.

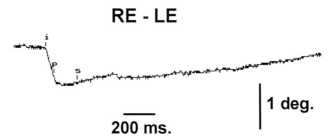

Fig. 8. Vertical vergence change when performing a vertical saccade to a target close to and in front of the left eye (compare fig. 5 left panel) but in this case the naturally occurring disparity was nullified by placing a neutralizing prism in front of the left eye. Note that the vertical vergence change is first similar to the the naturally occurring situation, but that the first rapid vertical vergence change is followed by a long period of a slow vertical vergence change when the eye drifts back to the correct position due to the prism.

Another way to check for the possible need for disparity to make disconjugate vertical saccades to near targets was to make the subjects produce disconjugate vertical saccades to the remembered location of near targets located closer to one eye. When making these saccades to the remembered position, there was no vertical disparity since the saccades were made in the dark and then after about 200 msec after the saccade the target light came on. All three subjects really did make disconjugate saccades to the remembered targets although there was no disparity (fig. 9).

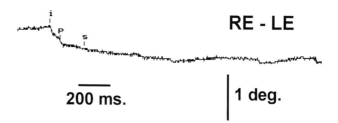

Fig. 9. Vertical vergence change to a remembered target close to and in front of the left eye. Note that there is a vergence change occurring although it is smaller than the one usually seen when doing saccades to naturally occurring targets in front of one eye (cf. Fig. 5). However, a vertical vergence does occur in this situation indicates that the vergence change occurs independently of the immediate presence of disparity cues.

Adaptive control of the vertical saccade yoking.

In a final set of experiments, two of the subjects wore a prism that covered the lower field in front of the left eye, sparing the lone of sight in the horizontal meridian, for periods of 8 and 20 hours respectively. For the lower visual field the prism nullified the naturally occurring disparity for near targets located close to the left eye, but also produced a disparity for targets on the midline and increased the disparity for targets located nearer to the right eye. Both subjects experienced diplopia in the lower visual field at the beginning of the adaptation period but the diplopia disappeared both at near and distance viewing after a few hours of wearing of the prism.

After the adaptation period, the change in ΔVA when making vertical saccades differed considerably between the upper and lower visual field. In the upper visual field (where no prism had been worn) the ΔVA during the saccadic pulse was still comparable to the ones seen without habitually wearing the prism. In the lower visual field however, the ΔVA during the pulse was about a 1/3, compared to the values when not wearing the prism. The ΔVA had thus been decreased in the lower visual field by wearing the nullifying prism. The subjects had thus developed an orbital-positioned dependent adaptive change in saccadic yoking (Sethi, 1986; Schor, Gleason & Horner, 1990; Schor, M, Gleason, Maxwell & Lunn, 1993).

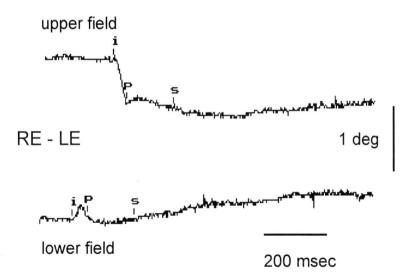

Fig. 10. Vertical vergence response after wearing a prism habitually over the lower visual field for 8 hours to nullify the naturally occurring disparity in that visual field for objects located close to and in front of the eye. The lower trace shows the adapted response with a small intrasaccadic change in vertical alignment during the pulse part of the saccade (about "i" - "p") and the following slow fusional movement to correct for the absent natural disparity. The upper trace shows the vertical vergence change in the upper visual field where no prism has been worn. Note the vertical vergence change similar to the one seen in the normal situation (cf. Fig. 5, right panel).

Conclusions

It can be concluded from the present experiments, that the vertical vergence change to a prism induced disparity is very slow and ineffective. If, however, the disparity instead is a natural one, such as if the targets are located in front of and close to one eye the vertical vergence change is much more rapid, in the order of several magnitudes over the prism induced response. It can also be concluded that this rapid alignment change does not seem to be dependent on immediate disparity cues since when trying to elicit prism-induced vertical vergence change in the midline we did not see such changes. Furthermore, the rapid change in vertical alignment i.e. vertical vergence does not seem to be dependent on the targets being close to one eye since when nullifying the naturally occurring disparity at near in front of and close to one eye, the rapid change in vertical alignment which is appropriate for targets close to and in front of one eye was still found. A further indication for this is that when the saccades were made to remembered targets, the rapid alignment change was still also there.

These results can probably best be interpreted by assuming that the vertical alignment change that has been shown does not depend so much on disparity per se. Instead, the results indicate that there exists a brain mechanism that automatically preprograms the vertical movements of each eye to be of the correct amplitudes as a function of where the target of interest is located with respect to the head or orbits. This mechanism is based on the equivalent of an internally-generated map of three-dimensional space. It thus seems that the brain performs calculations using the relative and absolute position in the orbit before the saccade and to where the eyes must point after the saccade. In other words the brain must calculate what the excursion of each eye in the orbit must be in order to ensure that the two eyes are correctly positioned for binocular fixation immediately after the next saccade. Because vertical disparities can occur naturally such a yoking mechanism is necessary to ensure that binocular foveal fixation is achieved as quickly as possible following vertical saccades.

After the adaptation, however, when making vertical saccades on the near array placed laterally in front of the left eye, the change in vertical alignment differed considerably between the upper and lower fields of vision.

In the upper visual field the intrasaccadic change in vertical alignment was similar to the ones found before the adaptation period. In the lower visual field, however, where the prism had been worn, the values for the change in vertical alignment was very different from the values obtained before the adaptation period (fig. 10).

There was a decrease of the change in vertical alignment or disconjugacy after wearing the prism in the lower visual field. However, there was no change in the relative excursions of the two eyes during saccades for viewing at distance on the midline. Thus in response to habitually wearing the prism, our subjects developed an orbital position dependent adaptive change in vertical saccade yoking.

Thus the earlier described proposed map of saccadic yoking seems to be under long-term adaptive control, probably driven by disparity or the attempt to overcome it. In this way the correct yoking of the eyes during vertical saccades can be maintained in the face of the demands of normal development and aging as well as disease or trauma.

Acknowledgements
I am very grateful to Dr David S. Zee who placed the research facilities available to me for making these experiments and for the fruitful collaboration. This study was supported by the Swedish Medical Research Council, the Bank of Sweden Tercentenary Foundation, the Bernadotte Foundation for Research in Pediatric Ophthalmology and the Swedish Society of Medicine.

References

Collewijn H. Vertical conjugacy: What coordinate system is appropriate? In Contemporary ocular motor and vestibular research: A tribute to David A. Robinson. Fuchs A.F., Brandt T., Buttner U. and Zee D.S (Eds) 1994, pp 296-303. Thieme, Stuttgart.

Collewijn H., Erkelens C.J. and Steinman R.M. Binocular co-ordination of human horizontal saccadic eye movements. *Journal of Physiology(London)* 1988, 404:157-182.

Ellerbrock V.J. Experimental investigation of vertical fusional movements. *American Journal of Optometry* 1949, 26:327-337.

Enright J.T. Changes in vergence mediated by saccades. *Journal of Physiology(London)* 1984, 350:9-31.

Enright J.T. Convergence during human vertical saccades: Probable causes and perceptual consequences *Journal of Physiology(London)* 1989, 410:45-65.

Erkelens C.J., Collewijn H. and Steinman R.M. Asymmetrical adaptation of human saccades to anisometropic spectacles. *Investigative Ophthalmology & Vision Science* 1989, 30:1132-1145.

Houtman W.A., Roze J.H. and Scheper W. Vertical motor fusion. *Documenta Ophthalmologica* 1977, 44:179-185.

Houtman W.A., Roze J.H. and Scheper W. Vertical vergence movements. *Documenta Ophthalmologica* 1981 ;51:199-207 1981,

Kertesz A.E. Vertical and cyclofusional disparity vergence. In Vergence eye movements, basic and clinical aspects. Schor C.M. and Ciuffreda K.J (Eds) 1983, pp 317-330. Butterworth, Boston.

Lemij H.G. and Collewijn H. Long-term nonconjugate adaptation of human saccades to anisometropic spectacles. *Vision Resarch* 1991a, 31:1939-1954.

Lemij H.G. and Collewijn H. Short-term nonconjugate adaptation of human saccades to anisometropic spectacles. *Vision Resarch* 1991b, 31:1955-1966.

Lemij H.G. and Collewijn H. Nonconjugate adaptation of human saccades to anisometropic spectacles: Meridian-specificity. *Vision Research* 1992, 32:453-464.

Mottier C.O. and Metz M.B. Vertical fusional vergence in patients with superior oblique muscle palsies. *American Orthoptic Journal* 1990, 40:88-93.

Oohira A. Vergence Eye Movements Facilitated by Saccades. *Japanese Journal of Ophthalmology* 1993, 37:400-413.

Oohira A., Zee D.S. and Guyton D.L. Disconjugate adaptation to long-standing, large-amplitude, spectacle-corrected anisometropia. *Investigative Ophthalmology & Vision Science* 1991, 32:1693-1703.

Peli E. Blink vergence in an antimetropic patient. *American Journal of Optometry & Physiological Optics* 1986, 63:981-984.

Schor C., M, Gleason G., Maxwell J. and Lunn R. Spatial aspects of vertical phoria adaptation. *Vision Research* 1993;33(1):73-84 1993,

Schor C.M., Gleason G. and Lunn R. Interactions between short-term vertical phoria adaptation and nonconjugate adaptation of vertical pursuits. *Vision Research* 1993, 33:55-63.

Schor C.M., Gleason J. and Horner D. Selective nonconjugate binocular adaptation of vertical saccades and pursuits. *Vision Research* 1990, 30:1827-1844.

Sethi B. Vergence adaptation: a review. *Documenta Ophthalmologica Documenta Ophthalmologica* 1986 Sep 30;63(3):247-63 1986, 63:247-63.

Sharma K. and Abdul-Rahim A.S. Vertical fusion amplitude in normal adults. *American Journal of Ophthalmology* 1992, 114:636-637.

Ygge J. and Zee D.S. Yoking of the eyes for vertical saccades. In *Contemporary Ocular Motor and Vestibular Research*: A Tribute to David A. Robinson. Fuchs A.F., Brandt T., Buttner U. and Zee D.S (Eds) 1994, pp 333-335. Thieme Medical Publishers, New York.

Ygge J. and Zee D.S. Control of vertical eye alignment in three-dimensional space. *Vision Research* 1995, 35:3169-3181.

Zee D.S., Fitzgibbon E.J. and Optican L.M. Saccade-vergence interactions in humans. *Journal of Neurophysiology* 1992, 68:1624-1641.

Zee D.S. and Levi L. Neurological aspects of vergence eye movements. *Revue Neurologique* (Paris) 1989, 145:613-620.

Accommodation and Vergence Mechanisms
in the Visual System
ed. by O. Franzén, H. Richter and L. Stark
© 2000 Birkhäuser Verlag Basel/Switzerland

Predicting accommodative performance in difficult conditions: A behavioral analysis of normal variations of accommodation

D. Alfred Owens,[1] Jeffrey T. Andre,[1] & Robert L. Owens[2]

1 Franklin & Marshall College, Whitely Psychology Laboratories, Lancaster, Pennsylvania, USA
2 Dr. Robert L. Owens & Associates Optometrists, New Holland, Pennsylvania, USA

Summary. The efficiency of accommodation varies widely and often deteriorates in difficult visual conditions. These variations degrade visual performance and, therefore, pose an interesting problem for clinical practice and for human factors engineering. In many situations, accommodation is biased toward an intermediate resting posture, operationally defined as the *dark focus.* Laboratory studies have revealed large inter-individual variations in dark focus that are closely related to variations in focusing accuracy and oculomotor adaptation to near tasks. New evidence also indicates that variations of performance for a given individual depend on autonomic arousal as well as optical conditions. Recent studies show that it is possible to assess these individual differences using standard clinical instrumentation, opening the way to systematic investigation of the practical benefits of assessment of accommodation in the interest of optimizing task performance.

Historical Background

In his seminal theory of vision, Descartes (1637) presented the first modern account of accommodation and binocular vergence. From basic optical principles and anatomical observations, he reasoned that these oculomotor adjustments are essential to attain both focus and fusion of the binocular sensory inputs, and to perceive the third dimension. In retrospect, the insight of Descartes' account is astonishing. He anticipated the Young-Helmholtz explanation of the mechanism of accommodation, and he launched the theory of distance "cues" as the means of three-dimensional visual perception (Boring, 1942). One could say without too much exaggeration that the following three centuries of research has been spent working out the details.

One of these details, which will be the focus of the present discussion, is the fact that the behavior of accommodation and vergence is highly variable. Although these mechanisms can be exceedingly precise and efficient, they are not always so. In fact, they are inclined to fail in a wide range of difficult, yet commonplace, conditions. The first evidence for such variations in efficiency came from observations of difficulties in challenging viewing conditions.

Lord Maskelyne (1789), the 4th Astronomer Royal at Greenwich, was probably the first to publish evidence of variation in the efficiency of accommodation when he reported that his nighttime observations were facilitated by a negative lens, which was of no benefit in bright conditions. Although he explained this phenomenon in terms of chromatic aberration ("the differential refrangibility of light"), later studies indicated something else was happening. The phenomenon of *twilight* or *night* myopia was rediscovered by several investigators over the next 175 years (Levene, 1965). While it was generally recognized that increased optical aberrations that accompany pupillary dilation may play a role, such aberrations could not account for the wide individual differences in the magnitude of night myopia (Wald & Griffin, 1947; Otero, 1951; Knoll, 1952; Mellerio, 1966). This became most evident when later workers found similar anomalous myopias also occur in bright viewing conditions as when viewing a bright empty sky (Whiteside, 1952) and when using optical instruments (Schober, 1954). As with night myopia, both "empty-field myopia" and "instrument myopia" exhibited unpredictable individual differences. And because they both occur in high luminance, the latter anomalous myopias could not be attributed to increased optical aberrations.

In 1954, Herbert Schober proposed that the anomalous myopias might be understood in terms of a simple process, the natural tendency of accommodation to return to an intermediate resting posture, *die Akkommodationsruhelage*, in the absence of adequate stimulation. Though appealing for its simplicity, this hypothesis was contrary to the long accepted theory, which had been advanced by Helmholtz (1909) and Donders (1864), that accommodation relaxes at the far point of its operating range. Yet the notion of an intermediate resting state was not contrary to the evidence; nor was the idea unprecedented. In fact, several noted theorists, including Weber (1855; cited in Cornelius, 1861), Cogan (1937), and Morgan (1946, 1957), had already advocated this viewpoint prior to Schober's paper. Perhaps for lack of empirical support, however, this hypothesis had not gained much attention.

In this context, one can find dual motivations for studying variations in the efficiency of accommodation and vergence. From a practical standpoint, variations in accuracy of focus and fixation could seriously impede visual detection, recognition, and localization. Such difficulties are often troublesome, as in the case of astronomers and microscopists, and they could be hazardous, as in the case of pilots or night drivers. Moreover, these practical difficulties reflected a gap in our basic understanding. A fuller explication of the normal variations in the efficiency of accommodation and vergence is fundamental to understanding oculomotor behavior; it could be integral to theories of visual perception, and it should provide key insights for uncovering the neural control of these systems.

Behavioral evidence for the intermediate resting state hypothesis

Interest in anomalous myopias emerged again in the 1970's as Herschel Leibowitz and his students began to investigate the visual problems of photointerpreters and night drivers from the perspective of human factors, which links basic behavioral science to engineering. Taking advantage of the availability of low-cost lasers, Hennessy and Leibowitz (1970, 1972) developed a new optometer, which offered several advantages for research on accommodation. First, its test pattern, generated by optical interference of a reflected laser beam, did not provide a "blur" stimulus and, therefore, could be used to assess the "open-loop" or resting state of accommodation. Second, unlike more complex infrared optometers, which provide relative measures of refractive power,

Figure 1. Herschel Leibowitz testing a student's accommodation using a laser optometer.

the laser-Badal optometer provides absolute measures of refractive power. Third, because of its simple optical design, the laser optometer could be configured to investigate accommodation of untrained subjects in a wide variety of viewing conditions, including viewing wide-field natural scenes and when using optical instruments.

A series of investigations with the laser optometer yielded new evidence in support of Schober's hypothesis that all the anomalous myopias result from a common process, namely the return of accommodation to an intermediate resting state. These studies also revealed unexpected individual differences. Figure 2 illustrates data from a sample of 220 college students, who had emmetropic or optically corrected vision (Leibowitz & Owens, 1978). Contrary to traditional theory, resting state measures, defined operationally as the *dark focus*

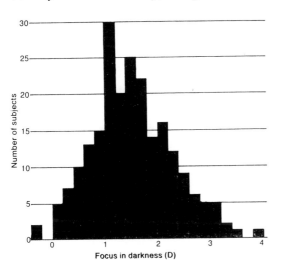

Figure 2. Dark focus distribution for 220 college-aged individuals. (From Owens, 1984; reprinted with permission.)

of accommodation, had an average value of 67 cm (1.5 Diopters, D), and ranged from low hyperopia to as near as 25 cm (-0.25 to 4 D).

The laser optometer was used to measure anomalous myopias of college students as they viewed a dim outdoor environment, an unstructured *Ganzfeld*, and a grating pattern imaged in a dissection microscope. Consistent with previous reports, the results showed wide individual differences in the magnitude of anomalous myopia, covering a range of about 3 D in all three conditions.

More important, comparison with measures of the dark focus for the same subjects showed that the anomalous myopia were highly correlated with the intermediate resting posture (Figure 3).

Other studies, using a wide variety of test conditions and subject populations, have reported evidence that when accommodation fails to respond accurately, the eyes' focus adopts an intermediate refractive state. This behavioral tendency appears to be pervasive (Table 1). This occurs, for example, with targets comprised of only low or high spatial frequency, with iso-luminant chromatic images, and with images created by optical interference that is independent of focus; it occurs when viewing a bright fixation point or retinoscope filament. It occurs when adults view attentively through pinhole pupils, which increase the eyes' depth of field, and when they are inattentive, "spacing-out" while viewing a complex field with normal pupils. It is interesting to note that young infants, who have not yet developed mature spatial vision or accommodation, tend to maintain an

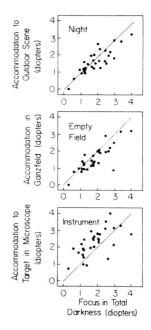

Figure 3. Relationship between dark focus and anomalous myopias, measured in dim light, a bright empty field, and a microscope. (From Leibowitz & Owens, 1975a; reprinted with permission.)

intermediate focus, which appears to depend more on autonomic arousal than on optical distance. Although not all of these accommodative biases have been correlated empirically with the dark focus, a consistent bias toward intermediate accommodation is evident. There is little, if any, behavioral evidence that accommodation "relaxes" at the far point. To the contrary, clinical assessment of refractive error demands special effort to induce negative accommodation to reach the far point.

From a behavioral standpoint, the evidence clearly supports the hypothesis of Weber, Cogan, Morgan, and Schober — accommodation returns to an intermediate resting posture whenever stimulation is degraded or the quality of the retinal image is independent of ocular focus. Much of the evidence for the intermediate resting state hypothesis derives from conditions of poor visibility, such as darkness, or artificially contrived targets, such as iso-luminant stimuli. One might question, therefore, whether the dark focus bias is relevant to more realistic visual tasks. At least two lines of investigation address this concern.

An interesting example of the dark-focus bias under naturalistic conditions was discovered by the ophthalmologist, Joseph Mandelbaum (1960). On a summer holiday, he observed

Table 1. Some of the situations in which accommodation is biased towards an intermediate resting position.	

Empty field	Whiteside, 1952; Westheimer, 1957; Leibowitz & Owens, 1975a; Benel & Benel, 1979
Fixation point	Owens & Leibowitz, 1975; Miller, 1980
Focus stabilization	Kruger et al., (1997)
Fogged vision (strong plus lenses)	Reese & Fry, 1941; Heath, 1956
Inattentive/"spacing-out"	Francis et al., 1989
Infancy	Haynes, White & Held, 1965
Intervening surfaces	Mandelbaum, 1960; Owens, 1979; Adams & Johnson, 1991
Iso-luminant contours	Wolfe & Owens, 1981
Low & high spatial frequencies	Owens, 1980; Mathews & Kruger, 1994
Low contrast	Luckiesh & Moss, 1940; Heath, 1956; Westheimer, 1957
Low luminance	Otero, 1951; Campbell & Primrose, 1953; Alpern, 1958; Leibowitz & Owens, 1975b
Monochromatic light	Kruger et al., (1997)
Optical instruments	Schober et al., 1970; Hennessy, 1975
Optical interference patterns	Hennessy & Leibowitz, 1970; Leibowitz & Owens, 1975b
Pinholes pupils	Hennessy et al., 1976; Ward & Charman, 1985
Retinoscope	Owens, Mohindra & Held, 1980
Reversed chromatic aberration	Kruger et al., (1997)
Visual fatigue	Östberg, 1980

that a distant sign was sometimes difficult to read from the screened-in porch of his cottage. This seemed odd because the sign was legible from outside the porch. Testing family and friends, he found the difficulty was experienced only by some individuals, and only when they were at certain distances from the screen. The difficulty was not experienced by older observers, nor by younger adults whose accommodation had been disabled by cycloplegia. Mandelbaum concluded that he had discovered a "phenomenon of accommodation."

Later experiments analyzed the Mandelbaum effect under controlled laboratory conditions and found that the key variable is the location of the intervening optical structure with respect to the observer's dark focus (Owens, 1979). When presented with objects superimposed at different distances, accommodation must "choose" which to focus. This choice is an interesting process that has received little scientific investigation. Under most circumstances, it is apparently accomplished by selective attention and binocular fixation. But in situations where an intervening unattended surface is located at the optical distance of dark focus, accommodation can be captured involuntarily by the irrelevant image. This interference is most troublesome when the head is stationary, precluding separation of the superimposed images through parallax, and view is monocular, minimizing the role of binocular vergence.

With practice, one can learn to overcome this unintentional bias. More important from the present viewpoint, the Mandelbaum Effect represents a strong behavioral bias toward the dark focus for young observers in bright, highly structured viewing conditions.

A second situation in which performance of complex tasks involves the dark focus is near work. Here the evidence points toward a general advantage for tasks that correspond to the optical distance of the dark focus. In his widely cited dissertation on variations of visual acuity as a function of distance, Chris Johnson (1976) found that accommodation is most accurate for stimuli that are located at the observer's dark focus. Previous research had shown that the slope of the accommodation response function declines with degradation of the stimulus (e.g., Heath, 1956; Toates, 1972). As illustrated in Figure 4, Johnson's study revealed the fact that these variations of slope pivot around the individual's dark focus. That is, accommodation for a stimulus at the dark focus was invariant and accurate, regardless of the quality of the target.

Johnson's finding has important implications for designing visual tasks that optimize visual performance (e.g., Kintz & Bowker, 1980; Jaschinski-Kruza, 1988) and it stimulated renewed interest in oculomotor behavior during stressful near work. A great deal of this work has focused on adaptive variations of the resting posture that are induced by near tasks (e.g., Östberg, 1980; Ebenholtz & Zander, 1987; Schor, Johnson & Post, 1984; Owens & Wolf-Kelly, 1986; Wolf, Ciuffreda & Jacobs, 1987). In this context, one should note evidence that even prolonged and difficult visual tasks produce little or no oculomotor adaptation when the visual display is located at the observers' resting posture (e.g., Miller et al., 1983; Ebenholtz, 1992).

In summary, evidence from a wide range of test conditions and subject populations has established that accommodative behavior is highly variable. Accurate accommodation appears to be unusual behavior. Variations in focusing accuracy are commonplace, even for healthy young adults with normal vision, and the errors of accommodation are systematic, tending to be myopic for distant targets and hyperopic for near targets. The evidence supports the theory that *accommodation behaves around an intermediate resting posture,* and active responses are required to achieve focus for farther as well as nearer

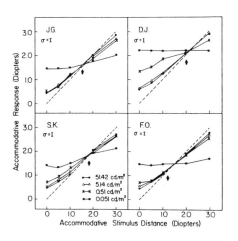

Figure 4. Accommodative response functions pivot near an individual's dark focus as luminance decreases. (From Johnson, 1976; reprinted with permission.)

distances. This resting posture has been extensively investigated through studies of the dark focus. Central to this work is the fact that the dark focus exhibits wide individual differences, which are highly correlated with focusing errors like the anomalous myopias and the Mandelbaum Effect. Conversely, the dark focus also predicts where accommodation will be most accurate and least susceptible to disturbance by stressful tasks.

Practical Implications

Performance of many visual tasks is limited by the quality of the retinal image. And the literature on variations in the accuracy of accommodation is linked repeatedly with problems of task performance. On one side, Lord Maskelyne's difficulty with night observations, like problems later encountered with optical instruments and high-altitude flight, are associated with the failure of accommodation to attain distant focus. On the other side, difficulties with near work, such as fatigue and asthenopia, have been associated with problems of accommodation in maintaining near focus (Owens, 1991). If the preceding theoretical approach is valid, then the intermediate resting state may prove useful in ameliorating these practical difficulties. As noted earlier, studies of the dark focus were initially motivated by efforts to optimize task performance. Those efforts have continued with emphasis on anomalous myopia and problems of near work.

The strong correlations illustrated in Figure 3 suggested that an individual's dark focus value could be used to predict and correct anomalous myopias. Several studies have investigated this hypothesis with encouraging results. Post et al. (1979) found that an optical correction based on the dark focus greatly improves detection of small targets presented in a bright empty field. A subsequent study by Luria (1980) showed that such correction of "empty-field myopia" is most beneficial for improving visibility of small targets. Efforts to predict and correct night myopia have been less successful. An early study showed that an optical correction based on the observer's dark focus can significantly improve acuity under simulated conditions of night driving, and field testing confirmed that drivers will accept and appreciate such night corrections (Owens & Leibowitz, 1976). Unlike empty-field myopia, however, the optimal night driving correction was equal to one-half the value predicted by the individual's dark focus. One possible reason that weaker corrections are better for night myopia is that, unlike a *Ganzfeld*, the night driving environment still provides some stimuli for accommodation. Thus, the driver's accommodation may adopt a posture that represents a compromise between optimal focus and the dark focus, as seen in the intermediate response slopes of Johnson's data (Figure 4). Later studies showed that the severity of night myopia depends on binocular vergence as well as the dark focus (Leibowitz, Gish & Sheehy, 1988).

Recent research by Morse, Kotulak, and Rabin (1997), which aims to optimize night visual performance of military pilots, has found that the best correction for night myopia depends on both the individual's CA/C ratio and dark focus.

The potential importance of visual detection and recognition for common, yet dangerous, tasks such as night driving justifies further research on correction of night myopia. Among the objectives for such work should be (1) identification of a viable clinical approach to correction of night myopia, and (2) determination of the incidence of night myopia in the general population.

With regard to near work, it is possible that a better understanding of the resting state will be useful in addressing the difficulties of asthenopia. Since the early epidemiology of Ramazzini (1713), it has been widely speculated that accommodation contributes to visual fatigue and asthenopia. Consistent with this hypothesis, numerous modern studies found that near work induces a proximal shift of the resting states of accommodation and vergence (e.g., Östberg, 1980). One study of college students indicated that asthenopic symptoms are differentially associated with accommodation and vergence, with blurred distance vision accompanying changes of the dark focus and subjective discomfort accompanying adaptive changes of dark vergence (Owens & Wolf-Kelly, 1987).

Aside from the problems of fatigue, the quality of task performance could be *enhanced* through consideration of the dark focus. As noted earlier, accommodation is most accurate for displays located at the individual's dark focus. Other studies have shown that the dark focus varies as a function of gaze inclination, falling at nearer distances with downward gaze and at greater distances with elevated gaze (Heuer & Owens, 1989). These findings suggest that performance can be enhanced by configuring visual displays to place key information at an optical distance of the worker's dark focus. It is also plausible that an optical aid, based on the worker's dark focus, could be beneficial. For example, a recent study by Best, Littleton, Gramopadhye, and Tyrrell (1995) showed that performance of a proof reading task improved when the video display was located optically at the reader's dark focus. Weisz (1980) proposed that the dark focus could be useful in prescribing reading corrections in early presbyopia, although this hypothesis has not been investigated systematically. It is interesting to note that recent studies indicate that autonomic arousal modulates the effects of near tasks on accommodation and vergence (e.g., Tyrrell, Pearson & Thayer, 1997). This suggests that optimization of vision for near work will require consideration of more than optical variables!

Practical application of research requires availability of efficient and economical methods for implementation. Most of the research on accommodation has utilized instruments that are not widely available in clinical settings. Although investigators have pointed to the possibility of assessing the dark focus with a standard retinoscope (e.g., Owens, Sterling & Owens, 1982; Hope & Rubin, 1984), this approach has not been exploited until quite recently. During the past few years, Robert Owens, O.D., has used a technique called *dark retinoscopy* to measure the dark focus of patients as part of a normal exam. His method is similar to conventional static retinoscopy, except that the examination room is darkened, and the patient is instructed to fixate the retinoscope filament with one eye, while the other eye is occluded. Previous studies had shown that accommodation does not respond to a bright dazzling stimulus (Owens et al., 1980). Therefore, if the patient's optical correction for distance is in place, the dark retinoscopy technique can provide a clinical estimate of the patient's dark focus.

Dr. Robert Owens' findings, summarized in Figure 5, show some resemblance to, and interesting differences from, the early data obtained from college students with the laser optometer (Figure 2). Similar to the earlier results, these clinical data show that the dark focus is nearer than the far point, but at 0.5 D the mean is lower than that for the college students. Also similar to the original finding, the clinical measures show clear individual differences, although the range is smaller with dark focus values covering a range of about 1.5 D. A reason for the differences between college and clinical data cannot be specified with certainty. It may be due to a difference in method. There is evidence that dark focus measures are lower in passive subjects than in active subjects (Post, Johnson, & Owens, 1985). So, dark retinoscopy may give lower numbers because the patients are gazing passively rather than actively looking for a laser speckle pattern. The discrepancy could also reflect differences in the subject populations. The early sample was limited to students between the ages of 17 and 26, while the clinical sample included individuals from numerous occupations who ranged in age from 6 to 55 years. In any event, these clinical data represent the largest sample of dark focus values yet reported. They confirm that variations of the dark focus are characteristic of the general population, and they demonstrate that standard clinical equipment can be used to measure the dark focus. Further ex-

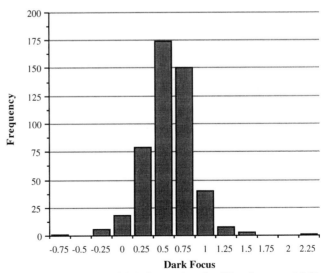

Figure 5. Distribution of dark focus values for 480 patients, aged 6-55, measured with dark retinoscopy.

periments are currently under way to investigate (1) the effects of task demands (active vs. passive) on measures of the dark focus, and (2) the utility of dark retinoscopy in prescribing corrections for night myopia.

Conclusion

The behavior of ocular accommodation is highly variable both across situations and across individuals. A growing body of evidence indicates that variations in the accuracy and efficiency of accommodation can be understood parsimoniously with reference to the intermediate resting posture. Accommodative responses are often biased toward the intermediate resting posture, giving rise to focusing errors for stimuli that fall farther or nearer than the resting focus, and resulting in consistently accurate focus for stimuli that correspond to the dark focus distance.

Much remains to be learned about the normal behavior of accommodation, and variations appear to be a key issue in this domain. We do not understand the basis of individual differences in the dark focus, nor the basis of intra-individual variations that are associated with attention, perception, and arousal. Past research on these questions was motivated by practical difficulties with challenging visual tasks. It seems likely that further insights regarding accommodative behavior will help to resolve such difficulties, and that continued attention to practical problems will help to advance our theoretical efforts.

Acknowledgments
Preparation of this manuscript was supported by Franklin & Marshall College and NIH grant EY06673-02.

References

Adams, C. W., & Johnson, C. A. (1991). Steady-state and dynamic response properties of the Mandelbaum effect. *Vision Research*, 31(4), 751-760.

Alpern, M. (1958). Variability of accommodation during steady fixation at various levels of luminance. *Journal of the Optical Society of America*, 48, 193-197.

Benel, R. A, & Benel, D. C. R. (1979). *Accommodation in untextured stimulus fields* (Eng Psy-79-1/AFOSR-79-1). Washington DC: Air Force Office of Scientific Research.

Best, P. S., Littleton, M. H., Gramopadhye, A. K. & Tyrrell, R. A. (1995). Relations between individual differences in oculomotor resting states and visual inspection performance. *Ergonomics*, 39(1), 35-40.

Boring, E. G. (1942). *Sensation and Perception in the History of Experimental Psychology*. New York: D. Appleton-Century.

Campbell, F. W., & Primrose (1953). The state of accommodation of the human eye in darkness. *Transactions of the Ophthalmological Societies of the United Kingdom*, LXXIII, 353-361.

Cogan, D. G. (1937). Accommodation and the autonomic nervous system. *Archives of Ophthalmology*, 18, 739-766.

Cornelius, C. S. (1861). *Die Theorie des Sehens und räumlichen Vorstellens* (pp. 283-285). Halle: H. W. Schmidt.

Decartes, R. (1637). *Discourse on method, optics, geometry, and meteorology*. English translation by P. J. Olscamp. Indianapolis IN: Bobbs-Merrill, 1965.

Donders, F. C. (1864). *On the anomalies of accommodation and refraction of the eye*. London: The Sydenham Society.

Ebenholtz, S. M., & Zander, P. A. (1987). Accommodative hysteresis: Influence on closed loop measures of far point and near point. *Investigative Ophthalmology & Visual Science*, 28, 1246-1249.

Ebenholtz, S. M. (1992). Accommodative hysteresis as a function of target-dark focus separation. *Vision Research*, 32(5), 925-929.

Francis, E. L., Jiang, B. C., Owens, D. A., Tyrrell, R. A., & Leibowitz, H. W. (1989). "Effort to see" affects accommodation and vergence but not their interactions. *Investigative Ophthalmology & Visual Science*, , 30(3), 135.

Heath, G. G. (1956). The influence of visual acuity on accommodative responses of the eye. *American Journal of Optometry*, 33, 513-524.

Haynes, H. M., White, B. L., & Held, R. (1965). Visual accommodation in human infants. *Science*, 148, 528-530.

Helmholtz, H. (1909). *Handbook of Physiological Optics* (Third Edition, Volume 1). Hamburg: Voss. English translation by J. P. C. Southall, Dover Publications, 1962.

Hennessy, R. T. & Leibowitz, H. (1970). Subjective measurement of accommodation with laser light. *Journal of the Optical Society of America*, 60, 1700-1701.

Hennessy, R. T. & Leibowitz, H. (1972). Laser optometer incorporating the Badal principle. *Behavioral Research Methods & Instrumentation*, 4, 237-239

Hennessy, R. T. (1975). Instrument myopia. *Journal of the Optical Society of America*, 65, 1114-1120.

Hennessy, R. T., Iida, T., Shiina, K., & Leibowitz, H. W. (1976). The effect of pupil size on accommodation. *Vision Research*, 16, 587-589.

Heuer, H., & Owens, D. A. (1989). Vertical gaze direction and the resting posture of the eyes. *Perception*, 18, 363-377.

Hope, G. M., & Rubin, M. L. (1984). Night myopia. *Survey of Ophthalmology*, 29, 129-136.

Jaschinski-Kruza, W. (1988). Visual strain during VDU work: the effect of viewing distance and dark focus. *Ergonomics*, 31(10), 1449-1465.

Johnson, C. A. (1976). Effects of luminance and stimulus distance on accommodation and visual resolution. *Journal of the Optical Society of America*, 66, 138-142.

Kintz, R. T., & Bowker, D. O. (1980). Visual accommodation during a prolonged search task. *Applied Ergonomics*, 13: 55-59.

Knoll, H. A. (1952). A brief history of "nocturnal myopia" and related phenomena. *American Journal of Optometry*, 29, 69-81.

Kruger, P. B., Mathews, S., Katz, M., Aggarwala, K. R., Nowbotsing, S. (1997). Accommodation without feedback suggests directional signals specify ocular focus. *Vision Research.*, in press.

Leibowitz, H. W. & Owens, D. A. (1978). New evidence for the intermediate position of relaxed accommodation. *Documenta Ophthalmologica*, 46, 133-147.

Leibowitz, H. W., Gish, K. W. & Sheehy, J. B. (1988). Role of vergence accommodation in correcting for night myopia. *American Journal of Optometry and Physiological Optics*, 65, 383-386.

Leibowitz, H. W. & Owens, D. A. (1975a). Anomalous myopias and the intermediate dark focus of accommodation. *Science*, 189, 646-648.

Leibowitz, H. W. & Owens, D. A. (1975b). Night myopia and the intermediate dark-focus of accommodation. *Journal of the Optical Society of America*, 65, 1121-1128.

Levene, J. R. (1965). Nevil Maskelyne, F.R.S. and the discovery of night myopia. *Notes and Records of the Royal Society of London*, 20, 100-108.

Luckiesh, M., & Moss, F. K. (1940). Functional adaptation to near vision. *Journal of Experimental Psychology*, 26, 352-356.

Luria, S. M. (1980). Target size and correction of empty-field myopia. *Journal of the Optical Society of America*, 70, 1153-1154.

Mandelbaum, J. (1960). An accommodation phenomenon. *American Medical Association Archives of Ophthalmology*, 63, 923-926.

Maskelyne, N. (1789). An attempt to explain a difficulty in the theory of vision, depending on the different refrangibility of light. *Philosophical Transactions of the Royal Society*, 19, 256-264.

Mathews, S., & Kruger, P. B. (1994). Spatiotemporal transfer function of human accommodation. *Vision Research*, 34(15), 1965-1980.

Mellerio, J. (1966). Ocular refraction at low luminance. *Vision Research*, 6, 217-237.

Miller, R. J. (1980). Ocular-vergence-induced accommodation and its relation to dark focus. *Perception & Psychophysics*, 28 (2), 125-132.

Miller, R. J., Pigion, R. G, Wesner, M. F. & Patterson, J. G. (1983). Accommodation fatigue and dark focus: The effects of accommodation-free visual work as assessed by two psychophysical methods. *Perception & Psychophysics*, 34, 532-540.

Morgan, M. W. (1946). A new theory for the control of accommodation. *American Journal of Optometry and Archives of American Academy of Optometry*, 23, 99-110.

Morgan, M. W. (1957). The resting state of accommodation. *American Journal of Optometry and Archives of American Academy of Optometry*, 34, 347-353.

Morse, S. E., Kotulak, J. C., Rabin, J. C. (1997). The Influence of oculomotor function on the optical correction for night myopia. This volume.

Östberg, O. (1980). Accommodation and visual fatigue in display work. In E. Grandjean & E. Vigliani (Eds.), *Ergonomic aspects of visual display terminals* (pp. 41-52). London: Taylor & Francis.

Otero, J. M. (1951). Influence of the state of accommodation on the visual performance of the human eye. *Journal of the Optical Society of America*, 41, 942-948.

Owens, D. A. & Leibowitz, H. W. (1975). The fixation point as a stimulus for accommodation. *Vision Research,* 15, 1161-1163.

Owens, D. A. & Leibowitz, H. W. (1976). Night myopia: Cause and a possible basis for amelioration. *American Journal of Optometry & Physiological Optics,* 53, 709-717.

Owens, D. A. & Wolf-Kelly, K. (1987). Near work, visual fatigue, and variations of oculomotor tonus. *Investigative Ophthalmology & Visual Science,* 28, 743-749.

Owens, D. A., Mohindra, I., & Held, R. (1980). The effectiveness of a retinoscope beam as an accommodative stimulus. *Investigative Ophthalmology & Visual Science,* 19, 942-949.

Owens, D. A. (1979). The Mandelbaum Effect: Evidence for an accommodative bias toward intermediate viewing distances. *Journal of the Optical Society of America,* 69, 646-652.

Owens, D. A. (1980). A comparison of accommodative responsiveness and contrast sensitivity for sinusoidal gratings. *Vision Research,* 20, 159-167.

Owens, D. A. (1984). The resting state of the eyes. *American Scientist,* 72, 378-387.

Owens, D. A. (1991). Near work, accommodative tonus, and myopia. In T. Grosvenor & M. Flom (Eds.), *Refractive anomalies: Research and clinical applications* (pp. 318-344). Boston: Butterworth-Heineman.

Owens, R. L., Sterling, R. H. & Owens, D. A. (1982, December). *The clinical determination of night driving corrections.* Paper presented at the Annual Meeting of the American Academy of Optometry. Philadelphia, PA.

Post, R. B., Owens, R. L., Owens, D. A., & Leibowitz, H. W. (1979). Correction of empty-field myopia on the basis of the dark focus of accommodation. *Journal of the Optical Society of America,* 69, 89-92.

Post, R. B., Johnson, C. A., & Owens, D. A. (1985). Does performance of tasks affect the resting focus of accommodation? *American Journal of Optometry & Physiological Optics,* 62(8), 533-537.

Ramazzini, B. (1713). *Diseases of workers.* English translation by W. C. Wright. Chicago: University of Chicago, 1940.

Reese, E. E. & Fry, G. A. (1941). The effect of fogging lenses on accommodation. *American Journal of Optometry and Archives of Optometry,* 18, 9-16.

Schober, H. (1954). Über die akkommodationsruhelage. *Optik,* 6, 282-290.

Schober, H., Dehler, H., & Kassel, R. (1970). Accommodation during observations with optical instruments. *Journal of the Optical Society of America,* 60, 103-107.

Schor, C. M., Johnson, C. A. & Post, R. B. (1984). Adaptation of tonic accommodation. *Ophthalmic & Physiological Optics,* 4, 133-137.

Toates, F. M. (1972). Accommodative function of the human eye. *Physiological Reviews,* 52, 828-863.

Tyrrell, R. A., Pearson, M. A. & Thayer, J. F. (1997). Behavioral links between the oculomotor and cardiovascular systems. This volume.

Wald, G. & Griffin, D. R. (1947). The change in refractive power of the human eye in dim and bright light. *Journal of the Optical Society of America,* 37, 321-336.

Ward, P. A., & Charman, W. N. (1985). Effect of pupil size on steady state accommodation. *Vision Research,* 25(9), 1317-1326.

Weisz, C. L. (1980). The accommodative resting state: Theoretical implications of lens prescriptions. *Review of Optometry,* 117, (7), 60-70.

Westheimer, G. (1957). Accommodation measurements in empty visual fields. *Journal of the Optical Society of America,* 47, 714-718.

Whiteside, T. C. D. (1952). Accommodation of the human eye in a bright and empty visual field. *Journal of Physiology (London),* 118, 65-66.

Wolf, K. S., Ciuffreda, K. J., & Jacobs, S. E. (1987). Time course and decay of effects of near work on tonic accommodation and tonic vergence. *Ophthalmic and Physiological Optics,* 7, 131-135.

Wolfe, J., & Owens, D. A. (1981). Is accommodation colorblind? Focusing chromatic contours. *Perception,* 10, 53-62.

The influence of oculomotor function on the optical correction for night myopia

S.E. Morse, J.C. Kotulak[1] and J.C. Rabin[1]

University of Houston College of Optometry, Houston, TX 77204
[1] *U.S.Army Aeromedical Research Laboratory, Ft. Rucker, AL 36362*

Summary. Night myopia is the increase in ocular refraction occurring in low levels of illumination. Although night myopia has been known for centuries, there is still no effective method for prescribing the optimal lens to correct the condition. One difficulty is that the tendency of accommodation to seek its dark focus at low luminance may be opposed by convergence. For example, if there is enough luminance to sustain binocular vision, accommodation is influenced by fusional vergence. Therefore, the best optical compensation for night myopia should depend not only on the dark focus of accommodation, but also on the influence that vergence has on accommodation. The degree to which fusional vergence influences accommodation is known as the convergence accommodation to convergence (CA/C) ratio. Our PURPOSE was to perform an experiment under realistic night myopia conditions (binocular viewing and natural pupils) which would allow us to relate the power of the best optical correction to individual differences in the dark focus of accommodation and CA/C ratio. METHODS. The optimal optical compensation for night myopia was determined by measuring visual acuity under binocular conditions across three luminances (0.04, 0.4, and 4.0 cd/m2) and four lens powers (0.0, -0.5, -1.0, and -1.5 diopters). The dark focus was measured with an infrared optometer, and CA/C ratio was measured with an infrared optometer and eyetracker. Sixteen subjects (mean age 25.4 years) with unaided visual acuity of 20/20 were tested. RESULTS. Regardless of CA/C ratio, the optimal lens power was significantly correlated with dark focus. However, the slope of the regression line relating lens power to dark focus was much steeper for subjects with low CA/C ratios than with high ones. CONCLUSIONS. It appears that variability in night myopia can be explained by intersubject differences in dark focus and CA/C ratio. We found that a lens equal to approximately the full dark focus value optimized vision in subjects with low CA/C ratios, and that a one-half dark focus correction did the same for subjects with high CA/C ratios. In addition, we also found evidence for a previously unknown natural mechanism that appears to limit the amount of night myopia through a non-random association of dark focus and CA/C. This process we termed "night emmetropization" because it appears to result in less night myopia than would be expected if there were only a chance association between CA/C and dark focus.

Introduction

Night myopia is a common condition that has defied effective treatment since its discovery two centuries ago (leibowitz et al, 1987, 1988). In night myopia, individuals with no refractive error under ordinary illumination become nearsighted under low light levels and suffer the same visual loss as those with true myopia. The search for an effective optical correction for night myopia has been frustrated by a lack of understanding of its physiologic mechanism. The first breakthrough came when it was learned that excessive accommodation was a major etiologic factor, which led to the view that night myopia is a specific tendency of accommodation to regress to its resting position under reduced illumination or degraded stimulus conditions (Leibowitz and Owens, 1975). Accommodation refers to the eye's ability to change focus such that its dioptric power matches viewing distance. However, as luminance decreases, this ability to adjust focus diminishes until accommodation reaches a fixed position known as the "dark focus," which is the level of accommodation in complete darkness. The result is that a myopic state develops, because in the absence of light, the eye is typically focused at about 1 M (Leibowitz and Owens, 1975).

Unfortunately, knowledge that accommodation tends to change toward the position of the dark focus under low luminance conditions has not been sufficient to provide a satisfactory correction for night myopia. This is because accommodation behaves in a peculiar way when lighting is dim but not completely dark, e.g., in some individuals, accommodation remains near the dark focus, while in others, it is closer to the target of interest. Leibowitz and colleagues (1987, 1988) accounted for this dichotomous behavior of accommodation by postulating a role of "fusional vergence" in night myopia. The primary function of fusional vergence is to converge or diverge the eyes to maintain single vision; the secondary function is that it is linked to and adjusts accommodation in such a way that the eyes tend to be focused and aligned for the same point in space. The accommodation that results from fusional vergence (vergence accommodation) varies among individuals according to the convergence accommodation to convergence (CA/C) ratio (Fincham and Walton, 1957). In those individuals with high CA/C ratios, the influence of fusional vergence is great and accommodation seeks the target rather than the dark focus under low light levels. Conversely, in individuals with low CA/C ratios, the influence of fusional vergence is small and accommodation regresses to the dark focus.

Our original contribution was to be the first to put Leibowitz's hypothesis to an empirical test (Kotulak et al, 1995). In this paper, we describe that experiment, interpret its scientific and applied significance, and offer suggestions for further research. We also describe and discuss the unexpected result which seems to demonstrate the existence of a natural process which limits the magnitude of night myopia in individuals who would otherwise be highly susceptible, i.e., those with low CA/C ratios. We refer to this process as "night emmetropization" after its apparent function of minimizing (emmetropizing) or producing less night myopia than expected in this group of individuals.

Methods

Optimal Lens

The methods, reported in detail elsewhere (Kotulak et al, 1995), are summarized below. There were sixteen subjects (mean age 25.4 years) with unaided visual acuity of at least 20/20 Snellen. The optimal night myopia corrective lens for each subject was determined by measuring visual performance at three different light levels using four different lens powers. Measurements were made binocularly to resemble natural viewing, and while pupils were dilated for optometer measures, artificial pupils were imaged at the entrance pupil of the eye which matched natural pupil size for each individual in each condition – again, to resemble natural viewing. Visual acuity (VA) and contrast sensitivity (CS) charts were computer-generated and displayed on a high-resolution video monitor. For VA, letter size varied by 0.1 log steps for each row (20/15 to 20/126), and for CS, letter size varied with luminance (20/126 at 0.04 cd/m^2, 20/64 at 0.4 cd/m^2, and 20/32 at 4.0 cd/m^2) while contrast varied by 0.1 log steps (93% to 5%). A different CS letter size was used at each luminance such that CS always was measured with a letter size slightly larger than the VA threshold to ensure that CS would provide a sensitive index of optical defocus (Rabin, 1994). A 3X4 block design was used with three luminance levels (0.04, 0.4, and 4.0 cd/m^2) and four lens powers (0.0, -0.5, -1.0, and -1.5 diopters (D). Luminance was presented in ascending order to save adaptation time and discourage learning effects, and lens powers were counterbalanced at each luminance level. VA and CS were measured at each luminance and with each lens power.

Oculomotor Responses

Oculomotor responses were measured while subjects viewed the letter charts binocularly. Accommodation was recorded with an infrared optometer that was integrated into a binocular, dual Purkinje image infrared eyetracker that measured eye position (vergence). CA/C ratios were determined by opening the accommodative feedback loop by imaging a 0.75 mm diameter artificial pupil into the entrance pupil of each eye and simulating fusional vergence by changing vergence demand from 0 to 6 prism diopters base out. The measured change in accommodation was divided by the change in vergence to determine the CA/C ratio. The infrared optometer required dilated pupil. Since this conflicted with our desire to use natural pupil size, we conducted experiments in two stages. In stage one, pupil size was measured at each light level without dilation using an infrared camera and monitor. In the second stage, visual performance and oculomotor data were recorded after the pupils were dilated with 2.5% phenylephrine, using artificial pupils that matched the natural pupil size in stage one.

Results

Oculomotor Responses

Mean $\pm$ SD dark focus and CA/C ratio were 0.84 $\pm$ 0.65 D and 0.59 $\pm$ 0.50 D/MA, respectively. Figure 1 shows the relationship between accommodation (at the 0.04 cd/m^2 condition), dark focus, and CA/C ratio. Subjects with the eight lowest CA/C ratios were placed in the low CA/C group (Figure 1a), while subjects with the eight highest CA/C ratios were placed into the high CA/C group (Figure 1b). Accommodation was related to dark focus for both groups (low CA/C r=0.90, p<0.01; high CA/C r=0.97, p<0.001). However, the slope of the regression line relating accommodation to dark focus was steeper for the low group (0.94 compared to 0.67). A similar relation was found at the other two luminances. Mean accommodation for low, medium, and high luminance was 0.78, 0.68, and 0.77 D, respectively. There was no significant difference among the means (F (1.92, 26.88) = 2.08, p > 0.12).

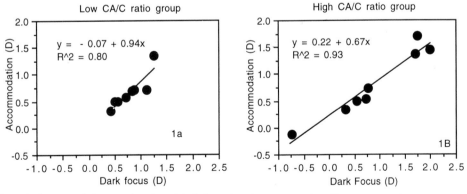

Figure 1. Accommodation as a function of dark focus. Luminance was 0.04 cd/m2, accommodative demand was 0.00 D, and vergence demand was 0.00 MA. Artificial pupils matched each subjects natural pupil size. Subjects were divided equally into low and high CA/C ratio groups. **A)** Low CA/C ratio group. **B)** High CA/C group.

Optimal Lens

Figure 2 shows VA and CS plotted as a function of lens power at each of the three light levels. Each data point represents the average performance of all 16 subjects. Both VA (Figure 2a) and CS (Figure 2b) peaked with the -0.50 D lens. However, this represents the performance of the group as a whole, and it should be noted that performance actually declined with this lens for some individuals. We defined the optimal lens as the one that provided best visual performance for each subject on an individual basis. In most cases, the lens that optimized VA also was the one that optimized CS. However, when a conflict occurred, the optimal lens was obtained by averaging the power of the best lens for VA with that of the best lens for CS.

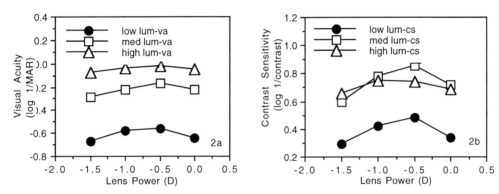

Figure 2. The effects of optical minus-lens overcorrection on visual performance. Each data point represents the mean for all subjects (N=16). No error bars are displayed as means are within-subject. **A** and **B** have equivalent units and ranges for ease of comparison. A) Visual Acuity. B) Contrast sensitivity.

Figure 3 shows the improvement in visual performance with the optimal lens over the control condition (0.00 D lens). Also shown is the improvement in visual performance over the control condition that would result if a -0.50 D overcorrection were given to all subjects. The latter is the best one-for-all lens for our sample of four lens powers. The optimal lens was significantly better than the control lens under all conditions, while the best one-for-all lens was no different than the control lens under four of the six conditions (see table 1). The average improvement in VA for the optimal lens was 0.09 log units (23%) compared to 0.053 log units (13%) for the best one-for-all lens. For CS, mean improvement with the optimal lens was 0.193 log units (56%) compare to 0.11 log units (29%) for the best one-for-all lens.

Luminance (cd/m^2)	Visual Acuity Increase	Visual Acuity P	Contrast Sensitivity Increase	Contrast Sensitivity P
Optimal Lens				
0.04	0.12	< 0.003 *	0.19	< 0.003 *
0.40	0.09	< 0.001 *	0.20	< 0.001 *
4.00	0.07	< 0.006 *	0.19	< 0.006 *
- 0.50 D Lens				
0.04	0.12	< 0.05	0.14	< 0.007 *
0.40	0.09	< 0.31	0.14	< 0.040 *
4.00	0.07	< 0.10	0.05	< 0.890

Table 3. Improvement in visual performance in Log units over the control condition (0.00 D) with the optimal lens and the best one-for-all lens (-0.50 D). Multiple comparisons were performed with paired t-tests and p values were adjusted for alpha inflation using the Bonferroni method. * Significant.

Figure 4 shows the relationship between the optimal lens power, dark focus, and CA/C ratio. The subjects were placed into low (n=7; Figure 4a), and high (n=9; Figure 4b), CA/C ratio groups depending on whether their individual CA/C ratios were lower or higher than 0.4 D/MA. This cutoff demarcated CA/C ratios which greatly limited regression of accommodation to the dark focus, from those that did not. It is evident from Figure 4 that optimal lens power is related to the dark focus for both groups. However, the slope was steeper for the low CA/C group (-1.12 compared to -0.36). This suggests that the optimal lens is stronger for a person with a low CA/C ration than for a person with a high CA/C ratio, given equal dark focuses.

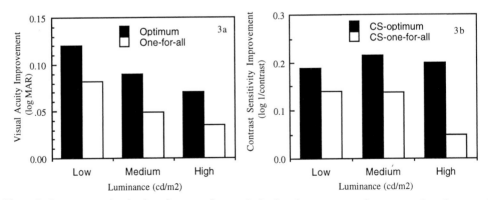

Figure 3. Improvement in visual performance from optical minus-lens overcorrection compared to the control condition of no lens (0.00 D lens). Error bars are not displayed as means are all within-subject. **A** and **B** have equivalent units and ranges for ease of comparison. **A**) Visual Acuity. **B**) Contrast sensitivity.

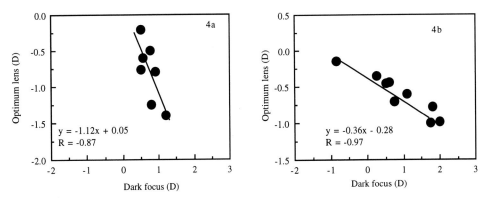

Figure 4. The effect of dark focus and CA/C ratio on optimal lens power. A) Subjects (N=7) with CA/C ratios lower than 0.4 D/MA. **B**) Subjects (N=9) with CA/C ratios higher than 0.4 D/MA.

Discussion

Our data confirm the hypothesis of Leibowitz and his coworkers (Leibowitz et al., 1987, 1988) that the optimal lens for night myopia is determined by both dark focus and CA/C ratio. For CA/C ratios less than 0.4 D/MA, a lens correction equalling the full dark focus produces the best result. For CA/C ratios above 0.4 D/MA, a correction equalling half the dark focus works best. However, if dark focus is small, say less than 0.50 D, then a full dark focus overcorrection was found to be preferable regardless of CA/C.

The visual performance improvements with the optimal night myopia lens are substantial, 23% for VA and 56% for CS. The latter measure is felt to be especially relevant to the mid and low spatial frequency detection tasks found under the degraded target conditions encountered by operators of automobiles and airplanes at night.

The present results indicate that it would be useful to determine the optimal lens correction for night myopia from clinical measures of CA/C and dark focus. It is possible to measure CA/C clinically, and this could be implemented as a standard clinical test. While there is presently no clinical test for measuring dark focus, this could be done with slight modifications to autorefractors.

We conclude by describing an unexpected finding which may have important theoretical implications. In Figure 1 and 4, it can be seen that the dark focus distributions appear to be different for the low and high CA/C groups. The range of dark focus valaues for the low group aappears to be narrower than for the high group. Indeed, this difference is statistically significant ($p < 0.05$ by the Levene test for variability). The more compact dark focus distribution of the low CA/C group could be beneficial in that this group has no mechanism to prevent extreme vaalues of dark focus from expressing themselves s extreme values of night myopia. A large saample size study needs to be done to determine whether the apparent difference in dark focus distributions between CA/C ratio groups is a general phenomenon. If so, it coud represent a night emmetropization process which limits the incidence of night myopia in the same way that the incidence of ordinary refractive error is held down by a non-random association of the eye's optical components, such as axial length and corneal curvature (Sorsby, 1979).

References

Fincham E.H., Walton J. (1957). The reciprocal actions of accommodation and vergence. *Journal of Physiology* (London); 137: 488-508.

Green D.G., and Campbell F.W. (1965) Effect of focus on the visual response to a sinusoidally modulated spatial stimulus. *Journal of the Optical Society of America*; 55:1154-1157.

Kotulak J.C. and Morse S.E. (1994) Relationship among accommodation, focus, and resolution with optical instruments. *Journal of the Optical Society of America*; 11: 71-79.

Kotulak J.C., Morse S.E., Wiley R.W. (1994) The effect of knowledge of object distance on accommodation during instrument viewing. *Perception*; 23: 671-679.

Kotulak J.C., and Schor C.M. (1987) The effects of optical vergence, contrast, and luminance on the accommodative response to spatially bandpass filtered targets. *Vision Research*; 27: 1797-1806.

Kotulak J.C., Morse S.E., Rabin J.C. (1995) Optical compensation for night myopia based on dark focus and CA/C ratio. *Investigative Ophthalmology and Vision Science*; 36: 1573-1580.

Leibowitz H.W., Gish K.W., Sheehy J.B. (1988) Role of vergence accommodation in correcting for night myopia. *American Journal of Optometry and Physiological Optics;* 65: 383-386.

Leibowitz H.W., Owens D.A. (1975) Anomalous myopias and the intermediate dark focus of accommodation. *Science*; 189: 646-648.

Leibowitz H.W., Sheehy J.B., Gish K.W. (1987) Correction of night myopia: The role of vergence accommodation. In: *Night Vision: Current Research and Future Directions.* Washington, DC: National Academy Press; 116-123.

Rabin J. (1994) Optical defocus: Differential effects on size and contrast letter recognition thresholds. *Investigative Ophthalmology and Vision Science*; 35: 646-648.

Sorsby A. (1979) Biology of the eye as an optical system. In: *Clinical Ophthalmology,* Duane TD, Ed, Volume 3. Hagerstown, MD: Harper and Row pp. 1-16.

Accommodation and Vergence Mechanisms
in the Visual System
ed. by O. Franzén, H. Richter and L. Stark
© 2000 Birkhäuser Verlag Basel/Switzerland

Effects of neck muscles proprioception on eye position and vergence movements

Ying Han and Gunnar Lennerstrand

Division of Ophthalmology, Department of Clinical Science, Huddinge University Hospital, Karolinska Institute, 141 86 Huddinge, Sweden

Summary. Perception of gaze direction can be influenced not only by visual input but also by extraretinal information from eye and neck muscle proprioception. Vibration of neck muscles with the head fixed is known to induce a visual illusory movement of a fixated object. A study of the effects of neck muscles proprioception on eye position and vergence movements has not been done before. Aims of this study are to investigate: 1) the effects of neck muscle proprioception on eye position during steady fixation; 2) the dynamics of the accommodative vergence movements driven by the dominant and non-dominant eye; 3) the effect of neck proprioception on accommodative vergence movements, driven by each eye. Subjects sat on a chair with the head fixed and one eye covered. Vibration was applied on three groups of neck muscles, two of which turn the head horizontally and the third group extend the head when the muscles contract. Accommodative stimulation was presented by a bar of LEDs lit in sequence. Eye movements were measured by an infrared system (Ober-2) during three stimulation conditions: 1). Vibration on neck muscles alone; 2). Accommodative stimulation alone in monocular viewing by each eye; 3). Vibration on neck muscles of the groups which turn the head horizontally, in combination with accommodative stimulation. Neck muscles vibration during fixation of a steady visual target could induce eye position changes which were related in directions to the muscles vibrated. During accommodative stimulation alone, the dynamics of vergence movements were different depending on whether the dominant or the non-dominant eye was driving the accommodation. The gain was higher when the non-dominant eye was fixating. Vibration on the neck muscles shortened the time constant of the accommodative vergence movement to square wave stimulation by the non-dominant eye. Extraretinal signals from neck proprioceptions could influence the static eye position, and the dynamics of accommodative vergence.

Introduction

The accommodative vergence system has both sensory and motor components, leading to adjustment of focus and alignment of the eyes with great precision. A previous study of dynamics of vergence movements, driven by pure accommodation, has shown that there were differences in the gain of the accommodative vergence movements under open-loop conditions (with one eye covered) depending which eye was driving accommodation.

Perception of gaze direction can be influenced not only by visual but also by extraretinal information from eye and neck muscle proprioception. Muscle spindles are stretch receptors and they can be selectively activated by vibration of the muscle tendon (Roll & Vedel 1982). The primary spindle endings are predominantly sensitive to dynamic stretching (Matthews 1982). It was reported from a microneurography study that the primary spindle endings were able to respond to each vibration cycle by one or several spikes in the vibration frequency range of 10 - 120 Hz (Roll et al 1989). Vibration of neck muscles with the head fixed is known to induce a visual illusion of movement of a fixated object (Biguer et al, 1988). This indicated that the afferent signal from muscle spindles would seem interpreted as if the muscle was pulled passively or stretched by the contraction of the antagonist muscles, and that neck muscle vibration would be interpreted as change of the head position.

A study of the effects of neck muscles proprioception on eye position and vergence movements has not been done before. Such investigations are of importance for understanding the background to disturbances of vergence movements in, e.g, different types of latent strabismus.

The aims of this study were to investigate: 1) The effects of neck muscle proprioception on eye position during steady fixation: 2) the dynamics of the accommodative vergence movements driven by the dominant and non-dominant eye: 3) the effect of neck proprioception on accommodative vergence movements, driven by each eye.

EXPERIMENT 1   The effect of neck proprioception on eye position

Material and methods

Eight normal subjects participated. They were from 28 to 36 years old and had normal vision and oculomotor function. The subject sat on a chair with the trunk straight, the neck slightly flexed. The head was fixed by a bite-bar on a head support mounted on a table. The dominant eye was fixating a light emitting diode at a distance of 3.8 meters. The non-dominant eye was covered. The eye movements were recorded by an infrared-system (Ober-2 system, Permobil Inc.), with goggles in which light emitting and sensing elements were placed. Horizontal and vertical movements of both eyes could be measured simultaneously but independently.

Vibration was applied to two muscles in each of 3 groups of neck muscles: (1) back muscles of the neck extending the neck; (2) left sterno-cleido-mastoid (SCM) and right splenius turning the head to the right; (3) right SCM and left splenius turning the head to the left. The amplitude of vibration was 1mm, the frequency was 70 Hz. During the experiment, the experimenter held the two vibrators on the skin over the muscle to be vibrated.

Results

Neck muscle vibration induced both an illusion of movement of the fixated object and an eye movement. A typical eye movement recording is shown in Fig. 1.

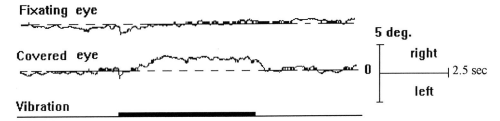

Fig. 1 An example of the effects of vibration on right SCM and left splenius in subject W. The covered eye deviated to the right during the vibration. The fixating eye remained stationary during vibration. Broken lines indicate baseline of each recording.

The directions of the deviation of the covered eye and the visual illusions induced by neck muscles vibration are shown in Table I.

Table I

Vibration	Left SCM & Right Splenius	Right SCM & Left Splenius	Back muscles of neck
Direction of visual illusion	Left	Right	Down
Direction of eye deviation	Left	Right	Down

SCM: sterno-cleido-mastoid

The directions were the same as if the head had been turned or bent by an activation of the antagonist muscles, stretching the proprioceptors of the vibrated muscles.

EXPERIMENT 2
The effect of neck proprioception on accommodative vergence movements

Material and Methods

Five normal subjects, who were 25-33 years old and had normal vision, normal stereopsis and were orthophoric, have participated in the experiment. Accommodation was elicited by a set of light emitting diodes (LEDs) as seen in Fig.2. The diods were lit in a sequence, either in a step-wise or a sine-wave fashion. Eye movement were recorded with the Ober-2 system.

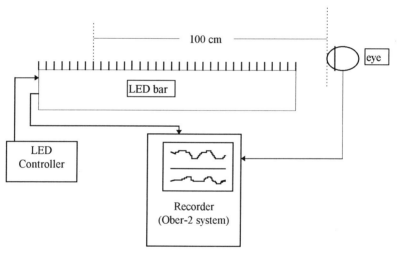

Fig. 2. A schematic picture of the LED bar and its controlling and recording system for accommodative stimulation of one eye and the recording of the eye accommodative vergence movements.

Each subject was tested on two separated occasions; first with the dominant eye fixating the visual target while the non-dominant eye was covered, and the second time *vice versa*. At each occasions recording were done without vibration or with the vibration of horizontally acting neck muscles.

The amplitude of the accommodative stimulation was from 1D to 5 D. The frequency of the accommodative stimulation was for the step changes 0.1 Hz and for sine wave changes 0.1, 0.2, 0.3, 0.5, 0.8, 1.0 and 1.2 Hz.

Results

Fig. 3 shows a typical eye movement response to step stimulation. In the response the latency and time constant (TC) were measured.

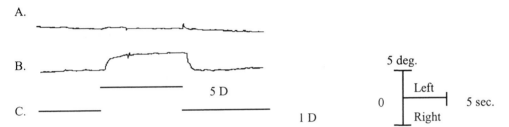

Fig. 3 Vergence response to step wave accommodative stimulation from subject LL. The amplitude of the stimulation was 1 D to 5 D. The duration of the stimulation was 10 sec. A. Recording from the fixating eye (Left); B. Recording from the covered eye (Right); C. The stimulation signal.

Fig. 4 depicts the values of the response latencies in the different testing condition. There were no significant differences in the values.

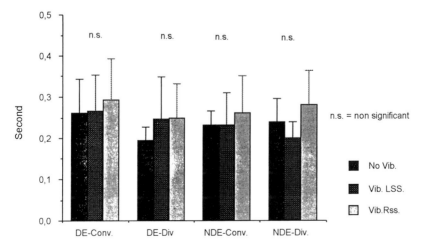

Fig. 4 Mean values of latencies of the vergence movement in the covered eye induced by square wave accommodative stimulation in four subjects. DE: Dominant eye; NDE: Non-dominant eye; Conv: Convergence movement; Div: Divergence movement.

A histogram graph of TC is shown in Fig. 5. Vibration of the neck muscles lowered significantly the TC-values of the non-dominant eye in both convergence and divergence.

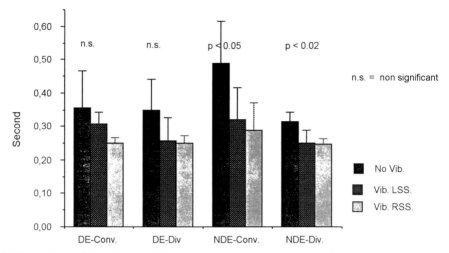

Fig. 5. Mean values of Time Costant of the vergence movements in the covered eye induced by square wave accommodative stimulation in four subjects. Abbreviation as in Fig 4.

In fig. 6 the typical vergence response to sine wave stimulation is shown. For a dynamic evaluation of the responses, measurements of gain and phase were done.

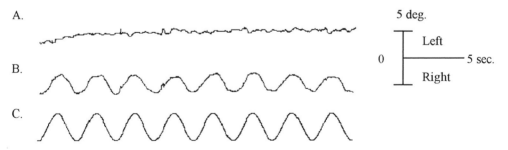

Fig. 6 Vergence responses to sine wave accommodative stimulation from subject YK. The amplitude of the stimulation was 1D to 5 D; the frequency of the stimulation was 0.3 Hz. A. Recording from the fixating eye (Left); B. Recording from the covered eye (Right); C. The stimulation signal.

The gain of the responses was lower in dominant than in non-dominant eye stimulation, as seen in Fig.7. There were no differences in gain depending on vibration of the neck muscles as shown in Fig. 8.

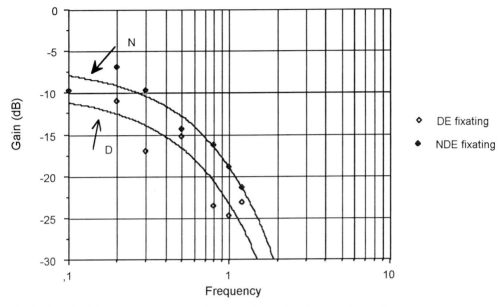

Fig. 7 The gain of the response to accommodative stimulation of the dominant (DE) and non-dominant (NDE) eye are presented. .

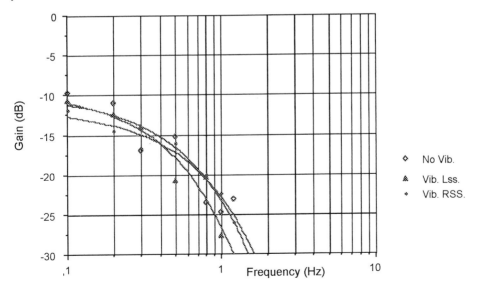

Fig. 8 The gain of responses to dominant eye stimulation without and with vibration of the horizontal neck muscles.

These results are also shown in table II with mean values of the gain and p-values from t-test and ANOVA in four subjects.

Table II

	Dominant Eye fixating	Non-dominant Eye fixating	p-value from t-test **
No Vib.	-17,522	-15,673	*0,0087*
Vib. LSS	-18,089	-17,187	**0.2926**
Vib. RSS	-17,905	-16,387	**0.1451**
p-value from ANOVA *	**0,9299**	**0,6478**	

No Vib.: accommodative stimulation without vibration; **Vib. LSS.**: accommodative stimulation with vibration on left sterno-cleido-mastoid and right splenius; **Vib. RSS.**: accommodative stimulation with vibration on right sterno-cleido-mastoid and left splenius; **DE**: Dominant eye; **NDE**: Non-dominant eye.* comparison between DE & NDE fixating; ** comparison between vibrations.

For the phase values there were no differences between responses of DE and NDE (see Fig. 9), or between tests with or without vibrations of the neck muscles.

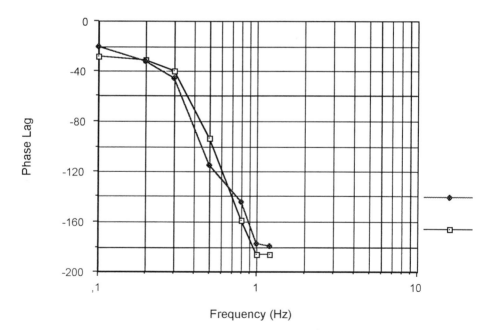

Fig. 9. The phase values of responses to sine wave stimulation of the dominant (DE) and the non-dominant eye (NDE).

Conclusions

1. Vibration of neck muscles is known to activate muscle spindles. This did induce eye position changes of a covered eye during steady fixation with the other eye. The direction of the position change was related to the muscles vibrated. Stimulation of the back muscles induce vertical movements, and stimulation of the side muscles horizontal movements.

2. During accommodative stimulation alone, the dynamics of vergence movements were different depending on whether the dominant or the non-dominant eye was driving accommodation. For example, the gain to sine wave stimulation was higher when the non-dominant eye was fixating.

3. Vibration on neck muscles (which turn the head in horizontal direction) shortened the time constant of the accommodative vergence movement to square wave stimulation driven by the non-dominant eye.

4. Vibration of the neck muscles during sine wave stimulation of accommodation did not influence the dynamics of the response, from either eye.

5. Thus, extraretinal signals from neck proprioceptors could influence the static eye position and the dynamics of accommodative vergence movements.

Acknowledgements
This study was supported by grants from the Medical Research Council (no, 4751), the Sigvard & Marianne Bernadotte Foundation and the Wenner-Gren Center Foundation.

References

Biguer B, Donaldson IHL, Hein A and Jeannerod M (1988) Neck muscle vibration modifies the representation of visual motion and direction in man. *Brain* 111:1405-1424.

Matthews PBC (1982) Where does Sherrington's 'muscular sense' originate? Muscles, joints, corollary discharges? *Annual Review of Neuroscience* 5:189-218.

Roll JP & Vedel JP (1982) Kinaesthetic role of muscle afferents in man, studied by tendon vibration and microneurography. *Experimental Brain Research* 47:177-190.

Roll JP, Vedel JP and Roll R (1989) Eye, head and skeletal muscle spindle feedback in the elaboration of body references. *In*: Allum JHJ & Hulliger M (eds): *Afferent control of posture and locomotion.* Elsevier, Amsterdam, pp.113-123.

Accommodation/vergence/fixation disparity and synergism of head, neck and shoulders

Ivar Lie, Reidulf Watten and Knut Inge Fostervold

Vision Laboratory, Institute of Psychology University of Oslo
P.O. Box 1094, Blindern, N - 0317 Oslo, Norway

Summary: The integration of scanning eye movements with head and body movements as well as the synergetic interaction of accommodation and convergence have been known for a long time. Surprisingly, little interest, if any, has been devoted to the possible interaction between vergence/accommodation and other body muscles. Experimental evidence suggesting the existence of an eye-head-body focusing programme for maintenance of clear and single vision was published by Lie and Watten in 1987. In the present paper, these results are summarised and the two-step vergence model of Schor (1980) is extended to account for an ocular motor/body muscle interaction. New evidence of the existence of very large fixation disparities suggesting a dynamic interplay between motor and sensory compensation of angular misalignment is presented. Binocular eye movement registrations are performed while prism induced vergence capacity is measured. In extreme cases prism induced angular deviations up to 15 degrees are seen to be substituted in full by disparity changes. Possible implications of these findings for our vergence-body muscle interaction model is discussed and some preliminary EMG results are mentioned.

Introduction

Eye movements are closely integrated with head and body movements in visual scanning performance under normal conditions of every-day life. This probably reflects a centrally controlled eye-head-body motor programme, as suggested by Bizzi (1971). The first study offering functional evidence of a synergetic-like interaction between vergence muscles and head and neck muscles dates back to Simons et al's experiments in 1943 (Simons, Goodell and Wolf, 1943). They used vertical prisms to stress the oculomotor muscles while recording EMG from frontalis, trapezius and neck muscles. The EMG was shown to increase gradually with recording time, and at the end of the experiment, subjects reported headache, pain and stiffness in the neck. They obtained similar results by using spherical and/or cylindrical lenses to stress the ciliary muscles.

A replication of Simons et al's experiments was performed by Lie and Watten (1987), applying horizontal instead of vertical prisms. When using vertical prisms the change in EMG activity may reflect the trivial effect of head tilt made by the subject to compensate for the vertical displacement of the eyes. Such compensations are well known and are frequently observed among hyperphorics.

The following briefly summarises the Lie and Watten study. Subsequently some new data on the existence of large Panums areas will be presented, followed by a discussion of the implication of this finding as to vergence-body interaction.

The Lie and Watten study – Method and Procedure.

We used a within-subject multiple repeated measures design with baseline periods between the experimental conditions. Variation of oculomotor stress was obtained by combining alteration of visual distance and application of optical lenses in the following four experimental conditions:

1. 6 m viewing distance without lenses
2. 40 cm viewing distance without lenses
3. 40 cm viewing distance with minus lenses
4. 40 cm viewing distance with base out prisms

Ten normally sighted students participated as subjects. They viewed a sample of letters arranged within a square subtending a visual angle of 1 degree, and they were asked to read the letters one by one in a given order. Letter size was individually selected for the visual acuity for each subject, so that the letters had approximately the same good legibility for all subjects at both viewing distances. Prism and lens powers were also individually selected on the basis of separate measurements of positive and negative relative convergence and accommodation, according to the procedure of The Optometric Extension Programme (Hofstetter, 1945). Reduced legibility of the letters was avoided by keeping the threshold criteria just below the blur or diplopia point.

The subjects were seated in an upright position for the reading test, for which the stimulus was placed at eye level for each subject. During baseline recordings, body and head posture were kept unchanged and the eyes were closed.

EMG was recorded bipolarly from the following six muscles on the right side of the body: frontalis, masseter, deltoid, middle trapezius, levator and upper trapezius and upper neck including splenius.

Results

The mean EMG-amplitudes from the six muscles across experimental and baseline conditions are shown in fig. 1. A two-way analysis of variance shows a significant, all-over increase in EMG-amplitudes across stimulation periods for all muscles ($p < 0.01$), except for masseter. The trend analysis shows a significant increase in EMG activity for all muscles when stimulation conditions are changed from 2 to 3 or 4 ($p < 0.01$). None of the muscles show significant EMG differences when stimulation is changed from 3 to 4. Deltoid, middle-trapezius and neck also show a significant EMG increase when viewing distance is reduced from 6 m to 40 cm ($p < 0.01$).

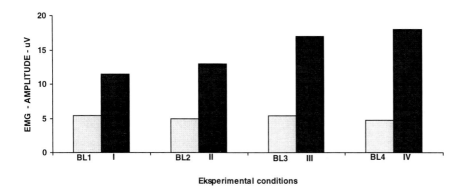

Figure 1. Mean EMG of six muscles over four experimental conditions (I = 6 m, II= 40 cm, III = 40 cm with minus lenses, IV = 40 cm with prism base out) and four baseline conditions (B1, B2, B3, B4). N = 10 (From Lie and Watten, 1987).

The results of Experiment 1 show that oculomotor stress may increase muscle activity in head, neck, shoulder and upper back muscles. This general conclusion was supported by EMG-results from a second experiment (Lie & Watten, 1987) showing positive effects of optical corrections of refractive errors and angular deviations among 31 heavily symptomatic subjects.

Discussion

These results suggest that the ocular muscles operate as an integrated part, not only of the eye-head visual scanning programme but also of *an eye-head-neck postural programme for maintaining clear and single vision.* A preliminary model of the latter programme is shown in figure 2.

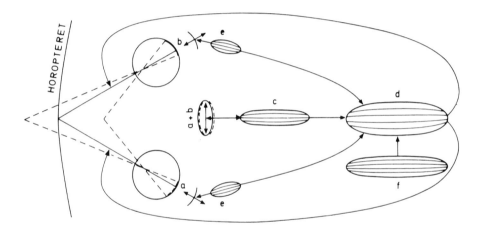

Figure 2.   Oculomotor interaction with other body muscles; a model based on C.Schor's theory of a two-step mechanism of vergence control. **a** and **b**= Panum's areas, **c** = disparity-driven fast-acting extraocular muscles, **d** = slow-acting extraocular muscles, **e** = ciliary muscles, **f** = other body muscles.

This model is based on Schor's theory of a two-step mechanism of vergence control; a fast-acting reflex mechanism stimulated by retinal image disparity and a slow-acting mechanism stimulated by the output of the fast mechanism (Schor, 1980). When orthophoria exists the activity of both mechanisms is minimal when inspecting distant objects. When focusing a near object, or having a phoria, the fast-acting mechanism must increase its activity in order to keep stimulation within the Panum's area. The new output of the fast mechanism stimulates the slow mechanism to the necessary extent in order to avoid fixation diplopia. Accordingly, vergence capacity is assumed to be dependent on both the efficiency of the fast-acting reflex mechanism and on the potentiation capacity of the slow mechanism. In this model we have assumed that the oculomotor interaction with other body muscles is linked to the slow-acting mechanism.

The results from our 1987 study do not, however, suggest any simple synergetic relationship between the slow mechanism and other body muscles. Large inter-individual differences exist as to which muscles are involved in the interaction pattern, indicating individual- specific interactions between oculomotor muscles and accessory body muscles. It is typical for any individual case that a significant increase of EMG is seen in one or two muscle groups only. While one subject may knit his brows, another subject may lift his shoulders.

Large Panuma's areas and the interplay of sensory and motor components of vergence.

Schor's two-step theory of vergence control is based on the conventional concept of a rather small rigid Panum's area changing its size mainly as a function of retinal eccentricity. A number of studies on retinal disparity have questioned this conventional concept of a rigid small Panum's area, by indicating some degree of plasticity of retinal correspondence (Fender & Julesz, 1967; H.-J. Haase, 1981; Wick, 1991; Watten, Lie & Klæboe, 1994). In his comprehensive review of the literature on flexibility of retinal correspondence, Wick concludes that "the correspondence of adult subjects with normal binocular vision is capable of small variation, particularly with strong sustained demands upon vergence, in order for binocular vision to be maintained".

During the last few years, we have carried out a series of experiments on dynamic aspects of fixation disparity, as measured by binocular eye movement recordings, using the Ober 2-instrument (Permobil AB, Timrå, Sweden) as well as a laboratory video recording system set-up. Results from a variety of experimental designs indicate a dynamic operation of surprisingly large Panum's areas.

In one of the experimental set-ups for video recordings of eye movements prism induced vergence was used to study motor and sensory compensation of angular deviations. The subject fixated the black ring of one of the binocular charts of the Polatest Zeiss (figure 3) when a prism bar was moved vertically in front of his left eye. The prism power increments were in steps of

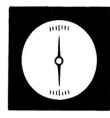

Figure 3. The Clock test chart of the Polatest (Zeiss). The pointer and the scale can be dissociated by use of a polaroid visor in order to control suppression during sustained fixation of the small central ring.

2 cm/m. The maximum power of base in and base out prisms avoiding diplopia was determined for each subject beforehand and used to define the individual fusional range to be applied throughout the experimental sessions. When maximum prism power was inserted the subject was asked to maintain fixation for 2 1/2 min. (total fixation time: 3 min). This experimental session was repeated four times for each subject with a rest period of 20 minutes between sessions. Measurements were made at both 5.7 m and 57 cm. Suppression was controlled by using orthogonal oriented polarisers allowing binocular dissociation of the pointer and the scale while the ring was seen binocularly. Subjects showing suppression tendencies were excluded from the study. Only normally sighted or optimally corrected subjects were allowed to participate in this study. A chin-head rest was used to stabilise head position. A wide-angle video camera was mounted in front of the chin-head rest slightly below and to the right of the fixation line in order to enable free camera view of both eyes behind the prism bar.

Results

A representative sample of individual results are shown in figures 4-6. The vergence curves show both over- and undercompensations representing very large fixation disparities, reaching maximum values of about 10 degrees in both directions. For other subjects disparities up to 15 degrees have been measured. Both inter- and intraindividual differences are observed when sessions are repeated as well as when fixation is shifted from far to near distance. No systematic parametric effects have been observed so far.

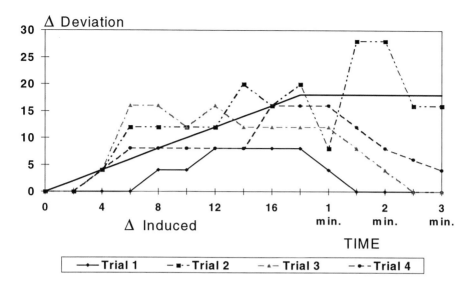

Figure 4. Vergence curves based on binocular eye-movement recordings during 3 minutes of prisme-induced exo deviation for subject 1. Fixation distance: 5.7m. Bold line represents full motor compensation. Full sensory compensation coincide with the X-axis.

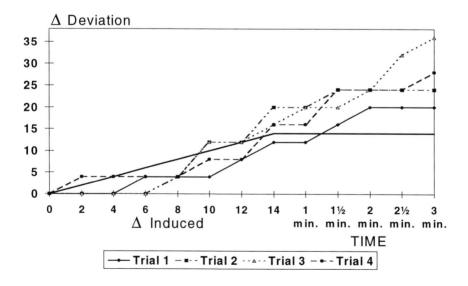

Figure 5. Vergence curves based on binocular eye-movement recordings during 3 minutes of prism-induced exo deviation for subject 2. Fixation distance: 5.7m. Bold line represents full motor compensation. Full sensory compensation coincide with the X-axis.

310

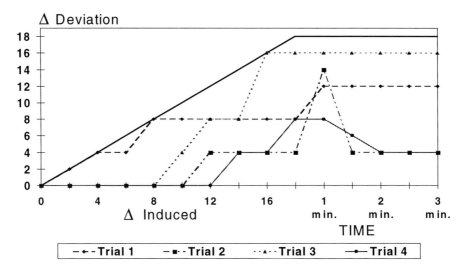

Figure 6. Vergence curves based on binocular eye-movement recordings during 3 minutes of prisme-induced exo deviation for subject 3. Fixation distance: 57cm. Bold line represents full motor compensation. Full sensory compensation coincide with the X-axis.

Discussion

The present results showing large size variations of fixation disparity suggests the existence of some kind of multiple neural mapping of binocular correspondence. Experimental evidence on disparity tuned cortical cells suggest that Panum's area constitute a cortical integration representing the extent to which receptive fields in the one eye, all with the same direction, are linked to fellow members of a pair in the other eye over a range of receptive field disparities.

Electrophysiological studies on animals show that many retinal receptive field sizes are represented at any site in the retina (e.g.Wiesel, 1960; Cleland et al, 1978). Psychophysical studies of the human contrast sensitivity using grating stimuli (Kelly, 1977) as well as single stimuli (Lie, 1980, 1981), is also likely to be interpreted on the basis of a multiple-channel theory as a composite reaction from several size-tuned receptive fields.

This multiple-channel concept of receptive field organisation may suggest a corresponding multiple-channel organisation of Panum's areas. It would not be surprising, therefore, to find concentrically overlapping Panum's areas of different sizes in any site of the retina. This pool of multiple-size Panum's areas may allow for a more or less free shift of pairs of corresponding Panum's areas to be brought into register upon demand.

How does the operation of large Panum's areas effect the vergence-body interaction?

One possibility would be to expect large fixation disparities operating in order to reduce the need for muscle support from the body, sustaining single vision with a minimum expenditure of motor energy. Alternatively, with reference to neurophysiological research on disparity detectors and their tuning functions (e.g. Barlow, Blakemore and Pettigrew ,1969: Pettigrew 1972), one may expect the operation of large disparities to reduce the disparity induced vergence reflex, and the demand for body support of the slow mechanism would be expected to increase.

Preliminary experiments on simultaneous recordings of eye movements and EMG-activity in the head, neck and upper back regions have only just begun . The results obtained so far suggest that body muscle support decreases when motor compensation is replaced by large fixation disparities. However, much more research is needed in order to reach any safe conclusions as to the effect of the operation of large fixation disparities on the vergence-body muscle interaction.

Acknowledgments
We thank Stine Vogt for assistance with the experiments and for valuable comments on the manuscript.

References

Barlow, H.B., Blakemore,C. and Pettigrew, J.D. (1967) The neural mechanisms of binocular depth discrimination. *Journal of Physiology* 193: 327-342 .

Bizzi, E., Kalil, R.E. and Tagaliasco, V. (1971) Eye-head coordination in monkeys : evidence for centrally patterned organization. *Science* 173: 452-454.

Cleland, B.G. and Enroth-Cugell ,C. (1968) Retinal sensitivity and summation. *Journal of Physiology* 198:17-38.

Fender, D. and Julesz, B. (1967) Extention of Panum's fusional area in binocularly stabilized vision *Journal of the Optical Society of America* 57: 819-830.

Fischer, B. (1973) Overlap of receptive field centers and representation of the visual field in the cat's optic. *Vision Research*13: 2113-2120.

Haase, H-J. (1980-84) Zur Fixationdisparation, 22 Fortsetzungen in *der Augenoptiker*, Hefte 3/80 bis 1/84.

Hofstetter, H.W. (1945) The zone of clear and single binocular vision. *Archives of the American Academy of Optometry* 22: 301-333.

Kelly, D.H. (1977) Visual contrast sensitivity. *Optica Acta* 24: 107-129.

Lie ,I. (1980) Visual detection and resolution as a function of retinal locus. *Vision Research* 20: 967-974.

312

Lie, I. (1981) Visual detection and resolution as a function of adaptation and glare. *Vision Research* 21: 1793-1797.

Lie , I. and Watten, R. (1987) Oculomotor factors in the aetiology of occupational cervicobrachial diseases (OCD). *European Journal of Applied Physiology* 56: 151-156.

Pettigrew, J.D. (1972) The neurophysiology of binocular vision. *Scientific American* 227: 84-95

Schor, C. (1980) Fixation disparity: A steady state error of disparity-induced vergence *American Journal of Optometry & Physiological Optics* 57: 618-631.

Simons, D.J., Day, E., Goodell, H. and Wollf, H.G . (1943) Experimental studies of headache: muscles of the scalp and neck as sources of pain. *Res Publica Association of Nervous and Mental Disease* 23: 228-244.

Watten, R., Lie, I. and Klæboe, R. (1994) Dynamische Vergencemessungen- eine neue Technik zur Einschätzung der visuellen Ermüdung. *Optometrie* 3: 88-93.

Wick, B. (1991) Stability of retinal correspondence in normal binocular vision. *Optometry and Vision Science* 68: 146-158.

Wiesel, T.N. (1960) Receptive fields of ganglion cells in the cat's retina. *Journal of Physiology* 153: 583-594.

Changes in accommodation and vergence following 2 hours of movie viewing through bi-ocular head-mounted display

Kazuhiko Ukai*, Hiroshi Oyamada** and Satoshi Ishikawa**

*Faculty of Social and Information Sciences, Nihon Fukushi University, Handa, Aichi 475, JAPAN
** Department of Physiology, School of Medicine, Niigata University, Niigata, Niigata 951, JAPAN
***Department of Ophthalmology, School of Medicine, Kitasato University, Sagamihara, Kanagawa 228, JAPAN

Summary: Head mounted display (HMD) units will be widely used as a visual device for virtual reality and personal video monitors. We attempted to examine the effect of these displays on human visual health. Three HMDs, which have 2 liquid crystal displays but no binocular parallax (bi-ocular condition), were used for viewing video movies. Several ophthalmic examinations were carried out in 42 normal subjects before and after viewing a movie for 2 hours. Examinations included refraction and visual acuity, accommodation (TA and velocity of step responses), vergence (distance and near heterophoria, heterophoria and gradient AC/A ratio, measured by alternative prism-cover test), intra-ocular pressure, cornea, tear secretion and fatigue complaint. Controls were asked to watch an ordinary, domestic video monitor. Further to this, refraction of seminar attendants were measured. In addition some subjects were given the task of watching 3 dimensional computer graphics. Results showed that refraction change in both sphere and astigmatism was not systematic but the occurrence of a change greater than 0.38 D was increased. The results also showed that accommodation did not change significantly, and that the eye position during fixating near target changed toward exophoria direction (AC/A ratio reduced). It has previously been proposed that three dimensional displays using binocular parallax such as HMDs for virtual reality causes conflict between the accommodation and convergence systems, resulting in fatigue or eye strain. Present results indicate that changes in the accommodative vergence system can be caused without conflict of either systems. Thus, another possible mechanism should be proposed, i.e. that visual stress may be a common origin of eye strain and changes in accommodative vergence.

Introduction

The head mounted display (HMD) usually has 2 displays and can present 3 dimensional (3D) images due to binocular disparity. These 3D HMDs are used to create virtual reality (VR). So far, VR is used only for a short time, maybe up to 15 minutes, because it involves many problems, such as eye strain and VR sickness which is similar to motion sickness (Mon-Williams *et al.*, 1993; Regan and Price, 1994). By contrast, 2 dimensional images presented by 2 displays, the so-called bi-ocular condition, (Rushton *et al.* 1994), may involve other problems. It will be used for relatively extended periods such as for personal monitor system.

Many people have experienced eye strain or fatigue after using a 3D display. The accommodation system has to focus on the display surface while vergence may vary due to the binocular disparity. This conflict between the accommodation and vergence systems may induce changes in the link between the two systems. causing eye strain and leading to general fatigue. However, in the 1980's, many problems, including eye strain and myopia development, appeared in VDU workers. These may have been caused by ergonomical factors such as displays and work style. It should be noted that eye strain and myopia development were reported even when using natural viewing 2D displays, when the visual distance was not so close.

Thus, even 2D displays may cause problems, especially when it requires near vision work (Ehrlich, 1987; Owens and Wolf-Kellry, 1987). Bi-ocular HMD may elicit some visual complaints from users. This study was undertaken to investigate the effects of bi-ocular HMD on the visual system after extended use.

Material and Methods

HMDs: The HMDs used in this study are listed in Table 1. They are: 2 Prototypes made by the Sony Corporation, Visortron 1 and 2 (Hara *et al.*, 1996; Hasebe *et al.*, 1996), and 1 prototype from the Sega Corporation, Virtuavisor. Both Visortron 1 and 2 have lens optical systems with 2 liquid crystal displays (LCDs). Sony released a low-cost model of HMD for consumers, Glasstron, in July 1996 but only in Japan. This model has about twice the pixels of the prototypes, and a mirror optical system without an adjusting mechanism for inter pupillary distance. However, this model was not used in this experiment. Virtuavisor is similar to Visortron 2 but has a mirror optical system. Table 1 shows specifications of the 3 models. Visortron 1 has a focusing mechanism which is user-adjustable and therefore renders the use of corrective lenses unnecessary.

Table 1: Specification of Head Mounted Displays

	Visortron 1	Visortron 2	VirtuaVisor
Pixels	473 * 218	473 * 218	473 * 218
Visual Angle (degree)	30 * 22	18 * 13	32 * 23
Weight (g)	250	400	400
Image Distance (m)	variable	2.0	2.0
Correction Glasses	No	acceptable	acceptable
Focusing	required	N.A.	N.A.
See Through	N.A.	Yes	Yes

Subjects

In the main experiment and in 3 additional experiments, 42 and 28 young subjects participated, respectively. About half of them were students of the department of image science and engineering and the others were medical staff or office workers in the hospital. Subjects who had a history of double vision were excluded. Their ages ranged from 19 to 43 years old (mean: 27.7; SD: 5.6).

Procedure

Subjects were asked to sit in a comfortable office chair in a well lit room and watch a video movie for about 2 hours. The see-through function of the HMD was set at minimum, if this was possible. Before and after the task, many ophthalmic examinations were carried out. The types of examinations and the number of subjects evaluated are listed in Table 2 (main experiment) and 3 (additional experiments). Fundamentally, pre- and post-task examinations were identical. These results were compared with the results obtained when an ordinary domestic TV monitor was used to watch a video movie.

Table 2: HMDs and Examinations in Main Experiments

	Visortron 1	Visortron 2	Virtua Visor	21" TV
number of subjects	20	7	6	9
Objective Refraction	all	all	all	all
Subjective Refraction	all	all	all	5
Visual Acuity	all	all	all	4
Cornea	all	all	all	all
Phoria, AC/A (Heterophoria)	18	5	all	7
Phoria, AC/A (Gradient)	2	all	all	5
Depth Perception (TNO)	0	1	5	5
Accommodation	all	all	all	2
I. O. P.	all	0	0	0
Tear Secretion	18	0	0	0
Questionnaire	all	all	all	all

Table 3: HMDs, Tasks and Examinations in Additional Experiments

	1	2		3	
HMD		Visor 2	Visor 2	Visor 1	Visor 1
task	Seminar	3D/CG	2D/CG	movie	
number of subjects	8	7	4	4	5*2
Objective Refraction	**all**	all	all	0	0
Subjective Refraction	0	6	0	0	0
Visual Acuity	0	6	0	0	0
Cornea	0	6	0	**all**	all*
Phoria, AC/A (Heterophoria)	0	**6**	0	0	0
Phoria, AC/A (Gradient)	0	**5**	0	0	0
Depth Perception (TNO)	0	**6**	4	0	0
Accommodation	0	6	0	0	0
I. O. P.	0	0	0	0	0
Tear Secretion	0	0	0	0	0
Questionnaire	0	all	0	0	0

* Humidity controlled (2 conditions)

Both general and specific ophthalmic examinations were performed including tear secretion, corneal state, intra ocular pressure (IOP), refraction (subjective and objective) and visual acuity, accommodation and vergence, and the process also included the completing of a questionnaire. Examination items varied from time to time. As the project lasted 3 years, new problems appeared which had to be solved. Furthermore, some examinations clearly affected others, for example, when measuring IOP, touching the cornea affected the corneal state. IOP and tear secretion were found to vary significantly in the initial stage of the experiment. IOP decreased, and tear secretion increased. Thus, we stopped examining subjects. The method used to measure the AC/A ratio was changed because no coincidence, which is better, was given among examiners, and finally both measurements were carried out for each subject.

As a reference for refraction change, the subjects were asked to attend a seminar for about 2 hours. This was the first additional experiment. In the second additional experiment, watching moving 3D computer graphics (CG) for 30 minutes was used as a task. In this case, we mainly examined heterophoria and vergence in the subjects. Depth perception was examined in only one subject and was found to decrease. Results were surprising. Thus, this examination was subsequently added to the list of methods. Measurements of accommodation may cause corneal epithelium erosion due to the slight heating effect of infrared light. That is why the corneal check was only performed in a few subjects.

Methods

Subjective refraction was evaluated but not included in the final analysis. Using an automated infrared refractometer, data were obtained under uniform conditions. Spherical equivalent (SE) and the degree of astigmatism before and after the task were compared. Since we did not have the opportunity of using instruments of high resolution, we decided to only count and include changes larger than 0,38 D.

Tonic accommodation (TA) and step responses were measured using an infrared optometer (Ukai *et al.*,1986). Tonic accommodation was actually measured as the averaged accommodation response to an empty field over 30 s. Step responses were measured to the accommodative stimuli which were moving between the subject's far point and the point above 3 D in a stepwise manner for 5 cycles. The target stayed in each position for 5 s. Maximum velocity was calculated mainly for far to near responses.

Eye position and accommodative convergence were evaluated by experienced orthoptists using the alternative prism cover test. There are many types of stimuli for distance and near fixation and they have different functions. Here two types of stimuli were employed. One was a real target placed at 5.0 m and at 0.3 m for the calculation of the distant/near heterophoria AC/A ratio. The other stimulus for near fixation was a combination a far target and a -3 D lens used to calculate the negative gradient AC/A ratio.

Depth perception was examined using TNO test charts. Examination of tear secretion was carried out according to Schirmer's method. Changes in corneal epithelium erosion were examined by the same ophthalmologist, before and after the task, using a slit lamp microscope. IOP was measured using a Goldmann aplanation tonometer.

Results

Refraction and Accommodation

Changes in SE are shown in Figure 1. In this figure, data from 3 HMDs are shown superimposed. Each line represents one eye. Bold lines indicate changes in refraction of more than 0.38 D. Upward lines show changes toward myopia. No statistically significant change was seen. The tendency of myopia development was not proved for the 2 hours' task. Occurrences of changes over 0.38 D were 7, 2, 2 and 2 eyes out of 40, 16, 12 and 18 in 3 HMD groups and the home-use TV group, respectively. Which, in percentage forms, are 15%, 12%, 16% and 11%, respectively. None of the seminar attendants were included.

Changes in astigmatism are shown in Figure 2. Bold lines indicate changes of more than 0.38 D. No systematic change was observed. Changes in the axis were not considered. Occurrences were 11, 4 and 0 in the HMD groups, 2 in the TV group, and zero for the seminar attendants. These occurrences were very simliar to changes in SE. The largest change was 1.25 D. Keratometry did not show such large changes in astigmatism, which may thus be caused by lens deformation, even though, in this subject, the change was normalized 1 or 2 hours later. These changes were transient.

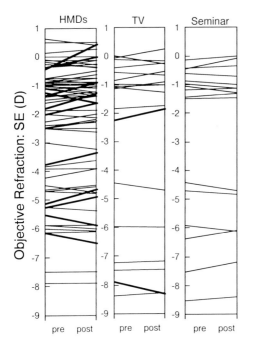

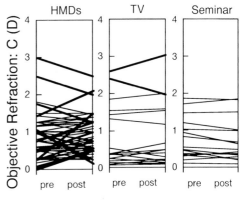

Figure 1 (Left): Changes in refraction (SE).
Figure 2 (Right): Changes in refraction (astigmatism).

Changes in TA are illustrated in Figure 3. Bold lines show the average values. No systematic change was seen. Some subjects showed very large TAs which were unstable. Only a few subjects with a small baseline showed large increases.

Figure 4 shows changes in velocities of far to near step responses. Bold lines show the average values. No statistical change was seen. In some cases, velocity varied between pre- and post-task. This may be due to the small number of trials. Each plot was calculated from 5 trials. We calculated the shift of far point but the results were unreliable due to the small number of cycles.

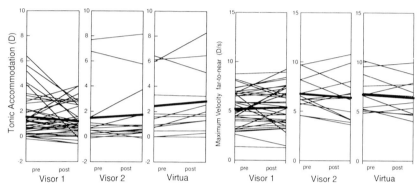

Figure 3. Changes in tonic accommodation.   Figure 4. Changes in accommodation velocity.

Eye Position and Accommodative Vergence

No significant difference was found with regard to distance heterophoria of pre- and post-task values as shown in Figure 5. Here subjects of 3 HMDs were superimposed. When the task was 3D CG (2nd additional experiment), heterophoria seemed to change towards the eso direction. But due to the small number of subjects, no statistical difference was seen.

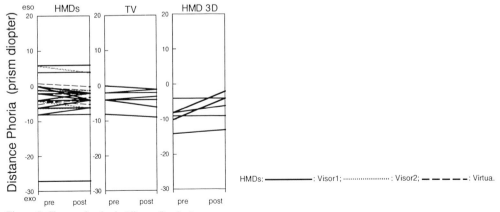

Figure 5. Changes in phoria (distant fixation).

Figure 6 shows changes in near phoria. A significant difference was found between pre- and post-task in the HMD video movie group (paired t-test: p<0.001) and in the HMD 3D CG group (paired t-test: p<0.05). Changes were towards the exo direction. Convergence was not measurable in these subjects. No significant difference was found when accommodative efforts were made by loading a negative lens for a distant target as shown in Figure 7.

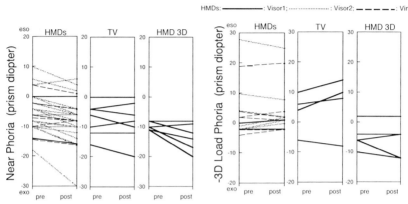

Figure 6. Changes in phoria (near fixation).   Figure 7. Changes in phoria (far target with minus lens).

AC/A ratios were calculated from the distant/near heterophoria value and shown in Figure 8. Note that none of the subjects in the HMD group showed an increased AC/A ratio. In all cases the AC/A ratio decreased or remained unchanged. Statistical tests (paired t-test) showed significant differences (2D HMD: p<0.001; 3D HMD: p<0.05). No statistical difference was found in the TV group. Figure 9 shows the results obtained from the eye position for the distant target with and without negative lens load, that is, the negative gradient AC/A ratio. In contrast to Figure 8, no significant difference was found in gradient AC/A ratio in the HMD video movie group while in the HMD 3D CG group a significant reduction (paired t-test: p<0.05) was noted.

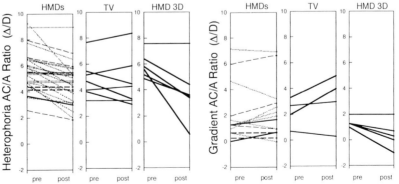

Figure 8. Changes in distant/near heterophoria AC/A ratio.   Figure 9. Changes in minus gradient AC/A ratio.

In the HMD group, changes in AC/A ratio varied according to the measurement method. In some subjects the AC/A ratio was examined by 2 methods. Figure 10 shows the relationship between the AC/A ratio measured by the heterophoria method and that measured by the gradient method. Arrows indicate pre- and post-task AC/A values. In general, the ratio tended to be higher when measured by the heterophoria method.

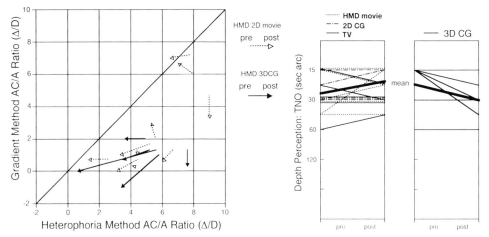

Figure 10. AC/A ratio (heterophoria vs gradient).   Figure 11. Changes in depth perception.

Other Examinations

Depth perception was examined in only a few subjects. The results are illustrated in Figure 11. All the subjects from the 3D CG group showed decreased or unchanged depth perception, while in the other groups, some improved and others worsened. This is surprising but the number of subjects (n=6) may be too small.

Other ophthalmic examinations show that IOP decreased significantly. Tear secretion increased and corneal state worsened significantly. Analysis of the questionnaire shows that fatigue complaint was typical when using the Visortron 1 and decreased when the later models were used.

Discussion

Refraction and Accommodation

According to the present results no systematic change was observed in refraction, but occurrence of changes of more than 0.38 D increased more in the HMD and TV groups than in the seminar group. Myopia development was not clear after two hours of experiments.

No systematic change was seen in TA or accommodation velocity. In the present study, TA was measured some minutes after the task, since many examinations were carried out. Accommodative adaptation will recover within a shorter period (Wolfe *et al.*, 1987). Thus, our measurements of TA are not involved with accommodative adaptation and further additional experiments will be required paying special attention to this point. Our method may have a relation to the lasting changes, as fatigue will not diminish within a few minutes. A similar situation may exist regarding changes in vergence.

Classically, the near point of accommodation moves further away when repeated measurements are made in asthenopia patients. Takagi *et al.* (1993) proposed that an accumulated shift to the closer direction of the far side of accommodation was observed in patients with asthenopia when sinusoidal stimulus was used. This supports our previous hypothesis that both near and far accommodation are affected in patients with asthenopia (Tsuchiya *et al.*, 1989). This hypothesis may also explain the relatively stable TA between pre- and post-task. We calculated the shift of the far point but results were unreliable due to the small number of cycles. One clear example of the shift of the far point was obtained in the present study, even though the repetitive cycle was very small. In future this may be useful as an index representing accommodative fatigue.

Analysis of the questionnaire showed that a complaint of fatigue was typical in the initial model of HMD and decreased in the later models. This finding suggested that the focusing mechanism, which is user adjustable, may cause fatigue after watching a video movie for 2 hours.

Eye Position and Accommodative Vergence
It is known that the AC/A ratio measured by gradient method is lower than that obtained by the heterophoria method (*e.g.*, Ciuffreda and Kenyon, 1983). This is attributed to the proximity when the closer target is used. The difference between both measurements is called the component of 'proximal vergence'. We have also confirmed this phenomenon using an infrared video refraction unit (Uematsu *et al.*, 1996). By using a video refraction unit it is possible to measure accommodation and eye position under relatively natural viewing conditions. An infrared filter can occlude one eye while the binocular eye position and accommodation are measured. Accommodative and vergence responses were measured for 5 accommodative stimuli and the slope was calculated for 10 normal subjects. Stimulus- and response-AC/A ratios measured by

the video refraction technique were 0.79 /D and 0.99 /D lower when using the gradient method compared with the heterophoria method, respectively.

Near exophoria increased after watching a video movie through HMDs. The AC/A ratio decreased, but in accordance with the examination method. The present work revealed that not only the AC/A ratio itself but also the changes in the AC/A ratio may vary in accordance with the measurement method.

Conventionally, it is known that if accommodation and vergence responses are in conflict, there will be an imbalance of the accommodation/vergence system, causing fatigue. However, we cannot neglect another possible mechanism, since not only 3D images but also the bi-ocular condition of the HMD affects the AC/A ratio. This change may be caused by the visual and general stress-strain relationship. Of course, the stress caused by the 3D display may be greater than that caused by the 2D display due to the above-mentioned conflict. Thus the effect of 3D displays is typical. However, working with a VDU, playing video games and maybe watching TV have milder but similar influences.

Other Examinations

Changes in depth perception may be caused by exophoria. However, the possibility remained that spatial vision decreased after the task. Depth perception may be influenced by the small changes in spatial vision which cannot be detected by conventional vision tests. Sheehy and Wilkinson (1989) have described changes in depth perception after prolonged use of night vision goggles.

Other ophthalmic examinations showed that the IOP decreased significantly. This has also been described in VDU workers (Araki, 1994) and discussed in relation to the development of myopia or continuing accommodative stress (Armaly and Jepson, 1962). Furthermore, we reported previously (Tsuchiya *et al.*, 1989) that the pupil constricted in the patients with asthenopia. Pupillary constriction reduces the IOP. These phenomena may be related.

Tear secretion (significantly increased) and corneal state (significantly worsened) may not be related to the accommodation and vergence system.

Conclusion

Bi-ocular HMD used for about 2 hours caused non-systematic but transient changes in refraction and vergence. Adequate rest should be essential when using this equipment. Not only 3D images but also the bi-ocular condition of HMDs cause changes in the AC/A ratio. Thus, the conventional theory that eye strain is attributed to the conflict between accommodation and vergence in binocular disparity stimuli, may not apply. The other possibility, as shown here, was that changes in accommodative convergence as well as eye-strain may be caused by visual and general stress-strain relationships. All visual tasks including viewing 3D HMD and bi-ocular HMD, working with VDU, playing video games, and maybe even watching TV may produce such stress, though the degree may vary.

Users cannot adjust the focusing mechanism adequately, and it is desirable to avoid such a mechanism. We have not examined how its use for an extended and longer period would affect visual health. We have not examined its use in infants and in the aged. Furthermore, there is a large population with considerable heterophoria and intermittent tropia. Future study is expected in these cases.

References

Araki, M. (1994) Changes in refraction and intraocular pressure in visual display terminal operators: A ten-year follow-up study (in Japanese). *Atarasii Ganka (Journal of the Eye)*, **11**: 255-257.

Armaly, M. A. and Jepson, N. C. (1962) Accommodation and the dynamics of the steady-state intraocular pressure. *Investigative Ophthalmology*, **1**: 480-483.

Ciuffreda, K. J. and Kenyon, R. V. (1983) Accommodative vergence and accommodation in normals, amblyopia, and strabismics. *In:* C. M. Schor and K. J. Ciuffreda (eds.): *Vergence eye movements: Basic and clinical aspects.* Butterworths, Boston, pp. 101-173.

Ehrlich, D. (1987) Near vision stress: Vergence adaptation and accommodative fatigue. *Ophthalmic and Physiological Optics,* **7**: 353-357.

Hara, N., Ukai, K., Ishikawa, S., Takagi, M., Bando, T. and Oyamada, H. (1996) Effects on visual functions following several hours' usage of head mounted display (in Japanese). *Nihon Ganka Gakkai Zasshi (Journal of Japanese Ophthalmological Society)*, **100**: 535-540.

Hasebe, H., Oyamada, H., Ukai, K., Toda, H. and Bando, T. (1996) Changes in oculomotor functions before and after loading a 3D visually-guided task by using a head-mounted display. *Ergonomics*, **39**: 1330-1343.

Mon-Williams, M., Wann, J. P. and Rushton, S. (1993) Binocular vision in a virtual world: Visual deficits following the wearing of a head-mounted display. *Ophthalmic and Physiological Optics*, **13**: 387-391.

Owens, D.A. and Wolf-Kellry, K. (1987) Near work, visual fatigue, and variation of oculomotor tonus. *Investigative Ophthalmology and Visual Science*, **28**: 743-749.

Regan, E. C. and Price, K. R. (1994) Some side effects of immersion Virtual Reality. *Aviation, Space and Environmental Medicine*, **65**: 527-530.

Rushton, S., Mon-Williams, M. and Wann, J. (1994) Binocular vision in a bi-ocular world: new generation head-mounted displays avoid causing visual deficit. *Displays*, **15**: 255-260.

Sheehy, J. B. and Wilkinson, M. (1989) Depth perception after prolonged usage of night vision goggles. *Aviation, Space and Environmental Medicine*, **60**: 573-579.

Takagi, M., Abe, H., Toda, H. and Usui, T. (1993) Accommodative and pupillary responses to sinusoidal target depth movement. *Ophthalmic and Physiological Optics,* **13**: 253-257.

Tsuchiya, K., Ukai, K. and Ishikawa, S. (1989) A quasistatic study of pupil and accommodation aftereffects following near vision. *Ophthalmic and Physiological Optics*, **9**: 385-391.

Tsuchiya, K., Aoki, S., Ukai, K. and Ishikawa S. (1992) Residual effect on sympathetic system by visual display terminal work. *In:* K. Shimizu (ed.): *Current Aspects in Ophthalmology, Volume 2*. Elsevier, Amsterdam, pp. 1713-1716.

Uematsu, T., Ukai, K. and Ishikawa, S. (1996) Measurement of accommodative convergence using video refraction unit (in Japanese). *Ganka Rinsho Iho (Japanese Review of Clinical Ophthalmology)*, **90**: 1341-1345.

Ukai, K., Ishii, M. and Ishikawa, S. (1986) A quasistatic study of accommodation in amblyopia. *Ophthalmic and Physiological Optics*, **6**: 287-295.

Wolfe, J. M., Ciuffeda, K. and Jacobs, S. (1987) Time course and decay of effects of near work on tonic accommodation and tonic vergence. *Investigative Ophthalmology and Visual Science*, **28**: 992-996.

Development of myopia due to environmental problems.
A possible interaction of anti-cholinesterase compounds examined by accommodative adaptation.

Satoshi Ishikawa, Kunihiko Tsuchiya, Norika Otsuka ,
Tatsuto Namba, and Kazuhiko Ukai

Department of Ophthalmology, School of Medicine, Kitasato, University, Sagamihara Kanagawa 228, Japan.

Summary: The method of measuring a tonic accommodation in an empty field was employed, and accommodative after-effect, defined as accommodation shift after 2 minutes viewing of a 4D near task, was recorded using an infrared electric optometer and the accommodation after-effect was measured. Subjects were classified into three groups according to the refraction. They are; 12 emmetropes [EMM], 12 early onset myopes [EOM], and 12 late onset myopes [LOM], respectively. Their age was approximately matched.

There was no statistical difference in the pre-task tonic accommodation among the three subject groups although it was slightly higher in the late onset myopes. Accommodative adaptation differed in the three groups during the 3 min post-task period. The average accommodative adaptation of the LOM and EOM groups was significantly higher than in the EMM group. Furthermore, the accommodative adaptation in the EMM group returned to its previous level but was significantly lower than in the EOM and the LOM groups. The relationship between accommodative adaptation and blood levels of butylcholinesterase (serum) as well as acetylcholin esterase (erythrocytes) activities were studied. Only the acetylcholinesterase activity showed a correlation with the slope of the after-effects in LOM.

Two possible aetiologies of myopia are discussed in tis study: one for the group which shows low accommodative aftereffect that has no reduction of acetylcholineterase activity and another for the LOM group which shows the highest accommodative after-effect related to reduced acetylcholinesterase activity. The after-effect of the LOM group is a potentiating chemical effect, possibly due to environmental expsoure to anticholinesterase chemicals, mainly at the synapses of the ciliary muscle. The effects of environmental factors on the development of myopia have been discussed.

Introduction

A tremendous increase in cases of myopia in children several years after World war II is one of the most serious public health problems in Japan and possibly in the countries of South West Asia. Until 1970, its etiology had never been analyzed. Most of the reports described the relationship of myopia with near-work, but this is a matter of speculation based mainly on statistics. There are reports of myopia development due to visual deprivation after birth in certain primates. We have been studying the correlation between the myopia and organophosphate pesticides (OP). The first report was made by Ishikawa who studied the outbreak of cases in the Saku district where massive amounts of mainly malathion were sprayed by helicopter (Ishikawa, 1971).

Thereafter, detailed epidemiological, clinical, and experimental investigations were conducted in the district by a governmental task force that had recommended specific diagnostic criteria (1973). The patients with myopia with vertical astigmatismus as a rule, differed from ordinary myopes in that they had a mild complication of the central, peripheral and autonomic nervous systems. In addition, all patients had minor abnormalities in the biochemical profile of enzymes including cholinesterases (Ishikawa, 1973). We will introduce our recent study of accommodation after-effect in young myopes, a possible relationship to OP may be demonstrated through reviews of previous literature.

Review of ocular toxicity

Saku disease and experimental myopia
The following is a brief review of the studies related to the organophosphate pesticides performed previously. Generally, organophosphate compounds produce an intense cholinomimetic response with an accumulation of acetylcholin at the synapses of the cholinergically-innervated organ(Ishikawa1971, 1973). The ocular toxicity of OP occurs through mechanisms of action other than the inhibition of cholinesterase enzymes. Evidence for the existence of these other mechanisms of action includes the acute electroretinographical changes reported (Miata, Imai et al., 1973) at dose levels far below those inhibiting acetylcholinesterase, and the degeneration of the retina following chronic treatment/exposure which occurs in regions of the retina that do not possess cholinergic synapses. Retinas from rats treated with an OP (fenthion) showed decreases in carbachol-stimulated release of inositol phosphate, an indicator of cholinergically-mediated intracellular second messenger systems (Boyes et al., 1994). Non-cholinergic actions of OP in the retinal neurons of cultured tissue showed a generation of superoxide by a low dosage of OP (Depterex) producing an inflow of Ca 2^+ ions into the cells (Ishikawa, 1996). These studies may explain impairments of nervous tissues caused by OP at even extremely low-dosages. This may produce damage to the cereb-

rum, brainstem, spinal cord and the peripheral nervous system. The delayed or chronic effects of mainly the OP malathion, which has been implicated in the outbreak of Saku Disease, has already been studied in a monograph by the Stockholm International Peace Research Institute (Chemical Warfare Agents-Delayed toxic effect) (Barnaby, 1975). Unfortunately, Japan appears to use more pesticides, especially organophosphate compounds, per unit of cultivated area than almost any other country in the world(1993). Therefore, the following description of a high incidence of visual disturbances in school children may be related to OP use. The development of myopia in young Beagle dogs following chronic, low dosage, oral administration of one potent (ethyl-thiometon) and one less potent (fenitrithion) neurotoxic pesticide have been reported(Suzuki, 1974). Ishikawa et al clearly mentioned that development of myopia by OP administration can only be induced after a period of at least one year (Ishikawa, 1980, 1993). It is unfortunate that a six month exposure to a potent organophosphate will show a negative development of myopia. The experiment should be continued for at least two years, according to the findings of the previous author, or the myopia will not be seen (Akinson and Bolte et al., 1994).

Accommodation after-effect

In the human study, we will introduce a relatively new concept of a tonic accommodation or accommodation after-effect.(Woung, 1993).A tonic accommodation represents the resting position of accommodative function which can be measured in an empty field, dark field or defocus-independent visual field (Wolfe, 1987 and Rosenfield, 1989). A myopic shift of tonic accommodation can be introduced following the stimulation of a near task. This is called accommodative hysteresis (Ebenholtz, 1983 and Shor, 1986), or adaptation of tonic accommodation. Post-task accommodative adaptation or the after-effect of accommodation has been investigated following long or short periods of near stimuli at various distances. (Ebenholz 1988, 1991). The rate of decay of accommodative adaptation appears to depend on the pre-task level of tonic accommodation and the visual circumstances operating after the near task. Subject groups of differing refractive error i.e. hypermetropia, emmetropia (EMM), early onset myopia (EOM) and late onset myopia (LOM), usually display different accommodative functions, for example, tonic accommodation lag to near target stimuli (Gilmartin 1991). Accommodative adaptation following a near task also differs among hypermetropia, myopia, and emmetropia. The rate of decay of accommodative adaptation returning to pretask tonic base line levels is significantly more rapid in EMM and EOM than in LOM. McBrien and Millodot (1988) observed accommodative adaptation following near accommodative stimuli for 15 minutes at 3D and 5D.(1988). The result revealed that the adaptation in LOM is greater than in EMM and EOM. Similarly Gilmartin and Bullimore, using the same accommodative stimuli for a shorter period (10 min), demonstrated that adaptation in LOM is greater than in EMM. The myopic population in Japan is greater than that in western countries. We will introduce a statistical analysis of children with myopia taking into consideration background factors from an environmental perspective.

Subjects and methods

The subjects used in our study were classified according to their refractive error and the age of myopia onset as in previous investigations (Woung, 1993). Because the myopic population in Japan is greater than that in western countries, our results may reveal different aetiological possibilities. In addition, blood analyses, measuring cholinesterase levels were performed, and a correlative study between the accommodative after-effect and the cholinesterase enzyme levels were compared in the EOM, EMM and LOM groups.

The subjects were 36 university students and hospital employes with an average age of 27.3±6.1 years (range 18-45). Each group had 12 subjects (Table I). Subject's age, cycloplegic refraction, cholinesterase acitivities of erythrocyte and serum as well as statistical analysis (t⁻Test) are shown.

Table I. Subject's Age, Refraction and Cholinesterase Activity.

mean ± SD (Min - Max)	Emmetrope	Late Onset Myopia	Early Onset Myopia
Cycloplegic Refraction	27.9 ± 5.6 (20 - 37)	26.8 ± 4.9 (20 - 36)	27.1 ± 7.7 (18 - 45)
Cycloplegic Refraction	** + 0.05 ± 0.28 (-0.50 -+ 0.50)	** -1.34 ± 0.56 (-3.00 -- 0.75)	** 5.01 ± 2.0 (-9.50 -- 2.25)
Acetylcholinesterase mmol/ml/min	** 1.76 ± 0.19 (1.5 - 2.1)	1.55 ± 0.12 (1.3 - 1.7)	* 1.72 ± 0.19 (1.5 - 2.0)
Butylcholin-esterase mmol/ml/min	5.88 ± 1.26 (4.0 - 8.9)	5.56 ± 0.81 (4.4 - 6.9)	5.88 ± 1.34 (3.5 - 8.8)

$* p < 0.05$ $** p < 0.01$

In the EMM group, healthy volunteers were selected. They had a refraction range of -0.50D to +0.50D in spherical equivalent, with the mean refraction of +0.05±0.28 diopters (D). Subjects with refractions stronger than -0.50D where excluded from this study. Their average age was 27.9±5.6 years. The myopes who knew the onset age of their myopia were subdivided into two groups;

1. The subjects whose myopia onset was at 14 years of age or less (n=12), the onset age ranging from 5-13 years with an average of 11.3 years.
2. The LOM group where 15 years older at myopia onset. The onset age ranged from 15 to 21 years with an average of 16.9 years.

Their average age was 26.8±4.9. The age distribution in each group shows no statistically significant difference (t-test, P > 0.1). The mean refraction for the EMM, EOM and LOM groups was +0.05±0.28D and -5.01±2.00D, -1.34±0.56D, respectively. The magnitude of myopia differed between the EOM and LOM groups (t-test, P<0.001). None of the subjects had ocular symptoms, reduced visual acuity or signs including fixation disparity or vergence abnormality.

We measured the accommodation after-effect. It was calculated from the stored data using custom made software. The line was determined by least square linear regression ($y = ax + b$) and the slope a was calculated. The slope of the after-effect was plotted against acetylcholinesterase. There is a correlation between the slope and ChE in the LOM group ($r=0.94$, $p=0.002$). No correlation existed in the EOM group.

Apparatus and Procedures

The accommodative state was monitored by a modified objective infrared optometer (Nidek AR-1100). By switching this equipment can measure accommodation continuously. The measurement range capability is from +13 to -19D. The accommodative target is built into the machine. Its movement is controlled by a personal computer. Basic target movements are already preset. Under these conditions, rapid and flexible optical alignment is possible by using a video monitor. Using an additional apparatus (Nidek IC -1100), the pupil area can also be measured. (Ukai, 1983, 1986 ,1989 & Tsuchiya 1989).

Before the experiment, the subjects were requested not to undertake any concentrated visual tasks, such as VDU work for at least 120 minutes. In addition, the detailed experimental procedures were explained to all the subjects. All measurements started from the right eye followed by the left eye. Only the data from right eye were studied. At first , each subject saw the internal asterisk-shape target placed at 8 Diopters below far point. This level of blur was considered to represent a bright empty field, so that the accommodative state under this condition represents tonic accommodation. After the 500 second settlement period the tonic accommodation was recorded for 60 seconds. Then, the target was moved to 4D above the subject's far point. The transition took 6 seconds due to the limitation of the control software. This near task continued for 120 seconds. At 200 seconds, the target returned to the first position to measure the accommodative adaptation lasting 500 seconds. This duration was selected for the adaptation period owing to the time limitation of the computer program. The accommodative state and target position were digitized with a sampling rate of 12.5 Hz before, during, and after the adapting stimulus, and all these data were stored on a disk. If necessary, the pupil area was also measured.

Visual acuity of children in Tokyo

The rate of poor visual acuity due to myopia has been increasing yearly among school children in Japan. The report was issued by the Ministry of Education-Health and included the vision, body weight and the height. The distribution of children with poor visual acuity has been reported annualy and is based on examinations of elementary and middle school pupils from 6 to 15 years of age. According to the 1993 report based on 1.5 milion pupils in elementary, middle, and high schools, the rate of pupils with vision less than 20/20 was 23.8%, 47.3%, and 61.9%, respectiviely. The 1994 report involving a total number of 655,600 pupils in elementary and middle school throughout Japan, showed the same tendency. Approximately 10 % are residents of the Tokyo bay area. Since this number is too high, the visual disturbance of 11 year old females in the Tokyo bay area, which has a much higher rate of children with reduced vision, was selected for reasons of managability.

Results

Tonic accommodation

Two examples of the accommodation recording obtained from an a 16 year old EMM subject (on the left) and a 18 year old LOM subject (on the right) are shown in Figure 1. The ordinate shows the change in accommodation, and the abscissa shows the time in seconds. Horizontally, the recording is divided into three sections; tonic accommodation before the near task(see empty field), accommodative response during near task and post-task accommodative adaptation, with respect to time. No after-effect can be seen. Conversely the LOM group showed an obvious after-effect following the near task.

Fig. 1

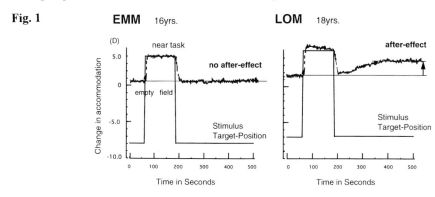

334

Fig. 2

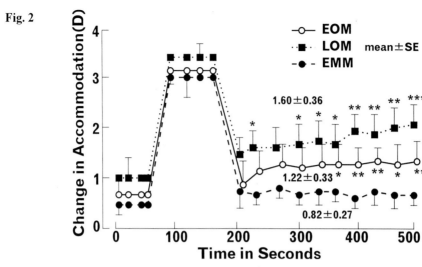

significance against EMM   *:P<0.05 **:<0.01 ***:<0.001
no significance between EOM and LOM

Accommodative after-effect

The average time course for each group is plotted in Figure 2. In this Figure, we used the normalized values of the accommodative responses. By shifting the accommodative far point to 0 Dioptor(D) normalized, i.e. ametropia, corrected data are shown. The average and S.D. of pre-task tonic accommodation in the EMM, EOM and LOM groups were 0.45±0.55D, 0.56±0.5D and 1.09±0.91 D, respectively. The EOM and LOM groups have a slightly higher tonic accommodation than the EMM group but the difference is not statistically significant. The post-task period starts from 200 seconds on the scale. For the entire 300 seconds of after-effect, the averages and S.D. for the EMM, EOM and LOM groups are 0.82±0.27D, 1.22±0.33D, 1.60±0.36D, respectively.

The accommodative after-effect in the EMM group is the lowest, and the highest is in the LOM group, the EOM group was in the intermediate place of the two groups throughout the entire period. A significant difference exists between the LOM and EMM groups (p<0.01).

Acetylcholinesterase activity

Topical administration of organophosphorus compounds such as phospholine iodide or DFP reduces AC/A and potentiates accommodative response. We wanted to ascertain the difference in cholinesterase activity betwen the 3 groups, since this activity can be a sensitive indicator of the nature of the cholinergic mechanism. As shown in Table 1, there were no significant differences among the 3 groups regarding butylcholinesterase activity, however, acetylcholinesterase activity in the LOM group was significantly lower when compared with the EMM ($p<0.01$) and EOM ($p<0.05$) groups.

A correlation of the slope of after-effect in the ordinate and the activity of acetylcholinesterase in the abscissa of the EOM and LOM groups was plotted. The results are shown in Figure 3. A significant correlation was seen only in the LOM group ($p=0.02$).

Fig. 3

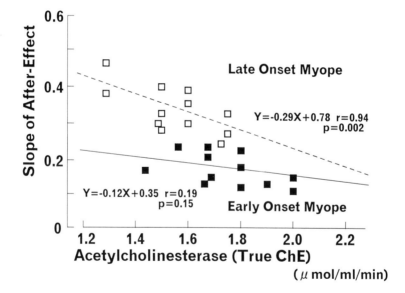

Children with reduced vision

Ueno and Masaki (1995) analyzed childrens' vision and related them to environmental factors. It was found that children who lived close to a reclaimed island of a garbage dump area showed different rates of distribution of reduced vision. The children who lived in the cities situated within 10 Km of the island: Koto, Taito, Churo and Choyoda, had a much higher incidence of reduced vision that the children who lived more than 20 Km west of the island in the cites of Kodaira, Murayama and Kunitachi (Figures 4A, 4B and 4 C).

Fig. 4A

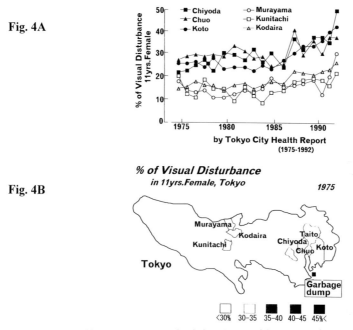

Fig. 4B

This marked difference was seen both in 1975 and in 1990. The average wind direction from the island expressed by an action vector showed a relationship to the organophosphorus pesticides sprayed on the island. The most prevalent wind directions from the island are: north-east, north and north-west. The incidence of visual disturbance in children is highest in Koto and decreases subsequently in Taito, Chiyoda and Chuo. In order to control the flies, the metropolitan authority sprayed the island with organophospherous pesticides including, among others, Diazinon, Depterex and Fenitrothion. The annual use of orthomethyl-parathion (fenitrothion) on the island is 24,500,000 liters. It is known that sprayed pesticides will not spread more than 20 Km from the sprayed area. Ueno and Masaki (1995) indicated that the organophosphate pesticides sprayed on the reclaimed island of garbage may be the cause of the visual disturbance found in the Tokyo bay.

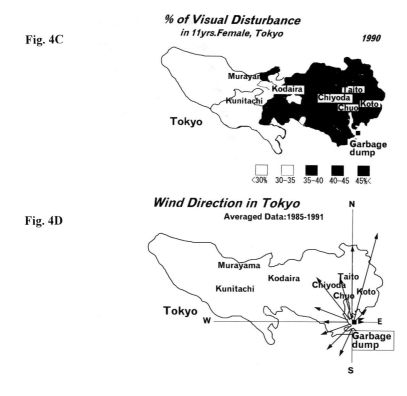

Fig. 4C

% of Visual Disturbance
in 11yrs.Female, Tokyo   *1990*

Murayan
Kodaira   Taito
Kunitachi   Chiyoda   Koto
                Chuo
Tokyo

Garbage
dump

<30%   30–35   35–40   40–45   45%<

Fig. 4D

Wind Direction in Tokyo
Averaged Data:1985-1991   N

Murayama
                Kodaira   Taito
Kunitachi   Chiyoda   Koto
                          Chuo
Tokyo   W   E

Garbage
dump

S

Discussion

Clear differences in accommodative after-effect following a near task are seen among the EMM, EOM and LOM groups. After the near task, the highest myopic shift is seen in the LOM group. It is difficult to explain why the after-effect is greater in the LOM group, smaller in EOM group and nonexistent in the EMM group. The degree of separation of the target from the dark focus, the task duration and the method of opening the accommodative loop may have the correlation between near task and accommodative after-effect adaptation and other factors (Ebenholz, 1983, 1991, Shor, 1986). There is a report that the near stimulus duration had no significant difference on either the magnitude of the adaptation or its rate of decay. (Fisher, 1990). According to the previous study, thr LOM group showed a significant myopic change in aftereffect (Gilmartin et al., 1991). In that study a ten minute near task was used, whereas in our study we used a two minutes near task, but the results are consistent with the previous studies.

The etiology of myopia may differ between the LOM and EOM groups. The possiblility should be kept in mind that the origin of myopia may differ according to race, geographical area or lifestyle. In Japan, severe competitive entrance examinations are set between the ages of 13-18 years. Many youngsters start to study excessively from their young teens and thus spend too much time in the near task of desk work. They also use computer displays for both study and amusement.

A hypothesis therefore was proposed that highly adaptive subjects will develop myopia due to these factors. Apart from heredity and near work, the difference in pathophysiology between the two types of myopia is still debatable. Firstly, it may derive from stress, tension and fatigue, and secondly, it may originate from an imbalance of tonic innervation from the autonomic nervous system. Cholinergic as well as reduced sympathetic innervation have also been suggested (McBrien, 1988, Bullimore, 1987, Woung, 1993, Gilmartin, 1987, 1986). We speculate that there might be a peripheral myoneural transmisssion anomaly inducing a long-term potentiating effect, which has been described by Rippus (1961) (Ishikawa, 1968, 1974) in relation to myopia following organophosphorus pesticide intoxication. In a systemic study of patients affected, they found that acetylcholinesterase in red blood cells decreased in the peripheral blood (Ishikawa, 1980). Moreover, the ratio of accommodative response to accommodative stimulus became decreased after the topical use of an anticholinesterase drug such as DFP or phosphorine iodide (Rippus, 1961, Oono, 1974). Recently, the presence of a soluble form of acetylcholinesterase in both the aqueous and vitreous humor of human ocular fluids has been reported (Appleyard, 1991). In the EOM and EMM groups, the myoneural junction may be much more stabilized than in the LOM group. The acetylcholine secreted in subjects belonging to these two groups is effectively hydrolyzed by normal concentrations of stable, abundant soluble acetylcholinesterase. However, in the LOM group, there may be reduced acetylcholinesterase due to a specific predisposition or to environmental exposure to anticholinesterases such as organophosphorus or carbamate pesticides. Therefore, acetyl-choline persists longer without being eliminated, neural transmission continues with potentiation as well as accumulation of acetylcholine and a myopic shift of tonic accommodation is induced.

In future work, we still need to clarify why we have an increasing ratio of visual disturbance with myopia especially among children in oriental countries. A possible etiological parameter arrived at, either through pharmacological analysis or physiological studies of accommodation, vergence and the pupil will solve these problems.

Acknowledgments
This work was in part supported by the Japanese Government of the Ministry of Health and Welfare "Chemical Sensitivity" 1996.

References

Akinson J.E., Bolte, H.F., Rubin, L.F. and Sonawane, M. (1994) Assessment of ocular toxicity in dogs during 6 months' exposure to a potent organophosphate. *Journal of Applied Toxicology,* 14:145-152.

Appleyard, M.E, McDonald, B. and Beniamin, L. (1991) Presence of a soluble form of acetylcholinesterase in human ocular fluids. *British Journal of Ophthalmology* 75:276-279.

Barnaby, F. (1975) Delayed toxic effects of chemical warfare agents. in Almqvist Wiksell: *Stockholm International Peace Research Institute (SIPRI),* Monograph, Stockholm, pp 1-57.

Boyes, W.K., Tandon, P., Barone, S.Jr. and Padillar, S. (1994) Effects of organophosphates on the visual system. *Journal of Applied Toxicology,* 14:135-143.

Bullimore, M.A. and Gilmartin, B. (1987) Aspects of tonic accommodation in emmetropia and late onset myopia. *American Journal of Optometry and Physiological Optics* 64.499-503 (1987)

Chin, Ishikawa, S., Lappin, Davidowitz (1968) Accommodation in monkeys induced by midbrain stimulation. *Investigative ophthalmology and vision science,* 7:386-396

Ebenholtz, S.M. (1983) Accommodative hysteresis: a precursor for myopia *IOVS,* 24:513-515.

Ebenholtz, S.M. (1988) Long term endurance of adaptive shifts in tonic accommodation. *Ophthalmic and Physiological Optics,* 8:427-431.

Ebenholtz, S.M. (1991) Accommodative hysteresis:fundamental asymmetry in decay rate after near and far focusing. *Investigative ophthalmology and vision science,* 32:148-153.

Gilmartin, B.(1986) A review of the role of sympathetic innervation of the ciliary muscle in ocular accommodation. *Ophthalmic and Physiological Optics,* 6, 23-27

Gilmartin, B. and Bullimore, M. A. (1986) Sustained near-vision augments inhibitory sympatheic innervation of the ciliary muscle. *Clinical Vision Science,* 1, 197-207

Gilmartin, B. and Bullimore, M. A. (1991) Adaptation of tonic accommodation sustained visual tasks in emmetropia and late onset myopia. *Optometry and Vision Science,* 68:22-26.

Fisher, S.K., Ciuffreda, K.J. and Bird, J.E. (1990) The effect of stimulus duration on tonic accommodation and tonic vergence. *Optometry and Vision Science,* 67,441-449.

Ishikawa, S. (1971) Eye injury by organic phosphorus insecticides. *Japanese Journal of Ophthalmology.* 15:60-66.

Ishikawa, S. (1973) Chronic optico-neuropathy due to environmental exposure of organophosphate pesticide (Saku Disease)--Clinical and experimental study. *Journal of Japanese Ophthalmological Society.* 77:1835-1886

Ishikawa, S. and Miyata, M. (1980) Development of myopia following chronic organophosphate pesticide intoxication: An epidemiological and experimental study, *In:* W.H. Merigan and B. Wise (ed.): *Neurotoxicity of the Visual System.* Raven Press, New York.

Ishikawa, S. and Miyata, M., Aoki, S. and Hani,Y. (1993) Chronic intoxication of organophosphorus pesticide and its treatment. *Folia Medica Cracoviensia,* 34:140-151.

Ishikawa, S. (1996) Ophthalmopathy due to environmental toxic substances especially intoxication by organophosphorus pesticides. *Journal of Japanese Ophthalmological Society,* 100:417-432

McBrien, N.A. and Milldot, M. (1988) Differences in adaptation of tonic accommodation with refractive state. *Investigative ophthalmology and vision science,* 29:460-469.

Oono, S. and Ishikawa, S. (1974) Pharmacological study on near reflex -the effects of parasympathomimetic drugs on accommodation and pupil (in Japanese). *Acta Societatis Ophthalmologicae Japonicae,* 78,516-523

Otsuka, N. Tsuchiya, K. Yoshitomi, T. Ukai, K. and Ishikawa, S. (1994) Adrenergic receptors affect accommodation by modulating cholinergic acitivity pp1281, *IOVS,* 35(suppl)

Ripps, H.Bereinin,G.M.and Baum,J.L. (1961)Accommodation in the cat. Transactions of the American Ophthalmological Society, 59, 176-780.

Rosenfield, M.(1989)Comparison of accommodative adaptation using laser and infrared optometers. *Ophthalmic and Physiological Optics,* 9:431-436.

Schor, C. M., Kotulak, J.C. and Tsuetaki, T. (1986) Adaptation of tonic accommodation reduces accommodative lag and is masked in darkness. *IOVS*, 27:820-827.

Suzuki, H. and Ishikawa, S. (1974) Ultrastructure of the ciliary muscle treated by organophosphate pesticide in beagle dogs. *British Journal of Ophthalmology,* 58:931-940.

Tamura, O. and Mitsui, Y. (1975) Organophosphorus pesticide as a cause of myopia in school children. *Japanese Journal of Ophthalmology,* 19:250-253.

Tsuchiya, K., Ukai, K. and Ishikawa, S. (1989) A quasistatic study of pupil and accommodation after-effects following near vision.*Ophthalmic and Physiological Optics*, 9:385-391.

Ueno, J. and Masaki, T. (1995) Poor visual acuity of school children and its environmental factors in Tokyo. *Japanese Journal of Clinical Ecology*, 4:77-81.

Ukai, K., Tanemoto, Y. and Ishikawa, S. (1986). Direct recording of accommdative responses versus accommodative stimulus. *In: Advances in Diagnostic Visual Optics*. G.M. Brenin, I.M. Ukai, K, Ishii, M and Ishikawa, S (1986).

Ukai, K., Ishii, M. and Ishikawa, S. (1986) A quasistatic study of accommodation in amblyopia. *Ophthalmic and Physiological Optics*, 6:287-295.

Ukai,K. and Ishikawa, S. (1989) Accommodative fluctuations in Adies syndrome. *Ophthalmic and Physiological Optics,* 9:76-78.

Wolfe, J.M. and O'Connell, K.M. (1987) Adaptation of the resting state of accommodation, dark and light field measures. *Investigative ophthalmology and vision science,* 28:992-996.

Woung, L.C., Ukai, K., Tshuchiya, K. and Ishikawa, S. (1993) Accommodative adaptation and age of onset of myopia. *Ophthalmic and Physiological Optics,* 13:366-370.

Accommodation and Vergence Mechanisms
in the Visual System
ed. by O. Franzén, H. Richter and L. Stark

Nearwork and visual well-being: A possible contribution of neuroscience

Bruno Piccoli

Department of Occupational Health - ICP Hospital, University of Milan, Milan, Italy

Summary. Work tasks require prolonged visual effort at near and thus cause asthenopia with an intensity and frequency rarely found in non-occupational activities. Complex processes with many contributing factors such as environmental conditions, type of work, psychosocial and physiopathological characteristics of the subject are involved. In this context the relationship between nearwork and overloading of the accommodation/vergence mechanisms surely represents an important issue to be cooperatively investigated by vision researchers, neuroscientists and occupational health doctors.

Evolution. For thousands of years human beings have used their visual functions mainly for distance vision --- hunting, orientation in open spaces, exploration and warfare. With social evolution, urbanization, and more complex and articulated social needs, requirements for visual functions have drastically changed. The development of arts and sciences and the diffusion of writing, reading, painting, sculpture, and music have increased the daily demands for detailed, or near, vision. We are concerned with the negative implications of these increased requirements.

Occupational Asthenopia. Indeed, Bernardino Ramazzini, writing the first book on occupational health, De Morbis Artificum Diatriba (Lesson about illnesses in handicraftsmen), in 1739 included a chapter dedicated to "the illness of people working with small objects", wherein disturbances caused by intense and prolonged occupational visual effort are described in great detail and breadth.

The advent of the Industrial Revolution and its factories, with repetitive and specialized work tasks, enforced near vision under artificial illumination (oil and gas lamps). In the present century, this trend has been accentuated with the enormous diffusion of computers and such optoelectronic displays as cathode tube oscilloscopes and liquid crystal and plasma displays. The major part of the average working day is spent looking at such displays, especially with air traffic controllers, film editing and montage operators, secretaries and professors.

The elevated prevalence of ocular and visual disturbances, occupational asthenopia, in many work sectors but particularly in office activities, is thus expected. Furthermore, we may maintain that such discomforts may be connected to the overloading of the accommodation and vergence mechanisms. Many investigators (Bergqvist, 1989; Berlinguet and Berthelette, 1990; Grieco, et al., 1995; Knave and Widebäck, 1987; Krueger, 1991; Luczack et al., 1993; Miyamoto et al., 1997), in fact, suggest positive correlations between asthenopia and nearwork, and raise concerns about medium and long term effects without, alas, clarifying the pathophysiological underlying mechanisms.

Recent proposals from the Scientific Committee of Work and Vision of the International Commission on Occupational Health (ICOH) suggest multifactorial causation of these largely subclinical disturbances, as involving: a) environment --- chemical, physical and microbiological agents; b) job content --- visual load and organizational climate; c) individual --- ophthalmological and psycho-emotional characteristics.

Accommodation/Vergence. The requirements for accommodation and vergence performance that depend on both the type of work as well as the functional or subclinical characteristics of the subject, have proven remarkably difficult for researchers in the occupational health and ophthalmological fields to establish.

Thus, I have enthusiastically listened to the papers presented during this conference, that demonstrate the ability of both long-standing and recent contributions of neuroscience to elucidate the problems of people who perform nearwork, that so far, can only be considered as subclinical from the ophthalmological point of view. The clinical approach fails with functional occupational asthenopia, and can only make inroads on chronic and consolidated alterations.

The chapters of the present volume have, on the other hand, shown that in neuroscience these themes have been the object of study for quite a long time and are able to furnish relevant information and knowledge for prevention and improvement of the well-being of nearwork operators. This is in contrast to the mainly clinical ophthalmological approach that is used for the protection of the integrity of the visual system in these working sectors.

Hence, we can hope for both theoretical and practical contributions of vision research and neuroscience to reduce and control occupational asthenopia and to eliminate possible long-term deterioration.

Epidemiology. Precise epidemiological information is not available regarding the number of nearwork operators world-wide, although we know these are very numerous (Table 1).

Table 1: Nearwork occupations.

SECTORS	TASK
AVIATION	air traffic controllers
BANK AND INSURANCE PRESS TELEPHONE COMPANIES AIRPORT SERVICE DESIGN ACTIVITIES IN INDUSTRY	VDUs
CLOTHING INDUSTRY	quality inspection and mendings
RESEARCH LABORATORIES MICROSURGERY ELECTRONIC COMPONENTS INDUSTRY	work with microscopes
PHARMACEUTICAL INDUSTRY FOOD INDUSTRY	impurity controls
MECHANIC INDUSTRY	precision mechanics welders
JEWELRY INDUSTRY	bench work
PHOTOLITOGRAPHIC INDUSTRY	photo compositors and binders map makers topographers
CRAFTSMEN (ARTISANS)	decorators sculptors engravers wood carvers lithographers
MOVIE INDUSTRY	film editing, splicing, cutting, montage

For these subjects, average viewing distance ranges from 50 to 60 cm, and task duration is over 5 hours/day (Krueger, 1991). According to research from the European Foundation for the Improvement of Living and Working Conditions (1998), based on a representative sample of

15.800 people, it was concluded that: a) the working population of the European Union (15 member states) is 147.000.000 workers; b) 38 % (55 million workers) make use of computers; c) 9 % (13.2 million workers) report ocular problems. Therefore, we may perhaps extrapolate and conclude that many hundreds of millions of workers are at risk from occupational asthenopia, and, possibly, tens of millions suffer from disturbances of accommodation and vergence.

The Future. We suggest that at least two topics should immediately be approached by ergo-ophthalmologists in collaboration with visual scientists. We need to set up criteria and methods to analyze the effects on the accommodation and vergence mechanisms in both normal subjects and in those already affected by asthenopia alterations. We have also to focus on the development of late-onset myopia, and search for a method to detect and protect hypersensitive subjects.

For a wider discussion in these two critical areas it is necessary to attain knowledge from an interdisciplinary approach involving researchers in visual neuroscience, occupational and environmental medicine, optometry, ophthalmology, and industrial hygiene, and to clarify the complex relationships between "Work and Vision". By reducing the risk of visual and occulomotor disturbances and possibly related pathologies, we could certainly contribute to a positive influence in providing a more productive and satisfactory working life.

Acknowledgements
I wish to thank Miss D. Casalati for her valuable contribution in preparing the bibliography.

References

Bergqvist, U. (1989). Possible health effects of working with VDUs. *British Journal of Industrial Medicine,* 46: 217-221.
Berlinguet, L. and Berthelette, D. (1990, eds.) *Selected Papers of the Second International Scientific Conference on Work with Display Units 89*. Montreal, Quebec, Canada, 11-14 September, 1989. North-Holland, Elsevier Science Publishers.
European Foundation for the Improvement of Living and Working Conditions: Working conditions in the European Union, 1998.
Grieco, A., Molteni, G., Occhipinti, E. and Piccoli, B. (1995, eds.) *Selected Papers of the Fourth International Scientific Conference on Work with Display Units 94*. Milan, Italy, 2-5 October, 1994. North-Holland, Elsevier Science Publishers.
Knave, B. and Widebäck, P.G. (1987, eds.) *Selected Papers of the First International Scientific Conference on Work with Display Units 86*. Stockholm, Sweden, May 12-15, 1986. North-Holland, Elsevier Science Publishers.
Krueger, H. (1991). Visual functions and monitor use. In Roufs J.A.J. (ed.): *The Man-Machine Interface. Vision and Visual Dysfunction*. Macmillan Press, London, pp 55-69.
Luczack, H., Cakir, A. and Cakir, G. (1993, eds.): *Selected papers of the Third International Scientific Conference on Work with Display Units 92*. Berlin, Germany, September 1-4, 1992. North-Holland, Elsevier Science Publishers.
Miyamoto, H., Saito, S., Kajiyama, M. and Koizumi, N. (1997, eds.): *Proceedings of the Fifth International Scientific Conference on Work With Display Units 97*, Tokyo, Japan, November 3-5, 1997.
Ramazzini, B. (1739) *De Morbis Artificum Diatriba* - Apud Paulum et Isaacum Vaillant, Londini. Chapter XXXVI. pp 94-96.

Subject Index